Springer Series in Reliability Engineering

Series Editor

Today's modern systems have become increasingly complex to design and build, while the demand for reliability and cost effective development continues. Reliability is one of the most important attributes in all these systems, including aerospace applications, real-time control, medical applications, defense systems, human decision-making, and home-security products. Growing international competition has increased the need for all designers, managers, practitioners, scientists and engineers to ensure a level of reliability of their product before release at the lowest cost. The interest in reliability has been growing in recent years and this trend will continue during the next decade and beyond.

The Springer Series in Reliability Engineering publishes books, monographs and edited volumes in important areas of current theoretical research development in reliability and in areas that attempt to bridge the gap between theory and application in areas of interest to practitioners in industry, laboratories, business, and government.

Now with 100 volumes!

Indexed in Scopus and EI Compendex

Interested authors should contact the series editor, Hoang Pham, Department of Industrial and Systems Engineering, Rutgers University, Piscataway, NJ 08854, USA. Email: **hopham@soe.rutgers.edu, or Anthony Doyle, Executive Editor, Springer, London. Email: anthony.doyle@springer.com.**

Kai Li · Xin Yuan · Wei Ni

Security and Resilience in Distributed Machine Learning

Challenges, Techniques, and Future Directions

Springer

Kai Li
Interdisciplinary Centre for Security,
Reliability and Trust (SnT)
University of Luxembourg
Luxembourg, Luxembourg

Xin Yuan
Data61 Business Unit
Commonwealth Scientific and Industrial
Research Organisation (CSIRO)
Sydney, NSW, Australia

Wei Ni
Data61 Business Unit
Commonwealth Scientific and Industrial
Research Organisation (CSIRO)
Sydney, NSW, Australia

ISSN 1614-7839 ISSN 2196-999X (electronic)
Springer Series in Reliability Engineering
ISBN 978-3-032-23958-7 ISBN 978-3-032-23959-4 (eBook)
https://doi.org/10.1007/978-3-032-23959-4

This Springer imprint is published by the registered company Springer Nature Switzerland AG
The registered company address is: Gewerbestrasse 11, 6330 Cham, Switzerland

May the book inspire you to look beyond algorithms and see the deeper interplay between intelligence, trust, and collaboration in distributed machine learning systems. Let every challenge described here remind the research community that resilience is not merely a goal but a continuous pursuit of balance between learning and adaptation. And may the book guide researchers to build distributed machine learning systems that are not only powerful, but also secure, fair, and worthy of the trust they seek to earn.

Preface I

This book, Security and Resilience in Distributed Machine Learning: Challenges, Techniques, and Future Directions, offers a structured and in-depth examination of the evolving field of distributed machine learning, such as federated learning (FL), with a particular focus on its resilience and security under real-world conditions. Divided into four comprehensive parts, the book begins by establishing the foundations of FL, highlighting its transformative role in decentralized, privacy-preserving machine learning across sectors such as healthcare, finance, and smart cities. It then outlines the growing threat landscape, including model poisoning, privacy leakage, and system exploitation, supported by case studies and graph-based modeling approaches.

Part II delves into cutting-edge adversarial attacks on distributed machine learning, introducing variational and autoencoder-based model poisoning strategies, as well as bias-inducing adversarial graph attention networks (AGAT). These chapters illustrate how malicious agents can exploit structural properties and update mechanisms to degrade global model performance and fairness. Part III shifts the focus to defense, presenting innovative countermeasures such as representational similarity analysis, visual explanation-based detection with GradCAM, and privacy-preserving perturbation techniques that navigate the trade-off between utility and protection.

In Part IV, the book explores future directions, including the integration of graph neural networks, the development of explainable distributed machine learning frameworks, and strategies for sustainable, energy-efficient learning systems aligned with global development goals. This book serves as a critical reference for researchers, engineers, and policymakers seeking to design secure, interpretable, and resilient FL systems in adversarial and resource-constrained environments.

Luxembourg, Luxembourg — Kai Li
Sydney, NSW, Australia — Xin Yuan
Sydney, NSW, Australia — Wei Ni
November 2025

Preface II

Resilience has become a cornerstone in the development and deployment of Distributed Machine Learning embodied by Federated Learning (FL), particularly as FL gains traction in real-world applications, such as autonomous driving, health care, and smart manufacturing. These deployments demand continuous model reliability within inherently heterogeneous, dynamic, and untrusted distributed environments. Yet, this growing reliance on FL has exposed it to a range of sophisticated adversarial threats that can severely compromise model integrity, training accuracy, and fairness. For instance, graph representation-based model poisoning and adversarial graph attention attacks have demonstrated the potential to manipulate global model convergence while violating fairness across clients. To counteract such threats, recent research has explored feature-oriented and interpretability-driven defenses, leveraging representational similarity analysis (RSA) and visual explanations to identify anomalous model updates. These efforts underscore the dual imperative of understanding both attacks and defenses to construct resilient FL frameworks capable of withstanding evolving adversarial behaviors in large-scale, decentralized environments.

Guided by a vision of trustworthy and sustainable distributed intelligence, the book extends beyond threat modeling to explore the broader architectural design of resilient learning systems. It integrates graph neural networks, explainable AI (XAI), energy-aware optimization, and privacy-preserving mechanisms to reconcile reliability, security, scalability, and regulatory compliance. By bridging theoretical foundations with practical design strategies, this book provides both a technical reference and a forward-looking roadmap for researchers, engineers, and policymakers aiming to shape the future of secure, distributed learning systems. Ultimately, it seeks to ensure the long-term viability of distributed intelligence, from edge learning to collaborative, reasoning-driven AI at scale.

While this book primarily focuses on Distributed Machine Learning, the concepts and mechanisms developed herein have broader applicability, particularly to Large Language Models (LLMs) and agentic AI systems. Both LLM ecosystems and agentic AI networks increasingly depend on distributed collaboration, continual learning, privacy-preserving coordination, and robustness against manipulation, i.e., challenges that closely parallel those faced in FL. The principles of trust-aware aggregation, decentralized optimization, explainable reasoning, and resilient communication discussed in this book can inform the evolution of LLM fine-tuning across federated data silos and the coordination of autonomous, self-adaptive agents operating under uncertain and adversarial conditions. These synergies highlight a converging trajectory where resilient distributed learning forms the foundation for scalable, trustworthy, and self-governing AI.

Luxembourg, Luxembourg — Kai Li
Sydney, NSW, Australia — Xin Yuan
Sydney, NSW, Australia — Wei Ni
November 2025

Acknowledgements The authors gratefully acknowledge the invaluable technical support and collaborative environment provided by several outstanding institutions that have shaped the development of this book. In particular, sincere appreciation is extended to the Research Centre in Real-Time and Embedded Computing Systems (CISTER) in Portugal, for their supportive research environment and academic resources; the Telecommunication Networks (TKN) Group at the School of Electrical Engineering and Computer Science, TU Berlin, Germany, for their insightful discussions on resilient networked intelligence; the Department of Electrical and Computer Engineering at Carnegie Mellon University (CMU), Pennsylvania, USA, for their guidance on large-scale distributed learning architectures; the Internet of Everything (IoE) Group in the Department of Engineering at the University of Cambridge, UK, for their stimulating research atmosphere and critical perspectives on federated and edge intelligence; and Commonwealth Scientific and Industrial Research Organisation (CSIRO) in Australia, for their sustained support, computational resources, and collaborative spirit that made this work possible.

Competing Interests The authors have no competing interests to declare that are relevant to the content of this manuscript.

Contents

Part II Adversarial Attacks on Federated Learning

Acronyms

AE	Autoencoder
AGAT	Adversarial graph attention network
AI	Artificial intelligence
AN-BA	Additive noise-based biasing attack
AR	Augmented reality
Async-FL	Asynchronous FL
BER	Bit error rate
BS	Base station
CDF	Cumulative distribution function
CFL	Centralized federated learning
CNN	Convolutional neural network
CSCG	Circularly symmetric complex Gaussian
CSI	Channel state information
DFL	Decentralized federated learning
DNN	Deep neural network
DP	Differential privacy
DRL	Deep reinforcement learning
F^2L	Flexible aggregation-based FL
FL	Federated learning
FLOPs	Floating-point operations
FMNIST	Fashion-MNIST
FPR	False positive rate
GAE	Graph autoencoder
GAN	Generative adversarial networks
GCN	Graph convolutional network
GFT	Graph Fourier transform
G-MPA	GAE-based model poisoning attack
GNN	Graph neural network
GradCAM	Gradient-weighted class activation mapping
GTSRB	German traffic sign recognition benchmark
IID	Independent and identically distributed

IoT	Internet of Things
IoV	Internet of Vehicles
KL	Kullback-Leibler divergence
LayerCAM	Layer class activation mapping
LDA	Latent Dirichlet allocation
LDP	Local differential privacy
LFA	Label flipping attack
LLMs	Large language models
LP	Linear program
LTE	Long-term evolution
MEC	Mobile edge computing
ML	Machine learning
MLP	Multilayer perceptron
MP	Model poisoning
MSE	Mean squared error
NOMA	Non-orthogonal multiple access
Non-IID	Non-independent and identically distributed
OFDMA	Orthogonal-frequency division multiple-access
OTA	Over-the-air
ReLU	Rectified linear unit
RHS	Right-hand side
RSA	Representational similarity analysis
SDG	Sustainable development goal
SGD	Stochastic gradient descent
SNR	Signal to noise ratio
SVD	Singular value decomposition
SVHN	Street view house numbers dataset
SVM	Support vector machines
Sync-FL	Synchronous FL
TEE	Trusted execution environment
TLS	Transport layer security protocol
UAV	Unmanned aerial vehicle
VGAE	Variational graph autoencoder
VGAE-MP	Variational graph autoencoder-based model poisoning
VR	Virtual reality
XAI	Explainable AI

Part I
Foundations of Federated Learning and Its Challenges

Chapter 1
Introduction to Federated Learning

Federated learning (FL) is a decentralized machine learning (ML) paradigm designed to train models collaboratively across a large number of distributed devices or organizations while keeping the raw data localized [23]. Instead of pooling data into a central server, clients compute model updates locally and only share parameters or gradients with a coordinating server [6]. This framework preserves data privacy, reduces communication overhead, and allows leveraging heterogeneous and sensitive datasets that are otherwise difficult to aggregate.

The motivation for FL extends beyond theoretical elegance; it arises from practical constraints and the urgent demand for privacy-preserving, distributed intelligence. Many industries generate sensitive data in large volumes but are limited by privacy, bandwidth, or regulatory barriers that prevent central collection. FL provides a framework to unlock the predictive power of data without exposing it.

1.1 Applications and Importance in Real-World Scenarios

One of the earliest and most visible domains is *mobile and personal computing*. Smartphone keyboards and voice assistants, for example, learn better next-word prediction and acoustic models by training on the private usage patterns of millions of devices while retaining the data locally. Google's Gboard and Apple's on-device learning are prominent instances of this approach. Similarly, wearable health devices can collaboratively train anomaly detection or early diagnosis models using physiological signals without exporting raw patient data.

In the *healthcare and medical imaging* sector, FL has become a powerful enabler of multi-institutional studies. Hospitals and research centers can collaboratively train deep neural networks (DNNs) for disease detection (e.g., cancer, diabetic retinopathy) or segmentation of radiological images while remaining compliant with data

K. Li et al., *Security and Resilience in Distributed Machine Learning*, Springer Series in Reliability Engineering, https://doi.org/10.1007/978-3-032-23959-4_1

protection regulations such as GDPR or HIPAA [1]. This capability is particularly valuable where patient data is siloed and ethically or legally non-shareable.

Industrial and IoT ecosystems also benefit substantially from FL. Sensors and edge devices in smart manufacturing, energy systems, human sensing systems [14], or predictive maintenance can collaboratively train models to forecast failures, balance energy demand, or detect anomalies without transmitting large volumes of raw data to the cloud. Beyond industrial settings, FL has become increasingly relevant across distributed domains, such as mobile edge computing [2, 11], personalized learnign [24], the Internet of Things (IoT) [6], autonomous driving [17], blockchain systems [18, 22], emerging quantum-native communication networks [26], and Internet of Drones [3, 10, 12]. In these environments, FL enables privacy-preserving collaboration among heterogeneous and geographically dispersed nodes. Blockchain can further enhance the security, transparency, and traceability of model updates, while quantum-native systems offer potential for ultra-secure and low-latency communication, together paving the way toward next-generation decentralized intelligence.

In the domain of *finance and cybersecurity*, banks and fintech platforms use FL to improve fraud detection, credit scoring, and anti-money-laundering systems without pooling highly sensitive customer data across organizations [21]. Similar strategies are adopted in collaborative intrusion detection or malware classification, where multiple entities share threat intelligence indirectly through model updates.

Autonomous and connected vehicles illustrate another growing area: individual cars generate enormous volumes of driving data, including camera images, LiDAR point clouds, and trajectory logs [8, 20]. FL enables fleet-level improvements in perception, prediction, and navigation models while avoiding the cost and risk of uploading raw driving logs.

The increasing convergence of FL with sustainable and energy-aware communication infrastructures, such as energy harvesting and smart grid-powered wireless networks [5, 13], has attracted growing research interest. In such environments, devices can opportunistically harvest ambient energy or leverage grid-integrated power management to support continuous learning and communication [7]. These developments align with recent advances in the modeling and analysis of energy harvesting and smart grid-powered wireless communication networks [9], which emphasize efficiency, reliability, and environmental sustainability.

Finally, FL supports *cross-border collaboration* in sensitive domains, such as national security, biosecurity, and environmental monitoring. Agencies or research institutions operating in different jurisdictions can cooperatively build detection and surveillance models (e.g., for invasive species, pests, or smuggling detection) while respecting local data governance and operational confidentiality.

These applications demonstrate why FL is more than a theoretical construct. It is a response to structural constraints of modern data ecosystems: privacy requirements, legal compliance, and the explosive growth of distributed data generation. Its ability to convert previously isolated data silos into collective intelligence without surrendering control makes FL strategically important for a wide range of scientific, industrial, and societal systems.

1.2 Overview of Federated Learning

Consider a set of N clients, each with a local dataset $\mathcal{D}_i$. The typical goal of FL is to optimize a global objective:

$$\min_{w \in \mathbb{R}^d} F(w) = \sum_{i=1}^{N} p_i F_i(w),$$

where $F_i(w)$ represents the local loss function at client i, and p_i denotes the relative weight proportional to $|\mathcal{D}_i|$. FL protocols such as Federated Averaging (FedAvg) coordinate multiple rounds of local updates and global aggregation to achieve convergence while minimizing direct data exchange [25].

FL is therefore not only a computational architecture but also a socio-technical solution, addressing concerns of privacy, data ownership, and regulatory compliance. The paradigm sits at the intersection of distributed optimization, secure multiparty computation, and DP.

1.3 Centralized Federated Learning

The most widely adopted instantiation of FL is the centralized or *server-based* approach. As depicted in Fig. 1.1, this architecture follows a star topology, where a single server orchestrates the collaborative training procedure. The server begins by initializing a global model $w^{(0)}$ and transmitting it to a selected subset of clients. Each client performs local training, typically running several epochs of stochastic gradient descent (SGD) on its own dataset. If client i receives model $w^{(t)}$ at round t, a standard local update is

$$w_i^{(t+1)} = w^{(t)} - \eta \nabla F_i(w^{(t)}), \quad (1.1)$$

where η is the learning rate. After local optimization, clients return their model updates $w_i^{(t+1)}$ to the server.

The server then aggregates these updates to form the next global model. A widely used aggregation rule is *Federated Averaging* (FedAvg) [19]:

$$w^{(t+1)} = \sum_{i=1}^{N} \frac{n_i}{n} w_i^{(t+1)}, \quad (1.2)$$

thus weighting each client by its local dataset size. This process repeats for multiple communication rounds until the model converges.

This centralized design is relatively simple to deploy and has strong software support (e.g., TensorFlow Federated, PySyft, Flower). It also facilitates the introduc-

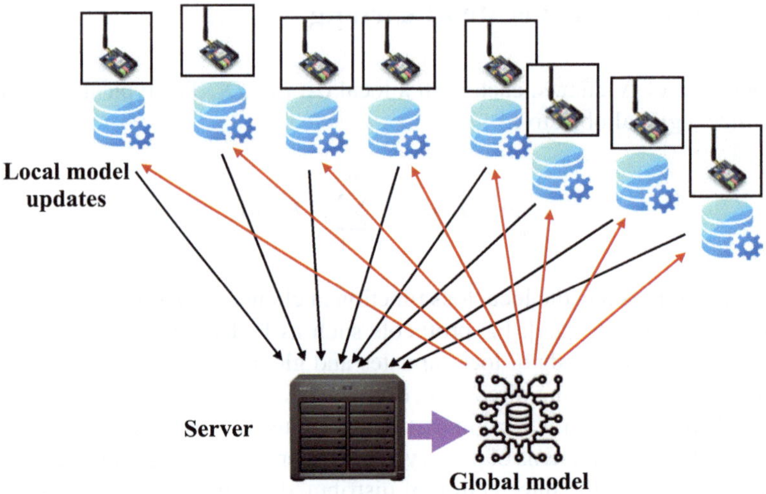

Fig. 1.1 Centralized federated learning (CFL): The server aggregates local model updates from the clients

tion of advanced optimization techniques such as adaptive learning rates, fairness weighting, or DP noise injection at the server side. Nevertheless, the central server introduces a single point of failure and a natural trust bottleneck. It can also become a scalability and bandwidth constraint if millions of clients attempt to communicate updates simultaneously.

1.4 Decentralized Federated Learning

An alternative to the server-centric paradigm is fully decentralized or *peer-to-peer* FL [15, 16], as shown in Fig. 1.2. Here, clients communicate directly with one another, forming a graph $\mathcal{G} = (\mathcal{V}, \mathcal{E})$ where vertices represent participants and edges represent communication links. Without a central aggregator, model consensus is achieved through local exchanges and distributed optimization techniques [4].

A common approach is gossip-based or consensus-based learning. Each client maintains its local model $w_i^{(t)}$ and, after performing a gradient step using its own data, mixes its model with those of its neighbors $\mathcal{N}(i)$:

$$w_i^{(t+1)} = \sum_{j \in \mathcal{N}(i) \cup \{i\}} \pi_{i,j} \Big(w_j^{(t)} - \eta \nabla F_j(w_j^{(t)}) \Big), \tag{1.3}$$

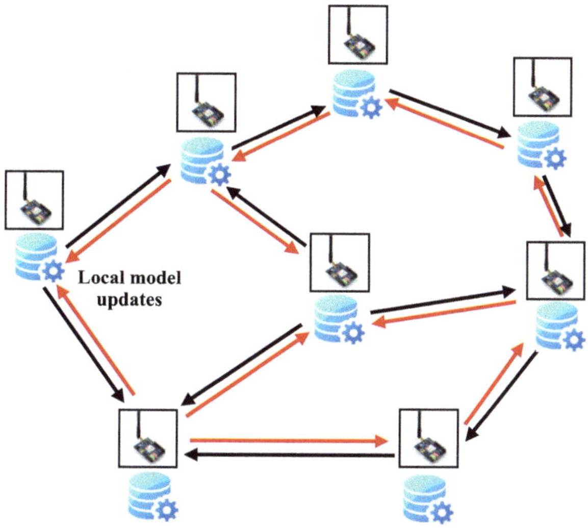

Fig. 1.2 Decentralized federated learning (DFL): clients communicate directly with one another, forming a graph, where vertices represent participants and edges represent communication links

where the coefficients $\pi_{i,j}$ are nonnegative and sum to one, reflecting the underlying topology. Over many rounds, and under mild connectivity assumptions on $\mathcal{G}$, the distributed models converge toward a common solution.

References

1. Ali M, Naeem F, Tariq M, Kaddoum G (2022) Federated learning for privacy preservation in smart healthcare systems: a comprehensive survey. IEEE J Biomed Health Inform 27(2):778–789
2. Deng D, Wu X, Zhang T, Tang X, Du H, Kang J, Liu J, Niyato D (2024) Fedasa: a personalized federated learning with adaptive model aggregation for heterogeneous mobile edge computing. IEEE Trans Mobile Comput
3. Guan Y, Zou S, Peng H, Ni W, Sun Y, Gao H (2024) Cooperative UAV trajectory design for disaster area emergency communications: a multiagent ppo method. IEEE Internet Things J 11(5):8848–8859. https://doi.org/10.1109/JIOT.2023.3320796
4. Hu S, Chen X, Ni W, Hossain E, Wang X (2021) Distributed machine learning for wireless communication networks: techniques, architectures, and applications. IEEE Commun Surv Tutorials 23(3):1458–1493. https://doi.org/10.1109/COMST.2021.3086014
5. Hu S, Chen X, Ni W, Wang X, Hossain E (2020) Modeling and analysis of energy harvesting and smart grid-powered wireless communication networks: a contemporary survey. IEEE Trans Green Commun Netw 4(2):461–496. https://doi.org/10.1109/TGCN.2020.2988270
6. Jiang Y, Ma B, Wang X, Yu G, Yu P, Wang Z, Ni W, Liu RP (2024) Blockchained federated learning for internet of things: a comprehensive survey. ACM Comput Surv 56(10) (2024). https://doi.org/10.1145/3659099
7. Li K, Ni W, Abolhasan M, Tovar E (2018) Reinforcement learning for scheduling wireless powered sensor communications. IEEE Trans Green Commun Netw 3(2):264–274
8. Li K, Ni W, Tovar E, Guizani M (2019) Optimal rate-adaptive data dissemination in vehicular platoons. IEEE Trans Intell Transp Syst 21(10):4241–4251

9. Li K, Ni W, Tovar E, Jamalipour A (2019) On-board deep q-network for uav-assisted online power transfer and data collection. IEEE Trans Veh Technol 68(12):12215–12226
10. Li K, Ni W, Wang X, Liu RP, Kanhere SS, Jha S (2015) Energy-efficient cooperative relaying for unmanned aerial vehicles. IEEE Trans Mob Comput 15(6):1377–1386
11. Li K, Ni W, Yuan X, Noor A, Jamalipour A (2023) Exploring graph neural networks for joint cruise control and task offloading in uav-enabled mobile edge computing. In: 2023 IEEE 97th vehicular technology conference (VTC2023-Spring). IEEE, pp 1–6
12. Li K, Noor A, Ni W, Tovar E, Fu X, Akan OB (2024) Poisoning federated learning with graph neural networks in internet of drones. In: IEEE International conference on computer communications and networks (ICCCN 2024)
13. Li K, Voicu RC, Kanhere SS, Ni W, Tovar E (2019) Energy efficient legitimate wireless surveillance of uav communications. IEEE Trans Veh Technol 68(3):2283–2293
14. Li K, Yuen C, Kanhere SS, Hu K, Zhang W, Jiang F, Liu X (2018) An experimental study for tracking crowd in smart cities. IEEE Syst J 13(3):2966–2977
15. Li W, Lv T, Ni W, Zhao J, Hossain E, Poor HV (2025) Route-and-aggregate decentralized federated learning under communication errors. IEEE Trans Neural Netw Learn Syst 36(9):16675–16691. https://doi.org/10.1109/TNNLS.2025.3552406
16. Li W, Lv T, Ni W, Zhao J, Hossain E, Vincent Poor H (2024) Decentralized federated learning over imperfect communication channels. IEEE Trans Commun 72(11):6973–6991. https://doi.org/10.1109/TCOMM.2024.3407208
17. Li X, Lu L, Ni W, Jamalipour A, Zhang D, Du H (2022) Federated multi-agent deep reinforcement learning for resource allocation of vehicle-to-vehicle communications. IEEE Trans Veh Technol 71(8):8810–8824. https://doi.org/10.1109/TVT.2022.3173057
18. Makhdoom I, Abolhasan M, Abbas H, Ni W (2019) Blockchain's adoption in iot: the challenges, and a way forward. J Netw Comput Appl 125:251–279. https://doi.org/10.1016/j.jnca.2018.10.019 https://www.sciencedirect.com/science/article/pii/S1084804518303473
19. McMahan B, Moore E, Ramage D, Hampson S, y Arcas BA (2017) Communication-efficient learning of deep networks from decentralized data. In: Proceedings of the 20th international conference on artificial intelligence and statistics, Proceedings of machine learning research, vol 54. PMLR, pp 1273–1282
20. Sun C, Ni W, Wang X (2021) Joint computation offloading and trajectory planning for UAV-assisted edge computing. IEEE Trans Wirel Commun 20(8):5343–5358. https://doi.org/10.1109/TWC.2021.3067163
21. Wang S, Asif M, Shahzad MF, Ashfaq M (2024) Data privacy and cybersecurity challenges in the digital transformation of the banking sector. Comput Secur 147:104051
22. Wang X, Zha X, Ni W, Liu RP, Guo YJ, Niu X, Zheng K (2019) Survey on blockchain for internet of things. Comput Commun 136:10–29. https://doi.org/10.1016/j.comcom.2019.01.006 https://www.sciencedirect.com/science/article/pii/S0140366418306881
23. Xu C, Qu Y, Ding M, Ni W, Gao L, Li J, Vincent Poor H (2025) What is the value of heuristic model selection and aggregation in federated learning? IEEE Netw 1. https://doi.org/10.1109/MNET.2025.3586566
24. Zhao X, Cui Q, Li W, Ni W, Hossain E, Sheng QZ, Tao X, Zhang P (2025) Convergence-privacy-fairness trade-off in personalized federated learning. IEEE Trans Mach Learn Commun Netw 3:246–262. https://doi.org/10.1109/TMLCN.2025.3528901
25. Zheng J, Li K, Mhaisen N, Ni W, Tovar E, Guizani M (2022) Exploring deep-reinforcement-learning-assisted federated learning for online resource allocation in privacy-preserving edgeiot. IEEE Internet Things J 9(21):21099–21110
26. Zhou X, Shen A, Hu S, Ni W, Wang X, Hossain E (2025) Towards quantum-native communication systems: state-of-the-art, trends, and challenges. IEEE Commun Surv Tutorials 1. https://doi.org/10.1109/COMST.2025.3575842

Chapter 2
Threat Landscape in Federated Learning

Although artificial intelligence (AI)-enabled Internet-of-Things (IoT) systems increasingly employ ML to provide localized intelligence, the rising heterogeneity and dynamic characteristics of IoT data highlight the limitations of task-specific models. FL with the ability to generalize across domains and tasks through large-scale pretraining is being widely adopted to support IoT applications, such as predictive maintenance, healthcare monitoring, and autonomous mobility. Moreover, FL facilitates cross-domain knowledge transfer, thereby reducing the need to train models from scratch on every device. However, integrating FL into IoT introduces unique security concerns: unlike conventional AI systems, FL operates in distributed and resource-constrained environments, often relying on federated or decentralized training across IoT nodes. This distributed interaction exposes FL to IoT-specific threats, including model poisoning (MP), inference attacks, and adversarial manipulations, particularly under non-independent and identically distributed (non-IID) data distributions and intermittent connectivity. Consequently, addressing the security challenges of FL within IoT ecosystems is crucial for enabling trustworthy AI and ensuring the resilience of IoT infrastructures.

2.1 Types of Threats: Poisoning, Privacy Breaches, and Model Exploitation

2.1.1 Data and Model Poisoning Attacks

Adversaries commonly exploit data poisoning and MP as key strategies against FL, seeking to inject malicious content into the training pipeline to compromise the integrity and reliability of the target model [26, 34, 35].

K. Li et al., *Security and Resilience in Distributed Machine Learning*, Springer Series in Reliability Engineering, https://doi.org/10.1007/978-3-032-23959-4_2

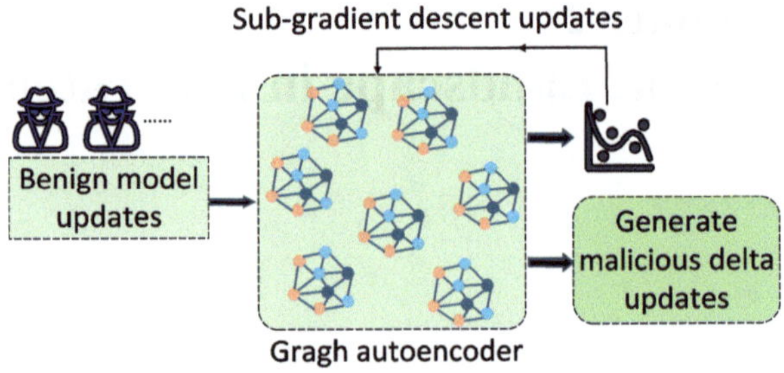

Fig. 2.1 An illustration of an adversarial VGAE-enabled MP attack [20]

Data poisoning attacks operate at the data level, where adversaries tamper with local training datasets to compromise the model's learning process. A typical approach involves injecting maliciously crafted samples, often mislabeled or subtly perturbed, into the data of compromised clients. Such samples are deliberately engineered to embed backdoors or distort the global model's decision boundaries [15, 28, 34, 37]. For instance, an attacker may insert inputs that associate a particular trigger pattern (e.g., a pixel patch in images or a phrase in text) with an incorrect label. Once updates from poisoned clients are aggregated, the global model learns to misclassify inputs containing that trigger, thereby embedding a covert attack [28]. These attacks are particularly severe in FL, where limited transparency into local client data and the absence of centralized control hinder effective monitoring and oversight. Furthermore, data poisoning tends to be stealthy and adaptive: malicious samples may be sparsely distributed, making them difficult to detect during aggregation, especially under non-IID data distributions that are prevalent in IoT environments.

MP attacks occur at the model update level, where adversaries manipulate gradients or parameters directly rather than altering training data. The objective is either to implant malicious behavior into the global model or to disrupt convergence by maximizing divergence during aggregation. In [20], a sophisticated threat model was introduced in which an adversary leverages an adversarial variational graph autoencoder (VGAE) to capture structural dependencies among benign local models, as illustrated in Fig. 2.1. These inferred relationships are then adversarially perturbed to craft malicious updates that remain statistically indistinguishable from benign ones. Importantly, this attack does not require access to private data, making it highly relevant to black-box FL scenarios such as IoT edge deployments and FL fine-tuning, where direct access to raw training data is restricted.

Another strategy, outlined in [21], allows adversaries to eavesdrop on both the global model and benign client updates in order to reconstruct the latent graph structure that connects model parameters with data features. By strategically perturbing these correlations, the attacker can degrade overall model performance or implant targeted malicious behaviors. This threat is particularly concerning in large-scale FL systems, where complex feature interdependency can be exploited without requiring direct access to local datasets. Complementary work in [3] highlights malicious user injection, wherein adversaries introduce fraudulent clients that contribute poisoned updates aligned with a predefined backdoor objective. This approach is inherently scalable and well-suited to FL environments such as IoT networks, where participant authentication mechanisms are often limited. Such vulnerabilities become especially critical in edge deployments, where even a small fraction of compromised clients can significantly influence the downstream behavior of the global model.

2.1.2 *Membership Inference Attacks*

Membership inference attacks in FL, as depicted in Fig. 2.2, leverage access to shared model updates or intermediate parameters to extract sensitive information, including user attributes, class distributions, or even raw training data [13]. Owing to FL's decentralized nature, adversarial clients can mount such attacks without requiring complete access to the global model, thereby enlarging the attack surface and making privacy preservation substantially more challenging.

In [11], the authors evaluated multiple secure aggregation protocols and showed that these mechanisms alone are insufficient to guarantee user data privacy. They introduced a differential selection attack combined with de-noising techniques, enabling adversaries to infer multi-label classification results from IoT node data, thereby breaching privacy in typical FL scenarios. Building on this, the authors of [38] proposed a poisoning-assisted inference attack, where malicious actors exploit benign model updates to extract sensitive feature information. Using a binary attack model, adversaries can uncover hidden data patterns that should remain private. Moreover, they devised a targeted poisoning strategy that manipulates training labels to shift the global model's decision boundaries, inadvertently inducing benign clients to disclose additional private feature information.

Inference attacks pose significant privacy risks in collaborative learning frameworks such as split learning and FL. In [10], both passive and active inference attacks were introduced: the passive approach employs semi-supervised learning on auxiliary datasets to infer labels, whereas the active variant manipulates the training process to increase the global model's dependence on the attacker's sub-model, thereby enhancing inference accuracy. A broader study in [24] demonstrated that data complexity strongly influences inference success, revealing a trade-off between model stealing and inference attack effectiveness. Using architectures, such as ResNet18 and datasets like CelebA and Fashion-MNIST (FMNIST), the authors highlighted

Fig. 2.2 Membership inference attacks in FL leverage access to shared model updates or intermediate parameters to extract sensitive information

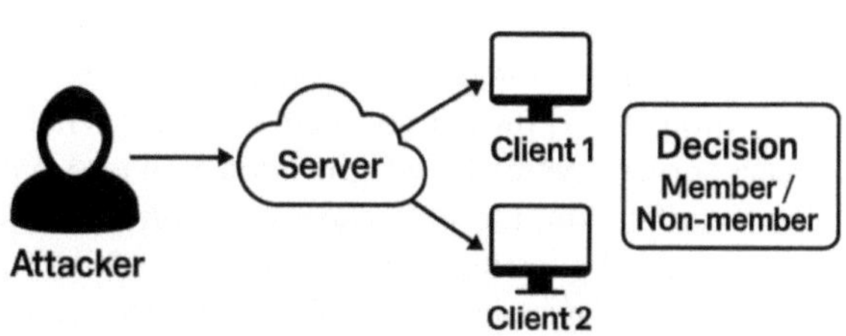

the sensitivity of inference outcomes to data characteristics. Complementary work in [23] investigated membership inference attacks based on prediction sensitivity, where attackers monitor changes in model predictions under slight input perturbations to determine whether a sample was part of the training set. Notably, this approach requires neither access to the training data nor detailed knowledge of the model, making it a powerful and pervasive privacy threat.

In [40], the authors proposed a hypothesis testing framework to strengthen membership inference attacks. By leveraging reference models, the framework improves the true positive rate while maintaining control over false positives. They further analyzed attacker uncertainty, showing that their method can reduce uncertainty to a single-bit secret, effectively determining whether a particular data sample was included in training. Nonetheless, this approach depends on the availability of well-calibrated reference models, which may not be feasible in practice, underscoring the need for future extensions to black-box settings with limited or unreliable references. Complementary work [25] investigated feature inference attacks at the model prediction stage under a strong adversarial assumption, where attackers have access only to the model and its outputs. By systematically analyzing model predictions across diverse architectures, they were able to infer private feature values. However, the attack's success depends on abundant prediction samples and tends to diminish under regularization or dynamic model updates. Future directions could include the development of adaptive defenses or noise-resilient attack strategies tailored to real-world deployment scenarios.

2.1.3 *Byzantine Failure Attacks*

As shown in Fig. 2.3, Byzantine failures in FL arise when compromised or unreliable clients submit erroneous or adversarial updates, ultimately hindering model convergence and degrading accuracy [39]. Such failures may stem from diverse sources, including random noise injection, corrupted training data, software faults, network delays, or skewed local distributions, posing significant difficulties for aggregation algorithms in separating honest updates from malicious ones [5].

Although certain FL frameworks are designed to tolerate Byzantine faults, they remain vulnerable to deliberate local MP attacks [9]. By tampering with the model updates generated on compromised IoT nodes, adversaries can substantially deteriorate training performance, effectively driving the optimization process away from its intended trajectory. While some defenses inspired by poisoning-attack countermeasures provide limited mitigation, their success is highly context-dependent. An investigation in [32] further revealed weaknesses in Byzantine-resilient aggregation. Because servers typically weight updates based on client-reported dataset sizes, without verification due to privacy constraints, malicious participants can exploit this mechanism by misreporting. Two manipulation strategies were analyzed: clients with small datasets exaggerating their sizes to match benign users, and clients with moderate datasets inflating them to amplify their influence during aggregation.

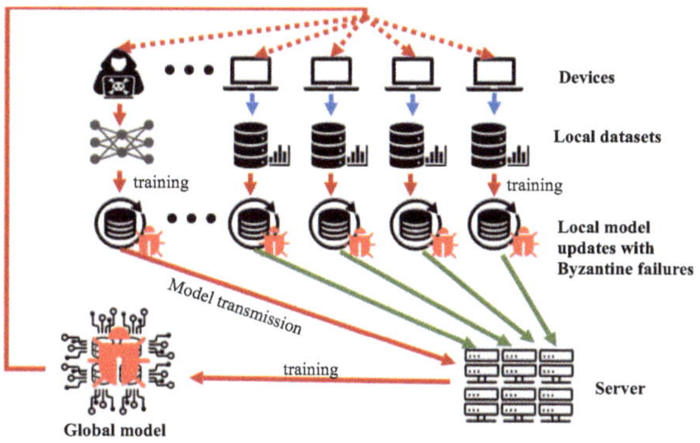

Fig. 2.3 Byzantine failure attacks against FL, ultimately hindering model convergence and degrading accuracy

In [2], it was demonstrated that linear aggregation strategies for IoT updates fail to withstand even a single Byzantine participant. A lone compromised client can coerce the system into adopting an arbitrary update, potentially characterized by extreme magnitude or misleading directionality. To address this, the study proposed a Byzantine-resilient framework that establishes sufficient conditions under which an aggregation rule can tolerate multiple adversarial devices. Building on this line of work, the authors of [6] introduced an FL framework that integrates both privacy preservation and Byzantine robustness. Their solution leverages three-party computation to perform secure aggregation while safeguarding local models. To improve scalability, the scheme incorporates a maliciously secure top-k protocol that reduces communication costs, along with an optimized secure shuffling mechanism, which is integral to the top-k procedure.

While the approaches developed in [2, 6] strengthen the security of federated systems, achieving a true zero-trust paradigm demands stricter assumptions, where all clients and intermediaries are regarded as potentially malicious. In this regard, the three-party computation and secure protocols proposed in [6] better reflect zero-trust principles by reducing dependency on individual parties and limiting data exposure. To fully realize a zero-trust architecture, such methods could be further enhanced with secure authentication mechanisms, trusted setup procedures, and continuous monitoring of client behavior.

2.1.4 Backdoor Attacks

Backdoor attacks represent a class of adversarial threats in which an attacker embeds a concealed trigger into an ML model during training, particularly in FL settings. As shown in Fig. 2.4, the trigger is crafted so that, when presented with specific inputs under the attacker's control, the model produces malicious or unintended outputs. Crucially, the model continues to perform normally on benign inputs, allowing the backdoor to remain stealthy and difficult to identify during routine evaluation.

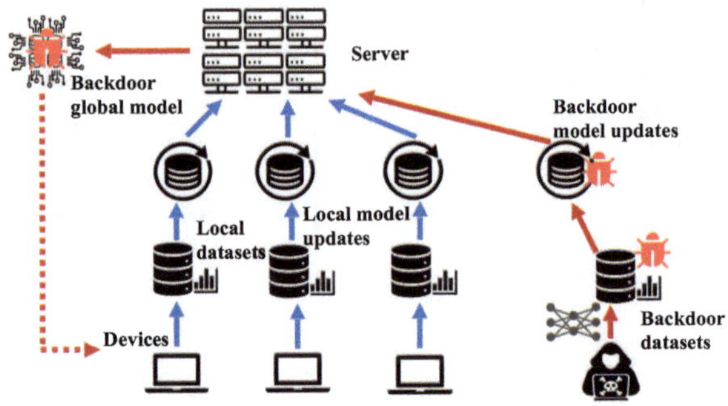

Fig. 2.4 Backdoor attacks enable an attacker to embed a concealed trigger into an ML model during training

Nguyen et al. [27] investigated the security vulnerabilities of FL under backdoor attacks, showing how malicious participants can covertly embed harmful behaviors into collaboratively trained models. Their study surveyed a range of attack strategies and outlined defense mechanisms, such as anomaly detection and robust aggregation to strengthen FL security. In a complementary line of work, Wang et al. [36] examined how backdoors can be introduced by exploiting tail distributions, i.e., data points that are rare or underrepresented. They demonstrated that even minor perturbations in these tails enable adversaries to implant hidden malicious behavior without degrading the model's overall accuracy. This finding highlights the susceptibility of FL to backdoor threats, particularly in scenarios with imbalanced datasets.

Gong et al. [12] investigated collaborative backdoor attacks in FL, where multiple adversaries work together to implant triggers into the global model. Their findings revealed that these triggers can be model-specific, exploiting architectural vulnerabilities that make the attack both more effective and more difficult to identify. This study underscores the heightened difficulty of defending against coordinated adversaries in federated settings. To counter such threats, Rieger et al. [30] introduced DeepSight, a detection framework that inspects the internal structure and output behavior of neural network updates to identify and filter out malicious contributions, thereby enhancing model robustness while maintaining accuracy on benign data.

2.2 Real-World Case Studies of Security and Resilience Challenges

In autonomous vehicular networks, FL enables collaborative learning of driving models from heterogeneous onboard sensor data without disclosing private information such as travel routes or passenger details. Hierarchical aggregation [4, 7, 33] plays a pivotal role in this context, allowing models trained on vehicles or roadside edge devices to first combine at regional servers before being synchronized globally. This layered aggregation has been shown to cut synchronization latency by as much as 40% compared with flat aggregation, while preserving accuracy under non-IID vehicular data distributions [31]. To ensure robustness in safety-critical scenarios, fault-tolerant mechanisms and anomaly detection are essential for identifying issues such as sensor failures, data spoofing, or adversarial attempts to influence automated driving decisions [16]. Empirical results indicate that resilient anomaly detection schemes can sustain convergence even with up to 25% of participants compromised, whereas conventional approaches fail to stabilize [14]. At the same time, feature-level defenses are vital for mitigating subtle manipulations of sensor inputs, such as misleading an object detection model to misclassify pedestrians. Finally, privacy-preserving defenses against inference and reconstruction attacks are crucial for safeguarding sensitive trajectory patterns and passenger behaviors, thereby ensuring both data confidentiality and compliance with regulatory requirements.

Smart healthcare scenarios, such as remote patient monitoring and medical image analysis, stand to gain significantly from FL, given the stringent privacy requirements in this domain [29]. Hierarchical aggregation enables the coordination of patient-specific models trained locally on heterogeneous data sources, including imaging, vital signs, and electronic health records, while adhering to institutional data silos and regulatory constraints. Experimental results indicate that hierarchical aggregation can cut communication overhead across hospital networks by nearly 28% compared with flat aggregation, while preserving accuracy within a 2% margin of baseline performance [22].

Ensuring fault tolerance and anomaly detection is particularly important, as compromised or corrupted medical inputs can lead to harmful misdiagnoses or inappropriate treatment [1, 8, 26]. For instance, anomaly-aware federated healthcare frameworks have sustained 58% diagnostic accuracy even when 10% of local updates were adversarial [42]. Feature-level defenses further counter more subtle manipulations, such as adversaries tampering with diagnostic imaging to bypass detection, thereby protecting the integrity of clinical outcomes. Equally crucial are privacy-preserving measures against inference and reconstruction attacks, which shield sensitive patient information from being recovered through shared model updates, maintaining both patient trust and compliance with healthcare data protection laws.

In immersive Metaverse applications, FL enables privacy preservation and computational efficiency by allowing models to be trained locally on user devices without sharing raw data [17, 18]. Hierarchical aggregation plays a central role by optimizing real-time, context-aware services, as edge nodes adapt models to regional interaction patterns and mitigate the challenges of highly non-IID data generated across diverse AR/VR interfaces. Studies show that edge-assisted hierarchical aggregation can achieve synchronization latencies below 200 ms, which is sufficient to support seamless (augmented reality) AR/(virtual reality) VR immersion [19]. Robust fault-tolerance mechanisms are equally important, ensuring uninterrupted user experiences by rapidly detecting compromised clients or malicious inputs designed to inject harmful content or destabilize interactions. Feature-level detection methods further safeguard against subtle manipulations, such as feature-based backdoors that distort user perception or alter in-world behavior. While defenses based on DP or secure aggregation strengthen confidentiality, they typically reduce AR/VR model accuracy by about 9%, illustrating the trade-off between privacy and service quality [41]. Finally, protection against inference and reconstruction attacks is essential to secure sensitive interaction traces and biometric information, ensuring that user-generated virtual content remains private, a prerequisite for trust and safety in shared Metaverse ecosystems.

References

1. Bi S, Yuan X, Hu S, Li K, Ni W, Hossain E, Wang X (2025) Resilience and failure analysis in next-generation communication networks: a contemporary survey. IEEE Trans Netw Sci Eng 1–29. https://doi.org/10.1109/TNSE.2025.3620950
2. Blanchard P, El Mhamdi EM, Guerraoui R, Stainer J (2017) Machine learning with adversaries: Byzantine tolerant gradient descent. In: Proceedings of NeurIPS, vol 30
3. Cao X, Gong NZ (2022) MPAF: model poisoning attacks to federated learning based on fake clients. In: 2022 IEEE/CVF conference on computer vision and pattern recognition workshops (CVPRW), pp 3395–3403. https://doi.org/10.1109/CVPRW56347.2022.00383
4. Chen X, Li Z, Ni W, Wang X, Zhang S, Sun Y, Xu S, Pei Q (2024) Toward dynamic resource allocation and client scheduling in hierarchical federated learning: a two-phase deep reinforcement learning approach. IEEE Trans Commun 72(12):7798–7813. https://doi.org/10.1109/TCOMM.2024.3420733
5. Distler T (2021) Byzantine fault-tolerant state-machine replication from a systems perspective. ACM Comput Surv (CSUR) 54(1):1–38
6. Dong C, Weng J et al (2024) Privacy-preserving and Byzantine-robust federated learning. IEEE Trans Dependable Secur Comput 21(2):889–904
7. Duan L, Huang H, Li C, Ni W, Cheng B (2025) A novel cross-chain hierarchical federated learning framework for enhancing service security and communication efficiency. IEEE Trans Serv Comput 18(3):1199–1212. https://doi.org/10.1109/TSC.2025.3562329
8. Duan L, Sun Y, Ni W, Ding W, Liu J, Wang W (2023) Attacks against cross-chain systems and defense approaches: a contemporary survey. IEEE/CAA J Automatica Sinica 10(8):1647–1667. https://doi.org/10.1109/JAS.2023.123642
9. Fang M, Cao X, Jia J, Gong N (2020) Local model poisoning attacks to Byzantine-Robust federated learning. In: 29th USENIX security symposium (USENIX Security 20), pp 1605–1622
10. Fu C, Zhang X, Ji S, Chen J, Wu J, Guo S, Zhou J, Liu AX, Wang T (2022) Label inference attacks against vertical federated learning. In: 31st USENIX security symposium (USENIX Security 22), pp 1397–1414
11. Gao J, Hou B, Guo X, Liu Z, Zhang Y, Chen K, Li J (2023) Secure aggregation is insecure: category inference attack on federated learning. IEEE Trans Dependable Secur Comput 20(1):147–160
12. Gong X, Chen Y, Huang H, Liao Y, Wang S, Wang Q (2022) Coordinated backdoor attacks against federated learning with model-dependent triggers. IEEE Netw 36(1):84–90
13. Hu H, Salcic Z, Sun L, Dobbie G, Yu PS, Zhang X (2022) Membership inference attacks on machine learning: a survey. ACM Comput Surv (CSUR) 54(11s):1–37
14. Huang SE, Feng Y, Liu HX (2021) A data-driven method for falsified vehicle trajectory identification by anomaly detection. Transp Res Part C Emerg Technol 128:103196
15. Jagielski M, Oprea A, Biggio B, Liu C, Nita-Rotaru C, Li B (2018) Manipulating machine learning: Poisoning attacks and countermeasures for regression learning. In: 2018 IEEE symposium on security and privacy (SP), pp 19–35. https://doi.org/10.1109/SP.2018.00057
16. Kurunathan H, Huang H, Li K, Ni W, Hossain E (2023) Machine learning-aided operations and communications of unmanned aerial vehicles: a contemporary survey. IEEE Commun Surv Tutorials 26(1):496–533

17. Li K, Cui Y, Li W, Lv T, Yuan X, Li S, Ni W, Simsek M, Dressler F (2023) When internet of things meets metaverse: convergence of physical and cyber worlds. IEEE Internet Things J 10(5):4148–4173. https://doi.org/10.1109/JIOT.2022.3232845
18. Li K, Lau BPL, Yuan X, Ni W, Guizani M, Yuen C (2023) Toward ubiquitous semantic metaverse: challenges, approaches, and opportunities. IEEE Internet Things J 10(24):21855–21872. https://doi.org/10.1109/JIOT.2023.3302159
19. Li K, Lau BPL, Yuan X, Ni W, Guizani M, Yuen C (2023) Toward ubiquitous semantic metaverse: challenges, approaches, and opportunities. IEEE Internet Things J 10(24):21855–21872
20. Li K, Yuan X, Zheng J, Ni W, Dressler F, Jamalipour A (2025) Leverage variational graph representation for model poisoning on federated learning. IEEE Trans Neural Netw Learn Syst 36(1):116–128
21. Li K, Zheng J, Yuan X, Ni W, Akan ÖB, Poor HV (2024) Data-agnostic model poisoning against federated learning: a graph autoencoder approach. IEEE Trans Inf Forensics Secur 19:3465–3480
22. Li W, Lv T, Zhao X, Yuan X, Ni W (2025) Free privacy protection for wireless federated learning: enjoy it or suffer from it? IEEE Trans Info Forensics Secur
23. Liu L, Wang Y, Liu G, Peng K, Wang C (2022) Membership inference attacks against machine learning models via prediction sensitivity. IEEE Trans Dependable Secure Comput 20(3):2341–2347
24. Liu Y, Wen R, He X, Salem A, Zhang Z, Backes M, Cristofaro ED, Fritz M, Zhang Y (2022) ML-Doctor: Holistic risk assessment of inference attacks against machine learning models. In: 31st USENIX Security symposium (USENIX Security 22), pp 4525–4542
25. Luo X, Wu Y, Xiao X, Ooi BC (2021) Feature inference attack on model predictions in vertical federated learning. In: 2021 IEEE 37th international conference on data engineering (ICDE). IEEE, pp 181–192
26. Makhdoom I, Abolhasan M, Lipman J, Liu RP, Ni W (2019) Anatomy of threats to the internet of things. IEEE Commun Surv Tutorials 21(2):1636–1675. https://doi.org/10.1109/COMST.2018.2874978
27. Nguyen TD, Nguyen T et al (2024) Backdoor attacks and defenses in federated learning: survey, challenges and future research directions. Eng Appl Artif Intell **127**(Part A), 107166
28. Nuding F, Mayer R (2022) Data poisoning in sequential and parallel federated learning. In: Proceedings of IWSPA@CODASPY, pp 24–34
29. Qayyum A, Ahmad K, Ahsan MA, Al-Fuqaha A, Qadir J (2022) Collaborative federated learning for healthcare: multi-modal covid-19 diagnosis at the edge. IEEE Open J Comput Soc 3:172–184
30. Rieger P, Nguyen TD, Miettinen M, Sadeghi A (2022) DeepSight: mitigating backdoor attacks in federated learning through deep model inspection. In: Proceedings of NDSS (2022)
31. Samarakoon S, Bennis M, Saad W, Debbah M (2019) Distributed federated learning for ultra-reliable low-latency vehicular communications. IEEE Trans Commun 68(2):1146–1159
32. Shi J, Wan W, Hu S, Lu J, Yu Zhang L (2022) Challenges and approaches for mitigating Byzantine attacks in federated learning. In: 2022 IEEE international conference on trust, security and privacy in computing and communications (TrustCom), pp 139–146. https://doi.org/10.1109/TrustCom56396.2022.00030
33. Sun P, Liu E, Ni W, Yu K, Qu X, Wang R, Bi Y, Zhang C, Jamalipour A (2025) Dual-segment clustering strategy for hierarchical federated learning in heterogeneous wireless environments. IEEE Wirel Commun Lett 14(6):1777–1781. https://doi.org/10.1109/LWC.2025.3556462
34. Tian Z, Cui L, Liang J, Yu S (2023) A comprehensive survey on poisoning attacks and countermeasures in machine learning. ACM Comput Surv 55(8):166:1–166:35
35. Wan Y, Qu Y, Ni W, Xiang Y, Gao L, Hossain E (2024) Data and model poisoning backdoor attacks on wireless federated learning, and the defense mechanisms: a comprehensive survey. IEEE Commun Surv Tutorials 26(3):1861–1897. https://doi.org/10.1109/COMST.2024.3361451
36. Wang H, Sreenivasan K, Rajput S et al (2020) Attack of the tails: yes, you really can backdoor federated learning. In: Proceedings of NeurIPS

37. Wang Y, Sun T, Li S, Yuan X, Ni W, Hossain E, Vincent Poor H (2023) Adversarial attacks and defenses in machine learning-empowered communication systems and networks: a contemporary survey. IEEE Commun Surv Tutorials 25(4):2245–2298. https://doi.org/10.1109/COMST.2023.3319492
38. Wang Z, Huang Y, Song M, Wu L, Xue F, Ren K (2023) Poisoning-assisted property inference attack against federated learning. IEEE Trans Dependable Secur Comput 20(4):3328–3340
39. Wu Y, Chen H, Wang X, Liu C, Nguyen P, Yesha Y (2021) Tolerating adversarial attacks and Byzantine faults in distributed machine learning. In: 2021 IEEE international conference on big data (Big Data), pp 3380–3389. https://doi.org/10.1109/BigData52589.2021.9671583
40. Ye J, Maddi A, Murakonda SK, Bindschaedler V, Shokri R (2022) Enhanced membership inference attacks against machine learning models. In: Proceedings of the 2022 ACM SIGSAC conference on computer and communications security, pp 3093–3106
41. Zhao X, Cui Q, Du Z, Ni W, Li W, Yu X, Zhang J, Tao X, Zhang P (2025) Enhancing convergence, privacy and fairness for wireless personalized federated learning: quantization-assisted min-max fair scheduling. IEEE Trans Mobile Comput
42. Zheng J, Li K, Tovar E, Guizani M (2021) Federated learning for energy-balanced client selection in mobile edge computing. In: 2021 international wireless communications and mobile computing (IWCMC). IEEE, pp 1942–1947

Chapter 3
Resilience and Security of Graph-Based Federated Learning

This chapter explores the role of graph-based methods in strengthening resilience and security within FL systems. Given the highly distributed and heterogeneous nature of FL, graph representations provide a powerful tool to model relationships among clients, data distributions, and model updates [10]. By capturing structural dependencies through graph neural networks (GNNs), VGAEs, and attention mechanisms, adversarial behaviors such as poisoning or inference attacks can be more effectively detected and mitigated. This chapter examines how graph-based modeling enhances the robustness of aggregation, supports anomaly detection, and facilitates secure knowledge transfer in dynamic and resource-constrained environments. Special attention is given to applications in IoT, edge computing, UAV networks [5, 9], and metaverse ecosystems [7], where client interactions are naturally graph-structured. Through this lens, graph-driven techniques are positioned as a promising foundation for advancing trustworthy and resilient FL architectures.

3.1 Graph Representations in Federated Learning Systems

In FL, users train local models on their private data and then share the resulting parameters either with neighboring peers in DFL or with a central edge server in conventional FL. These updates are subsequently aggregated to synchronize models across the network. This paradigm supports scalability, robustness, and trust, making it highly applicable to metaverse environments. Although FL mitigates privacy risks by ensuring data remains local, its distributed structure also exposes it to serious vulnerabilities, particularly MP.

A malicious participant can intentionally tamper with local parameters and disseminate corrupted updates to nearby nodes, thereby compromising model integrity and weakening system resilience. Adversaries may embed adversarial patterns or inject noise into their updates, enabling them to circumvent standard poison-

K. Li et al., *Security and Resilience in Distributed Machine Learning*, Springer Series in Reliability Engineering, https://doi.org/10.1007/978-3-032-23959-4_3

ing defenses and impair neighboring models. To counteract such threats, existing defenses rely on distance-based and ML-based filtering techniques, such as Euclidean distance, cosine similarity, support vector machines (SVM), and K-means clustering, to detect suspicious updates before aggregation. However, these methods face notable limitations. First, excessive filtering and the computational burden of analyzing high-dimensional updates can reduce efficiency and degrade model quality. Second, adversaries can imitate benign updates through eavesdropping, allowing them to bypass defenses, such as Krum [1], Trimmed-mean [17], and Median [17]. In addition, ML-based defenses are sensitive to parameter choices, where suboptimal tuning can severely limit effectiveness. Recent poisoning strategies have further demonstrated the ability to evade conventional distance-based defenses altogether.

Recent studies have placed growing emphasis on designing graph representation-based defenses against advanced MP threats in FL. Zheng et al. [19] proposed FedCAMAE, a defense framework that employs visual explanation techniques to move beyond traditional Euclidean distance—or ML-based detection schemes. The method combines Layer Class Activation Mapping (LayerCAM) with an autoencoder (AE) to generate detailed heat maps for each client update submitted to the server. These heat maps serve as fine-grained visual signatures, which the AE further enhances by revealing hidden patterns, thereby improving the system's ability to distinguish between benign and adversarial contributions.

Building on visual explanation-driven defenses, Zheng et al. [18] introduced another promising framework, illustrated in Fig. 3.1, which integrates Gradient-weighted Class Activation Mapping (GradCAM) with AE to enhance resilience against MP. By examining abnormal heat maps produced via GradCAM, the approach improves detection precision and strengthens FL security, enabling more effective identification and isolation of adversarial updates.

On the client side, each of the benign devices employs a DNN designed for image classification. The DNN extracts salient features from an input image (e.g., a bird) and maps them to the correct label. A typical DNN architecture consists of multiple layers, each serving a distinct purpose.

1. The initial stage is a convolutional layer, which applies filters to the input image to capture intrinsic patterns such as edges, corners, and textures. The resulting feature maps encode different structural aspects of the image.
2. Next, a pooling layer reduces the dimensionality of these feature maps while preserving critical information, for instance, distinctive traits like the bird's beak or feather patterns. Common pooling strategies include max pooling and average pooling, though some architectures deviate from this convention. Notably, networks, such as SqueezeNet [14], ResNet [3], DenseNet [15], and MobileNet [4] employ alternative mechanisms in place of standard pooling.
3. After several convolution and pooling stages, the aggregated features are passed through one or more fully connected layers, which ultimately perform the classification task.

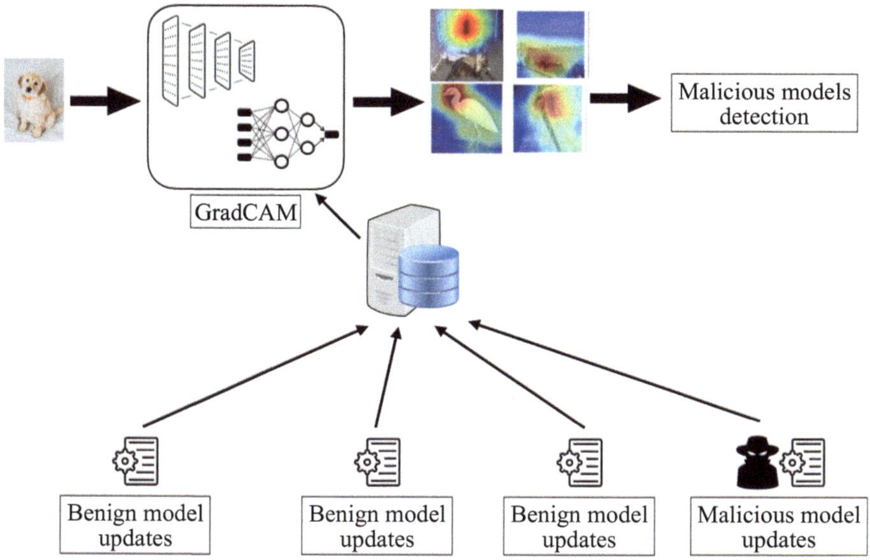

Fig. 3.1 Visual explanation-driven defenses which integrate GradCAM with AE

After receiving local updates from participating devices, the server aggregates them, though benign contributions may be mixed with malicious ones. To address this risk, a defense mechanism capable of identifying and filtering adversarial devices was developed. The design leverages GradCAM [12], which offers higher resolution and stronger class-discriminative capabilities compared with earlier visualization methods such as Class Activation Mapping (CAM) [20]. In this approach, the server selects a random image from the global test dataset, covering all categories represented in the devices' data, and forwards it through the convolutional layers, where the weights and biases are substituted with those from each submitted update. The resulting feature maps are then passed into a fully connected layer for classification, producing visualization patterns that help differentiate benign from malicious model updates.

3.2 Strengths and Motivations

The server can leverage a GNN constructed over an inter-client graph. The underlying assumption is that neighboring clients are likely to produce similar feature embeddings or local models. As in FL, the server begins by collecting model parameters from participating clients. These parameters are then represented as node features within the inter-client graph, on which a GNN is trained to guide the aggregation process. Once updated, the refined parameters are distributed back to the clients.

The inter-client graph may either be predefined or dynamically derived using a self-attention mechanism during training [16]. A key challenge, however, lies in the bilevel optimization problem-coordinating the simultaneous training of local client models and the server-side GNN.

3.2.1 Heterogeneity of Data Distribution and Model

Based on the underlying graph sources, clients can train GNN models to tackle two major challenges in FL: (a) data distribution heterogeneity and (b) model heterogeneity. In the first case, the implicit graph represents inter-client relationships, capturing dependencies among participants. In the second, the graph reflects intra-model structure, such as the connectivity between layers of a neural network. The overall paradigm still follows the standard FL workflow, where clients collaboratively train GNN models, upload their weights to the server, and receive aggregated updates for subsequent training rounds. Therefore, constructing different types of graphs within the clients enables addressing distinct issues, making graph design a critical component in enhancing FL performance and robustness.

Clients can go beyond FL by jointly training their local models and a GNN constructed on a global graph, enabling them to capture knowledge from other participants and mitigate data distribution heterogeneity. FedCG [2] implements this idea by forming a fully connected graph derived from the similarity of client model weights or feature patterns. Each client leverages this graph to train a GNN, producing a global embedding, which is then integrated with its local embedding through a trainable weighting mechanism. Conventional FL struggles in scenarios where client models differ in architecture. To address this, HAFL-GNN-based graph hypernetwork (HAFL-GHN) [11] employs GNNs as a solution. In this approach, each client's neural architecture is represented as a graph, where vertices correspond to parametric layers and edges denote computational flows between them. Node features are initialized using categorical one-hot vectors that specify the layer type. A GHN processes this graph representation and is trained to minimize the joint empirical risk across clients. The resulting latent node embeddings are then mapped back to layer weights for training the original network. By transforming neural networks into graph structures and learning through GHNs, heterogeneous client models can be indirectly aligned, as clients upload GHN weights to the server for aggregation instead of sharing raw model weights.

Training a GNN model on the FL server offers greater flexibility in aggregation, helping to mitigate non-IID challenges across clients by balancing personalization at the client level with generalization at the global level. However, this dual-objective optimization, namely, one for client models and another for the server-side GNN, makes convergence guarantees more difficult. Moreover, in large-scale deployments, maintaining and training a global inter-client graph incurs significant computational overhead, and inaccurate graph structures can further degrade performance. Alternatively, storing inter-client graphs locally and training GNNs on the client side can also alleviate non-IID issues, though this raises risks of exposing other clients' private information and still suffers from potential inaccuracies in graph construction. A different paradigm is serverless, graph-structured FL, where GNNs are designed for peer-to-peer settings in which clients communicate only with their neighbors. This setup enables personalized local models based on neighborhood interactions, yet introduces new challenges: communication costs grow due to direct model exchange, clients with high centrality become attractive attack targets, and the approach lacks continual learning, as models have to be retrained whenever new neighbors join the network.

3.2.2 Enhancing Resilience and Security of FL

Graph-based FL has emerged as a powerful paradigm for enhancing resilience and security, particularly in distributed and heterogeneous environments such as the IoT, mobile edge computing [6], and metaverse ecosystems [8]. One of its principal strengths lies in the expressive ability of graph representations to capture structural dependencies among clients, datasets, and model updates.

By modeling inter-client relationships through similarity graphs or attention-based mechanisms, GNNs can facilitate more robust aggregation that accounts for correlations and outliers. This enables the detection of anomalous patterns in local updates, improving defenses against poisoning, backdoor, or Byzantine failures. For instance, inter-client graphs can encode statistical similarities among local models, allowing the server to identify malicious updates that deviate significantly from benign neighborhoods.

Moreover, graph-based techniques provide a natural solution to personalization, striking a balance between global generalization and local adaptation. Clients connected in similar neighborhoods can exchange knowledge without being forced into a one-size-fits-all global model, which is especially valuable under non-IID data distributions. Graph hypernetworks extend this strength by enabling heterogeneous architectures across clients, mapping structural differences into a shared embedding space that facilitates indirect parameter alignment. In security-sensitive applications like healthcare, finance, and autonomous driving, this capacity to incorporate structural information translates to improved trustworthiness, fairness, and adaptability.

Another strength is scalability: while traditional defenses rely on distance metrics or clustering, GNN-based frameworks can efficiently process high-dimensional updates, leveraging message-passing mechanisms to aggregate information across the graph without requiring raw data access. This makes graph-based FL particularly suited to privacy-preserving contexts, where sensitive data has to remain local. In addition, the use of visual or feature-oriented graph explanations, such as GradCAM or LayerCAM heat maps embedded into graph structures, has enhanced transparency by enabling explainable anomaly detection. These capabilities collectively show that graph-based FL not only strengthens resilience against a wide spectrum of adversarial threats but also advances interpretability and trust in collaborative intelligence systems.

3.3 Challenges of Graph-Based FL

Despite these advantages, several challenges hinder the practical deployment of graph-based FL in adversarial environments. A central difficulty lies in the construction of reliable inter-client graphs. Graphs are often built using similarity measures between client updates or data distributions, yet in large-scale networks, these measures may be noisy, biased, or adversarially manipulated. An imprecise or poisoned inter-client graph can amplify vulnerabilities, allowing malicious users to mislead aggregation or disguise adversarial updates within trusted neighborhoods. This raises the paradox of resilience: while graphs can strengthen aggregation, they also introduce new attack surfaces. The bilevel optimization problem of training both local models and a server-side GNN further complicates convergence guarantees. Balancing client personalization with global generalization creates conflicting objectives that slow training and may destabilize learning under adversarial pressure.

3.3.1 Overheads and Parameter Sensitivity

Computational and communication overheads also pose challenges. Large-scale graphs with thousands of clients require extensive resources to store and process, and message passing across dense connections can overwhelm servers or edge nodes. In decentralized or serverless FL, where clients communicate directly with neighbors, communication costs rise significantly, and highly connected clients become hotspots vulnerable to targeted adversarial exploitation. Moreover, resilience in graph-based FL is not just a matter of filtering malicious updates; sophisticated attackers may exploit graph dynamics by injecting fraudulent nodes or manipulating edge weights to misrepresent relationships. This makes traditional defenses like Krum, Median, or Trimmed-mean ineffective when adversaries adapt to graph structures. Another challenge is privacy leakage: inter-client graphs themselves may inadvertently reveal sensitive information about client data distributions, particularly when similarity measures are derived from embeddings correlated with private data. Attackers could exploit these representations for inference or reconstruction attacks, undermining privacy guarantees.

ML-based defenses within graph contexts also inherit parameter sensitivity problems; poorly tuned hyperparameters in graph models may result in false positives that filter out benign clients, degrading overall performance. Finally, the non-stationarity of real-world environments adds another dimension: in dynamic networks, such as IoT or vehicular edge systems [13], clients frequently join or leave, and inter-client graphs need to be continuously updated. This raises concerns about continual learning, as retraining GNNs or graph hypernetworks from scratch is impractical and costly, while incremental updates risk propagating adversarial noise over time.

3.3.2 Potential Exploration of Graph-Based FL

The strengths and challenges of graph-based FL reveal its promise and complexity as a pathway toward resilient and secure collaborative learning. Its ability to exploit structural dependencies, personalize models, and enhance interpretability positions it as a superior alternative to purely statistical aggregation. Yet, realizing this potential requires addressing critical limitations. Graph-based FL can prioritize robust graph construction methods that are resistant to poisoning, for example, by integrating cryptographic assurances, consensus-based edge validation, or anomaly-aware graph learning.

Lightweight graph compression or hierarchical partitioning may help scale training to massive networks while reducing overhead. Privacy-preserving graph embedding techniques, such as DP applied to similarity metrics or secure multi-party computation for edge formation, are crucial to prevent leakage of sensitive information through graph structures. To strengthen resilience, defenses should move beyond static filtering to dynamic, adaptive strategies that can evolve with attacker behaviors. For instance, incorporating temporal graph models or attention mechanisms may allow systems to detect gradual poisoning trends over time rather than relying solely on snapshot detection.

Another frontier is explainability: while visual explanation methods like Grad-CAM or LayerCAM have improved interpretability, integrating them more deeply with graph models could provide richer, multi-modal insights into why updates are flagged as malicious, fostering user trust. From a system-level perspective, integrating graph-based FL with zero-trust architectures offers a promising direction, ensuring that no client, edge node, or server is implicitly trusted, and all communications and updates are continuously verified. However, implementing zero-trust principles in graph contexts requires innovations in secure authentication, real-time monitoring, and adversarially robust peer-to-peer protocols.

Lastly, resilience in graph-based FL should be benchmarked holistically: not only by measuring robustness to attacks, but also by evaluating scalability, fairness across clients, and the balance between privacy and utility. This requires standardized datasets, metrics, and evaluation frameworks tailored to graph-based settings. In conclusion, graph-driven techniques provide a transformative lens for advancing resilience and security in FL, but their long-term success depends on overcoming challenges in graph reliability, scalability, privacy preservation, and adaptive defense. Addressing these issues can motivate their full potential in powering trustworthy AI systems across IoT, healthcare, autonomous mobility, and immersive metaverse applications.

References

1. Blanchard P, El Mhamdi EM, Guerraoui R, Stainer J (2017) Machine learning with adversaries: Byzantine tolerant gradient descent. Adv Neural Inf Process Syst 30
2. Caldarola D, Mancini M, Galasso F, Ciccone M, Rodolà E, Caputo B (2021) Cluster-driven graph federated learning over multiple domains. In: Proceedings of the IEEE/CVF conference on computer vision and pattern recognition, pp 2749–2758
3. He F, Liu T, Tao D (2020) Why resnet works? residuals generalize. IEEE Trans Neural Netw Learn Syst 31(12):5349–5362
4. Howard AG, Zhu M, Chen B, Kalenichenko D, Wang W, Weyand T, Andreetto M, Adam H (2017) Mobilenets: efficient convolutional neural networks for mobile vision applications. arXiv:1704.04861
5. Kurunathan H, Huang H, Li K, Ni W, Hossain E (2023) Machine learning-aided operations and communications of unmanned aerial vehicles: a contemporary survey. IEEE Commun Surv Tutorials 26(1):496–533
6. Li K, Cui Y, Li W, Lv T, Yuan X, Li S, Ni W, Simsek M, Dressler F (2022) When internet of things meets metaverse: convergence of physical and cyber worlds. IEEE Internet Things J 10(5):4148–4173
7. Li K, Lau BPL, Yuan X, Ni W, Guizani M, Yuen C (2023) Toward ubiquitous semantic metaverse: challenges, approaches, and opportunities. IEEE Internet Things J 10(24):21855–21872
8. Li K, Li C, Yuan X, Li S, Zou S, Ahmed SS, Ni W, Niyato D, Jamalipour A, Dressler F et al (2025) Zero-trust foundation models: a new paradigm for secure and collaborative artificial intelligence for internet of things. IEEE Internet Things J
9. Li K, Ni W, Yuan X, Noor A, Jamalipour A (2022) Deep-graph-based reinforcement learning for joint cruise control and task offloading for aerial edge internet of things (edgeiot). IEEE Internet Things J 9(21):21676–21686
10. Li R, Yuan X, Radfar M, Marendy P, Ni W, O'Brien TJ, Casillas-Espinosa PM (2023) Graph signal processing, graph neural network and graph learning on biological data: a systematic review. IEEE Rev Biomed Eng 16:109–135. https://doi.org/10.1109/RBME.2021.3122522
11. Litany O, Maron H, Acuna D, Kautz J, Chechik G, Fidler S (2022) Federated learning with heterogeneous architectures using graph hypernetworks. arXiv:2201.08459
12. Selvaraju RR, Cogswell M, Das A, Vedantam R, Parikh D, Batra D (2017) Grad-cam: visual explanations from deep networks via gradient-based localization. In: Proceedings of the IEEE international conference on computer vision, pp 618–626
13. Tian C, Huang C, Wang Y, Chung E, Nguyen AT, Wong PK, Ni W, Jamalipour A, Li K, Huang H (2024) Recent estimation techniques of vehicle-road-pedestrian states for traffic safety: comprehensive review and future perspectives. IEEE Trans Intell Transp Syst
14. Tsivgoulis M, Papastergiou T, Megalooikonomou V (2022) An improved squeezenet model for the diagnosis of lung cancer in ct scans. Mach Learn Appl 10:100399
15. Wang SH, Zhang YD (2020) Densenet-201-based deep neural network with composite learning factor and precomputation for multiple sclerosis classification. ACM Trans Multimedia Comput Commun Appl (TOMM) 16(2s):1–19
16. Wu Z, Wu X, Long Y (2022) Multi-level federated graph learning and self-attention based personalized wi-fi indoor fingerprint localization. IEEE Commun Lett 26(8):1794–1798
17. Yin D, Chen Y, Kannan R, Bartlett P (2018) Byzantine-robust distributed learning: Towards optimal statistical rates. In: Proceedings of the 35th international conference on machine learning, proceedings of machine learning research, vol 80. PMLR, pp 5650–5659
18. Zheng J, Li K, Yuan X, Ni W, Tovar E (2024) Detecting poisoning attacks on federated learning using gradient-weighted class activation mapping. In: Companion proceedings of the ACM on web conference (WWW), pp 714–717

19. Zheng J, Li K, Yuan X, Ni W, Tovar E, Crowcroft J (2024) Exploring visual explanations for defending federated learning against poisoning attacks. In: Proceedings of the 30th annual international conference on mobile computing and networking, pp 1596–1598
20. Zhou B, Khosla A, Lapedriza A, Oliva A, Torralba A (2016) Learning deep features for discriminative localization. In: Proceedings of the IEEE conference on computer vision and pattern recognition, pp 2921–2929

Part II
Adversarial Attacks on Federated Learning

Chapter 4
Model Poisoning via Variational Graph Representations

FL has attracted significant attention recently, and emerged as a distributed deep learning paradigm. With FL, each user device trains its local model with its private data to generate local updates sent to the edge server without sharing the device's private data. The edge server then aggregates the local updates to train a global model, which is sent back to the user devices for the next round of FL training. Based on FL, individual data privacy is protected as no private data is shared [28].

Despite the fact that FL can protect user device's data privacy, it is still vulnerable to cyber-epidemic attacks in which attackers (i.e., malicious user devices) launch model or data poisoning attacks to manipulate the FL process and spread to other benign user devices, leading to training failure and compromised learning accuracy [19]. For the MP attacks, the attacker aims to manipulate the hyperparameters of the benign local model [29, 31]. In contrast, data poisoning attacks involve manipulating the training dataset of benign user devices. To launch effective model or data poisoning attacks [14], the attackers need to access the knowledge of the dataset used for FL training, which helps to minimize the detectability of malicious local models. FL could be manipulated if an attacker launches MP attacks based solely on the benign local and global models overheard without access to the data. Nevertheless, it is challenging for the attacker to achieve effectiveness and undetectability without knowledge of the data. This type of attack is new, has not yet been discussed in the existing literature, and requires further research to develop effective detection and prevention methods. This new attack underscores the importance of securing FL from local and global training threats.

K. Li et al., *Security and Resilience in Distributed Machine Learning*, Springer Series in Reliability Engineering, https://doi.org/10.1007/978-3-032-23959-4_4

4.1 Introduction of Adversarial VGAE-Based Model Poisoning

This chapter investigates a new adversarial VGAE-based MP (VGAE-MP) attack on FL. VGAE-MP is a new data-untethered cyber-epidemic attack, which extends the VGAE to create malicious local models based solely on the benign local models overheard without any access to the training data of FL. Specifically, the adversarial VGAE generates malicious local models by capturing the correlation of multiple data features in the benign local models. The VGAE-MP attack aims to maximize the FL training loss by regenerating the graph structure and using it to create manipulative, malicious local models that can effectively compromise the benign global and local models. The VGAE-MP attack first decouples the graph structure from the underlying data features of the local models, and then regenerates the graph using the VGAE. The regenerated graph structure is applied to the benign local models' data features to produce malicious local models that remain compatible with benign ones while effectively compromising the global model. As a result, the VGAE-MP attack is highly effective on FL while remaining reasonably undetectable on the server.

Since the user devices possessing large datasets could improve the learning accuracy of FL, the server selects a portion of the collected local models for the global aggregation. Likewise, the VGAE-MP, as a white-box attack, also selects the benign local models in the traini1 0ng of the VGAE. For example, the user device selection at the attacker ensures that the selected local models have sufficient data features for retrieving the correlation in the VGAE, while the generated malicious local model is within proximity to the global model in Euclidean distance.

The key contributions of the cyber-epidemic attack are summarized as follows:

- A new data-untethered MP attack, i.e., VGAE-MP, is proposed to manipulate the correlations of multiple data features in the selected benign local models and maintain the genuine data features substantiating the benign local models;
- A new adversarial VGAE, which is trained together with sub-gradient descent to regenerate the correlations of the local models manipulatively while keeping the malicious local models undetectable.
- The VGAE-MP attack is implemented in PyTorch, showing experimentally that VGAE-MP gradually reduces the accuracy of FL and bypasses the detection of existing poisoning defense mechanisms. This attack can propagate across all benign user devices, leading to an epidemic infection. The source code of the VGAE-MP attack is released on GitHub.

The remainder of this chapter is structured as follows. Section 4.2 introduces the background of adversarial attacks against wireless systems and FL. Section 4.3 investigates the FL system model with malicious agents. The proposed VGAE-MP attack is described in Sect. 4.4. Section 4.5 discusses the performance analysis. Table 4.1 lists the notation used in the paper.

Table 4.1 Notation of variables

Symbol	Description
$\beta'_{i,j}(t)$	Binary indicator for selecting benign local model weights
D	Total dataset of all I benign devices
$D_i(t)$	Dataset of benign device i at the tth communication round
$D'(t)$	Claimed data size of the attacker
I	Number of benign devices
J	Number of attackers
M	Total number of model parameters in $\boldsymbol{w}_i$
η_{loss}	Reconstruction loss of the VGAE decoder
$\tau_i(t)$	Training delay of $\boldsymbol{w}_i(t)$ at device i
$\lambda,\ \rho$	Lagrange dual variables
$\boldsymbol{\mathcal{A}}$	Adjacency matrix formulated by attacker
$\widehat{\boldsymbol{\mathcal{A}}}$	Reconstructed adjacency matrix at attacker
$\boldsymbol{\mathcal{F}}$	Feature matrix in the attacker's VGAE
$\widehat{\boldsymbol{\mathcal{F}}}$	Reconstructed feature matrix at attacker
$\boldsymbol{\mathcal{L}}$	Laplacian matrix derived from $\boldsymbol{\mathcal{A}}$
$\boldsymbol{\mathcal{L}}_k$	Rank-k SVD approximation of $\boldsymbol{\mathcal{L}}$
$\boldsymbol{w}_i(t)$	Local model weight parameters of benign device i
$\boldsymbol{w}_i^m(t)$	mth feature in $\boldsymbol{w}_i$
$\boldsymbol{w}'_j(t)$	Training parameters of the malicious model at attacker j
$\boldsymbol{w}'_G(t)$	Global FL model under attack

4.2 Landscape of Adversarial Attacks and Security Threats

This section reviews the existing adversarial attacks and security threats to FL, e.g., MP, data poisoning, inference, and backdoor attacks.

4.2.1 *Security Threats to FL*

The periodic model updates in FL bear a discriminative ability that reflects changes in data distribution, including sensitive properties, making it possible for an attacker to infer unintended information. In [33], the authors introduce a poisoning-assisted property inference attack, which injects malicious data into the training dataset to infer a targeted property of the FL model. The attack modifies the training data labels, thereby distorting the decision boundary of the shared global model in FL, resulting in the disclosure of sensitive property information by benign user devices. In [10], the authors present that the attacker can infer the presence or absence of a particular category in the data by carefully crafting a malicious training dataset, despite the secure aggregation methods. A category inference attack is developed, which iteratively generates malicious training data and utilizes them to update the global model in FL. The vulnerability of FL to label inference attacks is presented in [9], where a malicious user device can infer the private labels of other benign devices. With the observed aggregate model updates, three label inference attacks have been developed to infer private labels with the benign devices, including direct, passive, and active label inference attacks.

In [11], a coordinated backdoor attack on FL is designed using model-dependent triggers, where an attacker can inject a backdoor trigger into a target model and then train the models in FL to perform coordinated attacks. The trigger uses the model dependency in FL to activate the backdoor when the target client uses the compromised model. A distributed attacking algorithm is also provided to enable the attackers to select their respective backdoor models for a high attack success rate while maintaining a low impact on the overall training accuracy of FL.

The authors of [22] focus on data poisoning attacks in both sequential and parallel FL settings. These attacks could weaken the performance of trained models by injecting malicious data into the training datasets used in FL. Sequential FL involves user devices training successively, using the output model from the previous device. In contrast, parallel FL involves each user device simultaneously training a local model before sending updates to the server for aggregation. An attacker can later trigger malicious behavior during the prediction phase by modifying specific training inputs using a specific pattern. In [8], it is argued that FL based on weighted averaging and trimmed averaging for mitigating the Byzantine faults is still vulnerable to data poisoning attacks, leading to significant drops in the training accuracy. A data poisoning attack is studied, which targets the FL system designed to be robust against Byzantine attacks. The data poisoning attack can exploit the characteristics of FL and the Byzantine-robust mechanisms to insert malicious data into the system.

In [27], an MP attack is introduced, which accounts for the characteristics of FL, such as variability in the training data and randomness of the training process. The MP attack uses a transfer learning strategy to improve the attack efficiency. An MP attack on FL based on fake user devices added to the system and operating as legitimate devices is presented in [3]. This fake device can manipulate its data to influence the global model and potentially insert a backdoor or degrade the FL performance. In [6], the authors study a perception poisoning attack in which the attacker manipulates the FL model's perception by altering the training data. The attack can be captured by building a poison perception model for measuring a perception poisoning rate.

In [36], the authors focus on a generative poisoning attack against FL, which generates malicious data using generative adversarial networks (GAN) to target user devices in FL. The attack can introduce bias into the aggregated model by injecting poisoned data generated by the GANs. Another GAN-based poisoning attack against FL is presented in [37]. The GANs-based poisoning attack creates a set of malicious samples by generating poisoned data samples to attack the local models of benign user devices. To degrade the training accuracy of FL, the attacker deceives the aggregation process at the server by strategically altering the models of the benign user devices. The GAN-based poisoning attack is evaluated based on image classification datasets.

4.2.2 *Opportunities of the New VGAE-MP Attack*

The existing data or MP attacks against FL lack the description of the implicit relationship between different local models, which can be detected by recent poisoning defense frameworks based on the probabilistic graph model, e.g., [18, 25]. Additionally, convolutional layers at the aggregator can excessively smooth out the output features of the attacks, resulting in distinguishable discrepancies between the malicious local model and the benign ones. In contrast, the proposed VGAE-MP attack is a new attacking method for MP, which is independent of the data. The VGAE-MP attack manipulates the correlations among multiple data features in selected benign local models while preserving the genuine data features that support those models, thus keeping the malicious local models undetectable.

4.3 System Model and Problem Statement

This section presents the training of the local and global models of FL in mobile edge computing (MEC), for image classification as an example. Figure 4.1a presents an FL training process with I benign user devices. Each benign device $i \in [1, I]$ has $D_i(t)$ datasets at the tth communication round of FL. Let $x(d_i)$ denote a data sample captured at the ith benign device, and $y(d_i)$ the local model update trained at the ith benign device, where $d_i \in [1, D_i(t)]$.

The training loss function of a benign device i, denoted by $f(\boldsymbol{w}_i(t); x(d_i), y(d_i))$, measures approximation errors based on the inputs $x(d_i)$ and outputs $y(d_i)$ in the tth communication round, where $\boldsymbol{w}_i(t) \in \mathbb{R}^{1\times M}$ denotes the local model obtained in the communication round. For example, the loss function can be modeled as linear regression, i.e., $f(\boldsymbol{w}_i(t); x(d_i), y(d_i)) = \frac{1}{2}(\boldsymbol{w}_i^T(t)x(d_i) - y(d_i))^2$, or logistic regression, i.e., $f(\boldsymbol{w}_i(t); x(d_i), y(d_i)) = y(d_i)\log\Big(1 + \exp\big(-\boldsymbol{w}_i^T(t)x(d_i)\big)\Big) - (1 - y(d_i))\log\Big(1 - \frac{1}{1+\exp\left(-\boldsymbol{w}_i^T(t)x(d_i)\right)}\Big)$. Here, $(\cdot)^T$ denotes transpose.

Given $D_i(t)$, the local loss function of the FL at device i for the tth communication round is given by

$$F(\boldsymbol{w}_i(t))=\frac{1}{D_i(t)}\sum_{i=1}^{D_i(t)} f\left(\boldsymbol{w}_i(t);x(d_i),y(d_i)\right)+\alpha\zeta\left(\boldsymbol{w}_i(t)\right), \tag{4.1}$$

where $\zeta(\cdot)$ is a regularizer function capturing the effect of local training noise; $\alpha \in [0, 1]$ is a given coefficient [15].

With the learning rate μ, the local model of device i is updated for T_L local iterations throughout the tth communication round by

$$\boldsymbol{w}_i(t) \leftarrow \boldsymbol{w}_i(t) - \mu\nabla F(\boldsymbol{w}_i(t)), \tag{4.2}$$

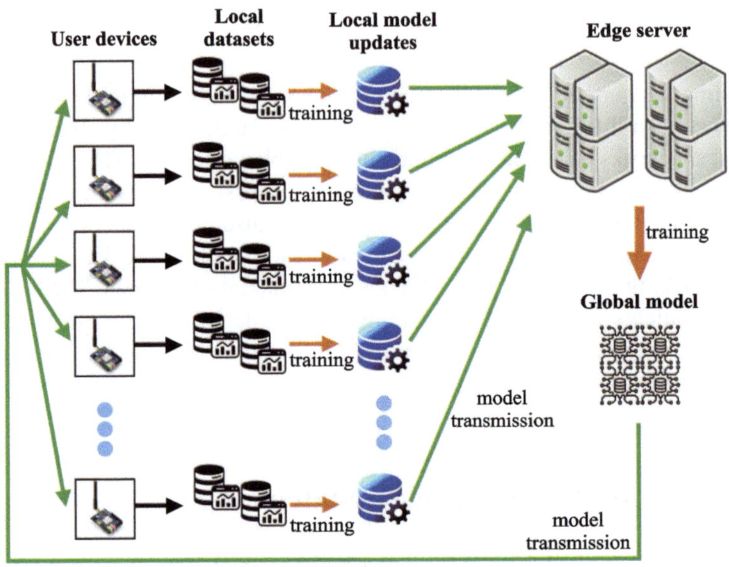

(a) FL with benign user devices

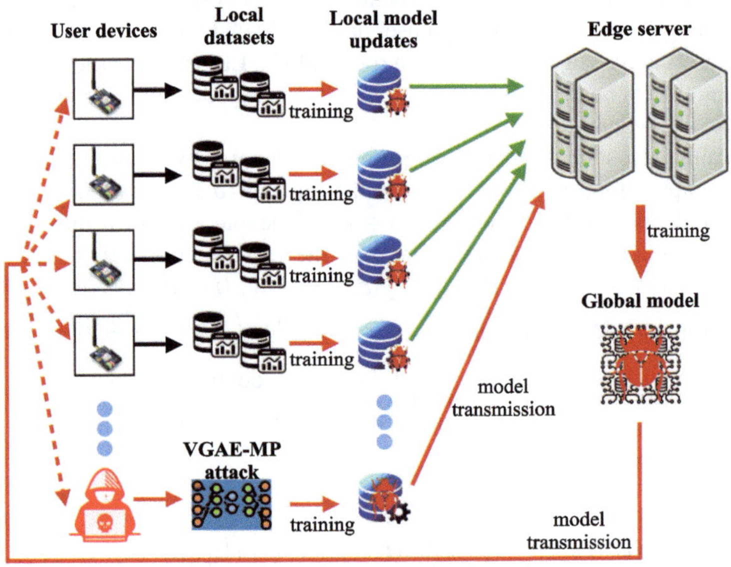

(b) The proposed VGAE-MP attack

Fig. 4.1 **a** Illustration of FL, where a local model update is trained at each benign user device based on its datasets. The edge server aggregates the benign local model updates to train a global model that will be broadcast to the user devices to update the training parameters of their local models. **b** By eavesdropping on the benign local model updates, the attacker performs the proposed VGAE-MP attack to create a malicious poisoning model that is sent to the server. The malicious model deviates from the FL in the opposite direction, thereby falsifying the local model updates of the devices

After the T_L local iterations, all devices upload their local models $\boldsymbol{w}_i(t), \forall i$ to the server. The server aggregates the local models to train a global model denoted by $\boldsymbol{w}_G(t)$ for the tth communication round. Then, $\boldsymbol{w}_G(t)$ is broadcast to all user devices for their training of $\boldsymbol{w}_i(t+1), \ \forall i$ in the $(t+1)$th communication round.

Figure 4.1b shows the FL of the benign user devices under the proposed VGAE-MP attack, where an attacker overhears $\boldsymbol{w}_i(t)$ uploaded by the benign devices. The attacker, who may appear as a legitimate device, can progressively contaminate the global model represented by $\boldsymbol{w}_G(t)$ and the local models of the benign users, i.e., $\boldsymbol{w}_i(t), \ \forall i \in [1, I]$, by creating and uploading malicious local models during each communication round t. The malicious local model at the attacker's device j is represented by $\boldsymbol{w}'_j(t)$. It is constructed based on the parameters of the benign local models overheard by the attacker during each communication round t. The server aggregates the local models of the user devices, including both benign and malicious models, without realizing the attacker's presence. This creates a contaminated global model, $\boldsymbol{\omega}'_G(t)$. The total size of the local training data reported to the server, $D(t)$, is calculated as the sum of the data size of all devices, $D_i(t)$, and the claimed data size of the attacker, $D'(t)$.

Some FL systems allow the server to select a portion of the collected $\boldsymbol{w}_i(t)$ to train $\boldsymbol{w}_G(t)$. For example, the authors of [38] considered a selection scheme in which the total bandwidth of the selected devices needs to be smaller than the bandwidth capacity. We define a binary indicator $\beta'_{i,j}(t)$ at the attacker. If $\boldsymbol{w}_i(t)$ is selected by the attacker to train its adversarial and contaminating local model, then $\beta'_{i,j}(t) = 1$; otherwise, $\beta'_{i,j}(t) = 0$. Thus, the contaminated global model can be written as

$$\boldsymbol{w}'_G(t) = \sum_{i=1}^{I} \frac{D_i(t)}{D(t)} \beta'_{i,j}(t) \boldsymbol{w}_i(t) + \frac{D'(t)}{D(t)} \boldsymbol{w}'_j(t), \tag{4.3}$$

where $\boldsymbol{w}'_j(t)$ is the weight parameter of the malicious model trained at attacker j. Then, the server broadcasts $\boldsymbol{w}'_G(t)$ to all I devices.

The FL trains the global model based on the local datasets of all user devices, including the non-existent dataset claimed by the attacker, by minimizing the global loss function:

$$\begin{aligned} \min_{\boldsymbol{w}'_G(t)} F(\boldsymbol{w}'_G(t)) = & \sum_{i=1}^{I} \frac{D_i(t)}{D(t)} \beta'_{i,j}(t) F_i(\boldsymbol{w}_i(t)) + \\ & \frac{D'(t)}{D(t)} F'_j(\boldsymbol{w}'_j(t)), \end{aligned} \tag{4.4}$$

where the attacker's claimed local loss function, represented by $F'_j(\cdot)$, is in accordance with (4.1).

The optimization of the VGAE-MP attack aims to construct an optimal $\boldsymbol{w}'_j(t)$ based on the overheard $\boldsymbol{w}_i(t)$ to maximize $F(\boldsymbol{w}'_G(t))$ in (4.4), while maintaining a reasonably small Euclidean distance between $\boldsymbol{w}'_j$ and $\boldsymbol{w}'_G$. This ensures that $\boldsymbol{w}'_j(t)$

remains undetectable by the server, which usually evaluates the similarities among all local models and eliminates those that are significantly different, for example, using algorithms such as Krum or multi-Krum [1]. Consequently, $\boldsymbol{w}'_G(t)$ deviates the most in the opposite direction that the benign global model would change in the absence of the attack.

The optimization of VGAE-MP launched by the attacker j, $\forall j \in [1, J]$, in the communication round t is formulated as

$$\max_{\boldsymbol{w}'_j(t), \beta'_{i,j}(t)} \quad F(\boldsymbol{w}'_G(t)) \tag{4.5a}$$

$$\text{s.t.} \quad d(\boldsymbol{w}'_j(t), \boldsymbol{w}'_G(t)) \leq d_T, \tag{4.5b}$$

$$\sum_{i=1}^{I} \beta'_{i,j}(t) d(\boldsymbol{w}_i(t), \bar{\boldsymbol{w}}(t)) \leq \Upsilon, \tag{4.5c}$$

$$\beta'_{i,j}(t) = \{0, 1\}, \tag{4.5d}$$

where $d(\cdot, \cdot)$ calculates the Euclidean distance between $\boldsymbol{w}'_j$ and $\boldsymbol{w}'_G$, d_T is a threshold of the Euclidean distance, $\bar{\boldsymbol{w}}(t) = \sum_{i=1}^{I} \frac{D_i(t)}{D(t)} \boldsymbol{w}_i(t)$, and Υ is a predefined upper bound of the overall distance between the selected local models and the aggregated model of the local models.

Constraint (4.5b) guarantees that the jth attacker's $\boldsymbol{w}'_j$ is in proximity to the global model in terms of Euclidean distance. Constraint (4.5c) describes that the overall distance between the selected local models and their aggregated model, i.e., $\sum_{i=1}^{I} \beta'_{i,j}(t) d(\boldsymbol{w}_i(t), \bar{\boldsymbol{w}}(t))$, has to be below the upper bound Υ. This is because the FL server may perform local model selection to rule out those dissimilar to the rest, e.g., using Krum or multi-Krum. Constraint (4.5d) defines $\beta'_{i,j}(t)$ as a binary indicator.

4.4 VGAE-Based MP

Due to a lack of correlation between the arbitrary features in $\boldsymbol{w}'_j(t)$ and $\boldsymbol{w}_i(t)$, the malicious local model $\boldsymbol{w}'_j(t)$ could be detected by the server. For example, recent GNN-based FL privacy protection schemes [30, 35] can classify the local model weights based on their features. To tackle this issue, we develop a new adversarial VGAE model in this section to generate $\boldsymbol{w}'_j(t)$ in such a way that the individual feature correlation in $\boldsymbol{w}_i(t)$, $\forall i$ is captured in $\boldsymbol{w}'_j(t)$. As a result, the server can hardly detect $\boldsymbol{w}'_j(t)$. For the brevity of notation, we omit the subscript $_j$ for the attacker in the following discussions.

The optimization of VGAE-MP in (4.5) is a non-convex combinatorial problem intractable for conventional optimization techniques. We decouple the VGAE-MP problem in (4.5) between the model attack and the bandwidth selection using the

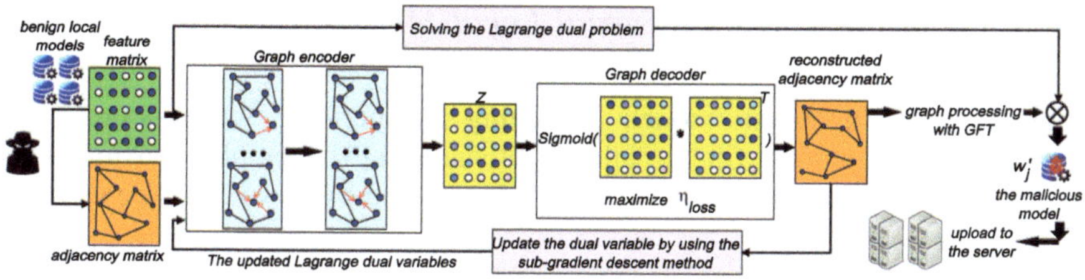

Fig. 4.2 The VGAE-MP attack creates $\boldsymbol{w}'_j(t)$ based on learning the correlation among the parameters of the models being trained in FL, i.e., $\boldsymbol{w}_i(t),\ \forall i$. A graph encoder trains $\mathcal{F}_j$ and $\mathcal{A}_j$ to build a feature representation matrix $\mathcal{Z}$. The output of the encoder is input to the decoder for the reconstruction of $\mathcal{A}_j$. The VGAE-MP attack is designed to adjust $\boldsymbol{w}'_j$ to maximize the reconstruction loss η_{loss}, according to (4.19)

Lagrangian-dual method [23]. A new approach is developed to iteratively optimize the adversarial local models by running a graph autoencoder and subgradient descent, as depicted in Fig. 4.2.

Let λ and ρ denote the dual variables. The Lagrange function of (4.5) is given by

$$\begin{aligned}\mathcal{L}(\beta'_{i,j}(t), \lambda, \rho) &= F(\boldsymbol{w}'_G(t)) + \lambda(d_T - d(\boldsymbol{w}'_j(t), \boldsymbol{w}'_G(t))) \\ &\quad + \rho\left(\Upsilon - \sum_{i=1}^{I} \beta'_{i,j}(t) d(\boldsymbol{w}_i(t), \bar{\boldsymbol{w}}(t))\right).\end{aligned} \tag{4.6}$$

The Lagrange dual function is given by

$$\mathcal{D}(\lambda, \rho) = \max_{\boldsymbol{w}'_j(t), \beta'_{i,j}(t)} \mathcal{L}(\beta'_{i,j}(t), \lambda, \rho). \tag{4.7}$$

The dual problem of the primary problem in (4.5) is given by

$$\min_{\lambda, \rho} \mathcal{D}(\lambda, \rho). \tag{4.8}$$

4.4.1 Client Selection

At communication round t, given $\lambda = \lambda(t)$ and $\rho = \rho(t)$, the primary variable $\beta'_{i,j}(t)$ of the bandwidth selection can be optimized by solving

$$\beta'_{i,j}(t)^* = \arg\min_{\beta'_{i,j}(t)} \left\{ \sum_{i=1}^{I} \beta'_{i,j}(t) d(\boldsymbol{w}_i(t), \bar{\boldsymbol{w}}(t)) \right\}, \text{ s.t. } (4.5d), \tag{4.9}$$

which is a standard 0/1 knapsack problem and solved using dynamic programming.

4.4.2 *Generation of Adversarial Local Models*

A new adversarial VGAE model, leveraging unsupervised learning on graph-structured data according to the variational autoencoder [5], is employed to maximize the Lagrange function in (4.6). Given $\beta'_{i,j}(t)^*$, $\lambda(t)$ and $\rho(t)$, we optimize $\boldsymbol{w}'_j(t)$ by solving

$$\begin{aligned}\boldsymbol{w}'_j(t)^* = \arg\max_{\boldsymbol{w}'_j(t)} \Bigg\{ & F(\boldsymbol{w}'_G(t)) - \lambda(t) d(\boldsymbol{w}'_j(t), \boldsymbol{w}'_G(t))) \\ & + \rho(t) \sum_{i=1}^{I} \beta'_{i,j}(t)^* d(\boldsymbol{w}_i(t), \bar{\boldsymbol{w}}(t)) \Bigg\}. \end{aligned} \tag{4.10}$$

An attacker, i.e., the jth attacker, can observe the local model parameters of the benign devices to establish the intrinsic correlation between the different parameters of the local models. A graph can be used to characterize the correlation. The graph is then regenerated manipulatively with the VGAE, and used to produce the malicious local model $\boldsymbol{w}'_j(t)$. By this means, we can maximize (4.10) while preventing the convergence of $\boldsymbol{w}'_G(t)$. Constraints (4.5b) and (4.5d) are satisfied by designing the decoder of the VGAE to reproduce the correlations. This approach reduces structural dissimilarity between $\boldsymbol{w}_i(t)$ and $\boldsymbol{w}'_j(t)$.

4.4.2.1 Graph Construction and Feature Extraction

As illustrated in Fig. 4.2, the graph represented by $\mathcal{G} = (\mathcal{V}, E, \mathcal{F})$ is utilized to characterize the correlations among the parameters of the models being trained in FL, i.e., $\boldsymbol{w}_i(t)$, $\forall i$. The vertices, edges, and feature matrix of the graph are represented by $\mathcal{V}$, E, and $\mathcal{F}$, respectively. The VGAE comprises a graph convolutional network (GCN) encoder and an inner product decoder. The encoder encodes the graph data using its features, and the decoder takes the encoded output as input to reconstruct the original graph $\mathcal{G} = (\mathcal{V}, E, \mathcal{F})$ [32].

Let $\boldsymbol{\mathcal{F}}(t) = [\boldsymbol{w}_1(t), \cdots, \boldsymbol{w}_I(t)]^T \in \mathbb{R}^{I\times M}$ be the feature matrix containing all I benign local models at communication round t, where M is the dimension of the local model. Let $\boldsymbol{\omega}^m(t) \in \mathbb{R}^{I\times 1}$ be the mth column of $\boldsymbol{\mathcal{F}}(t)$. We use $\delta_{m,m'}(t)$ to denote the *cosine similarity* between the $\boldsymbol{\omega}_m(t)$ and $\boldsymbol{\omega}_{m'}(t)$ at communication round t. $m, m' \in [1, M]$. $\delta_{m,m'}(t)$ is defined as [39]

$$\delta_{m,m'}(t) = \frac{(\boldsymbol{\omega}^m(t))^T \boldsymbol{\omega}^{m'}(t)}{\|\boldsymbol{\omega}^m(t)\| \cdot \|\boldsymbol{\omega}^m(t)\|}. \tag{4.11}$$

The adjacency matrix, denoted by $\boldsymbol{\mathcal{A}}(t) = [\delta_{m,m'}(t)] \in \mathbb{R}^{M\times M}$, is one of the inputs to the encoder of the VGAE model at the attacker. According to $\boldsymbol{\mathcal{A}}(t)$, the topological structure of the graph $\mathcal{G}$ can be constructed at the attacker. The feature matrix $\boldsymbol{\mathcal{F}}(t)$ is the other input to the encoder of the VGAE model at the attacker.

4.4.2.2 Encoder Design of VGAE Model

The encoder in the proposed VGAE maps $\mathcal{G}$ to a lower-dimensional representation. We build the encoder based on the GCN architecture, which learns a latent representation that captures the underlying features of $\mathcal{G}$. The encoded representation is then used as input to the decoder to reconstruct the original graph from the lower-dimensional representation to obtain the malicious local model $\boldsymbol{w}'_j(t)$ in (4.9). For the brevity of notation, we omit the index of communication rounds "t" in the following discussions.

A graph encoder is defined as

$$\mathcal{Z}_1 = f_{\text{relu}}(\mathcal{F}, \mathcal{A}, |\boldsymbol{W}_0); \tag{4.12}$$

$$\mathcal{Z}_2 = f_{\text{linear}}(\mathcal{Z}_1, \mathcal{A}|\boldsymbol{W}_1), \tag{4.13}$$

where $f_{\text{relu}}(\cdot)$ is the Rectified Linear Unit (ReLU) activation function employed for the first layer, while $f_{\text{linear}}(\cdot)$ is the Linear activation function used for the second layer; and $\boldsymbol{W}_l$ is the learnable parameters specific to the lth layer of the neural networks.

Since determining the probability distribution of the latent representation of vertices $\mathcal{Z}$ in $\mathcal{G}$ is difficult and intractable [12], we approximate the true posterior by using a Gaussian distribution $\mathcal{N}(\cdot)$, while the encoder takes $\mathcal{F}$ and $\mathcal{A}$ as its input to an inference model parameterized by a two-layer GCN. Thus, we have

$$q(\mathcal{Z}|\mathcal{A}, \mathcal{F}) = \Pi_{m=1}^{M} q(\boldsymbol{z}_m|\mathcal{A}, \mathcal{F}), \tag{4.14}$$

and

$$q(\boldsymbol{z}_m|\mathcal{A}, \mathcal{F}) = \mathcal{N}(\boldsymbol{z}_m|\boldsymbol{\mu}_m, \text{diag}(\boldsymbol{\sigma}^2)), \tag{4.15}$$

where $\boldsymbol{\mu} = \mathcal{Z}_2$ builds the matrix of mean vectors $\boldsymbol{\mu}_m$. Likewise, we have $\log \boldsymbol{\sigma} = f_{\text{linear}}(\mathcal{Z}_1, \mathcal{A}|\boldsymbol{W}_1)$ that shares the first-layer parameters $\boldsymbol{W}_0$.

With the identity matrix $\mathcal{I} \in \mathbb{R}^{M \times M}$, we define $\widetilde{\mathcal{A}} = \mathcal{A} + \mathcal{I}$ with the (m, m')th element $\widetilde{\mathcal{A}}_{m,m'}$, and the (diagonal) degree matrix $\mathcal{D}$ with the (m, m)th element $\mathcal{D}_{m,m} = \sum_{m'=1}^{M} \widetilde{\mathcal{A}}_{m,m'}$. Each layer of the GCN can be written as

$$f_{\mathcal{G}}(\mathcal{Z}_{l-1}, \mathcal{A}|\mathrm{W}_l) = \phi(\mathcal{D}^{-\frac{1}{2}} \widetilde{\mathcal{A}} \mathcal{D}^{-\frac{1}{2}} \mathcal{Z}_{l-1} \mathrm{W}_l), \tag{4.16}$$

where $\phi(\cdot)$ is the activation function such as relu$(\cdot)$.

4.4.2.3 Decoder Design of VGAE Model

The input to the decoder of the proposed VGAE model is $\mathcal{Z}$, which is the output of the GCN in the encoder. The decoder aims to reconstruct $\mathcal{A}$, denoted by $\widehat{\mathcal{A}}$, predicting whether there is a link between two vertices by an inner product between latent variables, which is designed as

$$p(\widehat{\mathcal{A}}|\mathcal{Z}) = \sum_{m=1}^{M}\sum_{m'=1}^{M} p(\hat{\delta}_{m,m'}|\mathbf{z}_m, \mathbf{z}_{m'}); \tag{4.17}$$

$$p(\hat{\delta}_{m,m'} = 1|\mathbf{z}_m, \mathbf{z}_{m'}) = \text{sigmoid}(\mathbf{z}_m^T \mathbf{z}_{m'}), \tag{4.18}$$

where $\mathbf{z}_m \in \mathbb{R}^{M\times 1}$ is the mth column of $\mathcal{Z}$, and sigmoid$(\cdot)$ is the logistic sigmoid function, i.e., sigmoid$(x) = 1/(1+\exp^{-x})$. Here, the larger the inner product $(\mathbf{z}_m^T \mathbf{z}_{m'})$ in the embedding, the more likely vertexes m and m' are connected in the graph, according to $\widehat{\mathcal{A}} = [\hat{\delta}_{m,m'}] \in \mathbb{R}^{M\times M}$ in the AE [24].

We can view (4.17) as the inverse operation of the encoder for constructing a reconstructed adjacency matrix $\widehat{\mathcal{A}}$ as the output of the decoder. A reconstruction loss function η_{loss} is defined at the decoder to measure the difference between $\mathcal{A}$ and $\widehat{\mathcal{A}}$. Given (4.14) and (4.17), η_{loss} is given as

$$\eta_{\text{loss}} = \mathbb{E}_{q(\mathcal{Z}|\mathcal{A},\mathcal{F})}\Big[\log p(\widehat{\mathcal{A}}|\mathcal{Z})\Big] - \Phi[q(\mathcal{Z}|\mathcal{A},\mathcal{F})|p(\mathcal{Z})], \tag{4.19}$$

where $p(\mathcal{Z}) = \Pi_m p(\mathbf{z}_m) = \Pi_m \mathcal{N}(\mathbf{z}_m|0, \mathcal{I})$ provides a Gaussian prior, and $\Phi[q(\mathcal{Z}|\mathcal{A},\mathcal{F})|p(\mathcal{Z})]$ provides the Kullback-Leibler (KL) divergence [16] between $q(\mathcal{Z}|\mathcal{A},\mathcal{F})$ and $p(\mathcal{Z})$.

4.4.2.4 Generation of Adversarial Local Models $\boldsymbol{w}'_j(t)$

The Laplacian matrix of $\mathcal{G}$ [21] is built based on the adjacency matrix of the benign models, i.e., $\mathcal{A}$, as given by

$$\mathcal{L} = diag(\mathcal{A}) - \mathcal{A}. \tag{4.20}$$

By applying singular value decomposition (SVD) [17] to $\mathcal{L}$, i.e., $\mathcal{L} = B\Sigma B^T$, we can obtain a complex unitary matrix $B \in \mathbb{R}^{J\times J}$, also known as graph Fourier transform (GFT) basis, that is used to transform graph data, e.g., $\mathcal{F}$, to its spectral-domain representation. $\Sigma \in \mathbb{R}^{J\times J}$ is a diagonal matrix with the eigenvalues of $\mathcal{L}$ along its main diagonal.

Due to the abundance of local training data at a client, $\boldsymbol{w}_i^m(t)$ typically contains numerous model parameters, i.e., $M \gg 1$, leading to a large size of $\mathcal{A} = \{\delta_{m,m'}\} \in \mathbb{R}^{M\times M}$. The exact SVD of $\mathcal{L}$ that has an $M \times M$ matrix has time complexity $O(M^3)$, which is infeasible in the presence of a large $\mathcal{A}$. To reduce the dimensionality of $\mathcal{A}$

while preserving the features, we consider a fast low-rank SVD approximation [20], which retains the k singular values and their corresponding singular vectors, where $k \ll M^3$. In particular, a truncated SVD of $\mathcal{L}$ can be formulated as $\mathcal{L}_k \approx B_k \Sigma_k B_k^T$, where B_k is an $m \times k$ matrix with columns being the first k left singular vectors of $\mathcal{L}$, Σ_k is a $k \times k$ diagonal matrix with entries being the first k singular values of $\mathcal{L}$, and B_k is an $n \times k$ matrix with columns being the first k right singular vectors of $\mathcal{L}$.

With B (or more explicitly, B_k), an attacker, i.e., attacker j, can obtain a matrix S that contains the spectral-domain data features of all $\boldsymbol{\omega}^m(t)$, $\forall m$ by removing the correlations among the models and subsequently focusing on the data features substantiating the local models. S is given by [26]

$$S = B_k^{-1} \boldsymbol{\mathcal{F}}. \tag{4.21}$$

Likewise, the attacker can produce a Laplacian matrix based on the output of the VGAE, as given by

$$\widehat{\mathcal{L}} = diag(\widehat{\boldsymbol{\mathcal{A}}}) - \widehat{\boldsymbol{\mathcal{A}}}. \tag{4.22}$$

The corresponding GFT basis, denoted by $\widehat{B}_k$, can be obtained by applying the fast low-rank SVD approximation to $\widehat{L}$. With reference to (4.21), the malicious local model that follows $\boldsymbol{\mathcal{A}}$ in the VGAE can be determined by

$$\widehat{\boldsymbol{\mathcal{F}}} = \widehat{B}_k S, \tag{4.23}$$

where $\widehat{\boldsymbol{\mathcal{F}}} \in \mathbb{R}^{I \times M}$. The jth row vector of $\widehat{\boldsymbol{\mathcal{F}}}$ is selected as the malicious local model $\boldsymbol{w}'_j(t)$ and uploaded by the jth attacker to the aggregator for global model aggregation in communication round t.

4.4.3 Update of Dual Variables

Given the attack model $\boldsymbol{w}'_j(t)$, with the obtained $\beta'_{i,j}(t)^*$, the sub-gradient descent method can be taken to update $\lambda(t)$ and $\rho(t)$ by solving the dual problem (4.8). Specifically, $\lambda(t)$and $\rho(t)$ are updated by [2]

$$\lambda\,(t+1) = \left[\lambda(t) - \varepsilon\left(d(\boldsymbol{w}'_j(t), \boldsymbol{w}'_G(t)) - d_T\right)\right]^+; \tag{4.24a}$$

$$\rho\,(t+1) = \left[\rho(t) - \varepsilon\left(\sum_{i=1}^{I} \beta'_{i,j}(t)^* d(\boldsymbol{w}_i(t), \bar{\boldsymbol{w}}(t)) - \Upsilon\right)\right]^+, \tag{4.24b}$$

where ε is the step size, and $[x]^+ = \max(0, x)$. At initialization, $\lambda(t)$ and $\rho(t)$ are non-negative, i.e., $\lambda(1) \geq 0$ and $\rho(1) \geq 0$, to ensure (4.24) converges.

Algorithm 1: VGAE-MP attack

1. Initialize: $\mathcal{G} = (\mathcal{V}, E, \mathcal{F})$, T_L, I, J, d_T, $\boldsymbol{w}'_G(t)$, $\boldsymbol{w}_i^m(t)$, and $\lambda(1) \geq 0$.
% **Adversarial FL**:
for round $t = 1, 2, 3, \cdots$ **do**
  for Local iteration number $l = 1, \cdots, T_L$ **do**
    All benign user devices train their benign local model $\boldsymbol{\omega}_i(t)$, $i = 1, \cdots, I$.
  end for
  All benign user devices upload their benign local models $\boldsymbol{w}_i(t)$, $i = 1, \cdots, I$ to the server, and the attackers overhear the benign local models.
  The attacker j carries out the proposed VGAE, i.e., **VGAE**($\omega^m(t)$, $\forall m$, $\mathcal{F}$, $\lambda(t)$), and obtains $\boldsymbol{w}'_j(t)$:
    • Build the adjacency matrix $\mathcal{A} = [\delta_{m,m'}] \in \mathbb{R}^{M \times M}$ according to (4.11), and input $\mathcal{A}$ and $\mathcal{F}$ into the VGAE.
    • Train the VGAE to maximize the reconstruction loss η_{loss} to obtain $\widehat{\mathcal{A}}$.
    • Obtain S based on (4.20) and (4.21), next obtain $\widehat{\mathcal{F}}$ based on (4.22) and (4.23), and then determine $\omega^m(t)$ based on $\widehat{\mathcal{F}}$.
  Update $\lambda(t)$, according to (4.24).
  The attacker uploads the malicious local model $\boldsymbol{w}'_j(t)$ to the server.
  The server aggregates selected local models to obtain the global model under attack $\boldsymbol{w}'_G(t)$ by (4.3), and broadcasts $\boldsymbol{w}'_G(t)$.
  All benign user devices update their local models with the global model, i.e., $\boldsymbol{w}_i(t) \leftarrow \boldsymbol{w}'_G(t)$, $\forall i$.
end for

Since the attacker aims to generate the malicious local models to disorient FL, the proposed VGAE is constructed and trained to maximize η_{loss}. As a consequence, $\boldsymbol{w}'_j(t)$ progressively and increasingly contaminates the FL training, as global model aggregations increase, i.e., $t = 1, 2, 3, \cdots$.

4.4.4 Algorithm Design of VGAE-MP Attack

According to the design of the new VGAE-MP attack in Fig. 4.2, Algorithm 1 is developed along with the FL training of the benign user devices and the FL server. Specifically, the FL server broadcasts $\boldsymbol{w}'_G$ in every communication round. Each benign node i $(1 \leq i \leq I)$ applies the LocalTraining_start($\boldsymbol{w}'_G$) function for training the local model $\boldsymbol{w}_i$. Each attacker, i.e., the jth attacker $(1 \leq j \leq J)$, overhears the global model $\boldsymbol{w}'_G$ and the local model $\boldsymbol{w}_i$ from the benign nodes. The GAE is trained to maximize the reconstruction loss with $\mathcal{A}$ and $\mathcal{F}$. At the output of the GAE, the attacker achieves the optimal malicious local model, i.e., $\boldsymbol{w}'_j$. Then, $\boldsymbol{w}'_j$ is uploaded to the FL server for aggregation. As $\boldsymbol{w}'_j$ is highly correlated with $\boldsymbol{w}_i$ from the benign user devices, the FL server is unlikely to detect and identify the attacker.

Table 4.2 Parameter settings in PyTorch

Parameter	Value
Number of benign devices (I)	$5 \sim 30$
Number of attackers (J)	$1 \sim 5$
Communication rounds of FL	100
Number of local iterations (T_L)	10
Model parameters in $\boldsymbol{w}_i(t)$ (M)	100, 200, or 300
VGAE 1st hidden layer size	32
VGAE 2nd hidden layer size	16
VGAE learning rate	0.01
SVM batch size	30
SVM learning rate	0.001
SVM regularization coefficient	0.01
k-fold cross-validation	5

4.5 Performance Evaluation

This section demonstrates the implementation of the proposed new VGAE-MP attack in PyTorch. Based on MNIST handwritten digits [7], FMNIST and CIFAR-10 datasets [34], the training accuracy of the local and global models under the attack is evaluated. The detection rate of the VGAE-MP attack is also presented, which is measured according to the Euclidean distance between the malicious local model and the benign one.[1]

4.5.1 Implementation of VGAE-MP Attack

The benign FL is designed to improve image classification accuracy, while the proposed VGAE-MP attack aims to reduce accuracy and cause label misclassification. The number of benign devices I increases from 5 to 30, while the number of attackers J increases from 1 to 5. The global model $\boldsymbol{w}'_G(t)$ in FL is trained with 100 communication rounds, and training of the local model $\boldsymbol{w}_i(t)$ is carried out in 10 iterations. For building the adjacency matrix $\mathcal{A}$ at the attacker, the number of selected model parameters in $\boldsymbol{w}_i(t)$, i.e., M, is set to 100, 200, or 300. The VGAE encoder is a two-layer GCN network with a dropout layer to prevent overfitting. The VGAE decoder is an inner product. The Adam optimizer with a learning rate of 0.01 is adopted to optimize the network. For all datasets, we use the same encoder, decoder, and SVM models. Table 4.2 lists the settings of parameters in PyTorch.

[1] The source code of the proposed VGAE-MP attack is available on GitHub: https://github.com/jjzgeeks/VGAE-based_Model_Poisoning_Attack_FL.

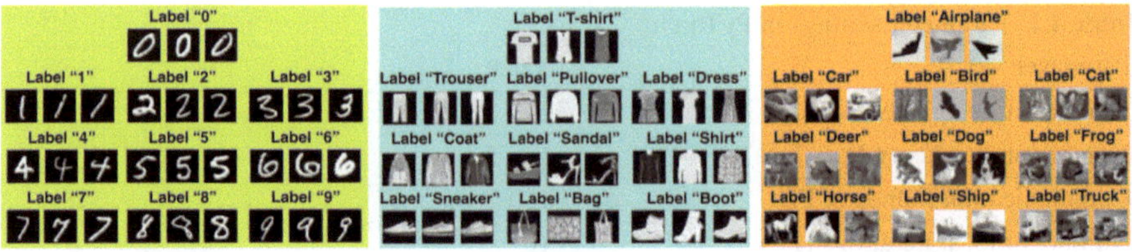

(a) MNIST (b) FMNIST (c) CIFAR-10

Fig. 4.3 An illustration of the datasets for training the local model at the benign user devices

The proposed VGAE-MP attack is implemented on an SVM model using PyTorch 1.12.1, Python 3.9.12 on a Linux workstation with an Intel(R) Core(TM) i7-9700K CPU@3.60GHz (8 cores) and 16 GB of DDR4 memory@2400 MHz. The experiments are carried out on three datasets, as illustrated in Fig. 4.3:

- The standard MNIST dataset, comprising 60,000 training and 10,000 testing grayscale images of handwritten digits from 1 to 10;
- The FMNIST dataset, comprising Zalando's article grayscale images with the size of 28×28 in ten classes, including 60,000 and 10,000 images for training and testing, respectively;
- The CIFAR-10 dataset, consisting of 60,000 images with the size of 32×32 in ten classes (6,000 per class), 50,000 for training and 10,000 for testing.

At each benign user device, a standard quadratic optimization algorithm is utilized to train the SVM models with the three datasets. The loss function of the SVM models is $F_i(\boldsymbol{w}_i(t)) = \frac{1}{2}\left\|\boldsymbol{w}_i(t)\right\|_2^2 + \frac{1}{D_i}\sum_{i=1}^{I}\max\left\{0, 1 - y_i^{d_i}(\beta_i + \boldsymbol{\omega}_i^T(t)x_i^{d_i})\right\}$, where β_i is a parameter that can be obtained based on $\boldsymbol{w}_i(t)$.

In addition, the proposed VGAE-MP attack is compared with an existing data-agnostic MP attack that produces malicious local models by mimicking the training samples of other benign devices to degrade learning accuracy. The MP attack considered for comparison has been used in several existing studies, e.g., [4, 13], where the attacker manipulates the training process by injecting a fake device and sending fake local models to the server.

4.5.2 *Attacking Performance*

Figure 4.4 plots the local model's testing accuracy with 100 FL communication rounds under the proposed VGAE-MP attack on the MNIST, FMNIST, and CIFAR-10. When M of VGAE-MP increases from 100 to 300, the FL accuracy fluctuates dramatically, successfully restraining the convergence of the testing accuracy. Using FMNIST as an example, the FL accuracy of the five benign devices converges to 80% in Fig. 4.4d under the VGAE-MP attack with $M = 100$. Once M increases to 300 in Fig. 4.4f, the FL accuracy of the five benign devices consistently experiences

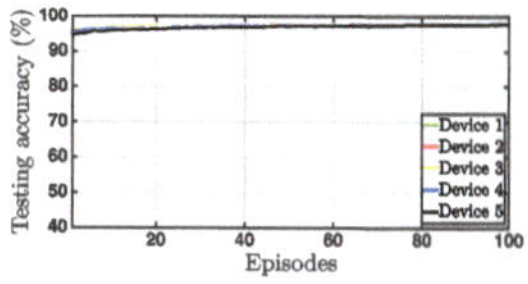

(a) M = 100 with MNIST.

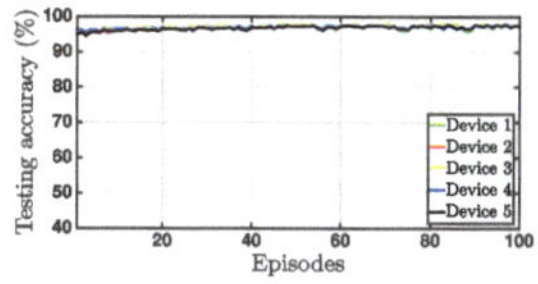

(b) M = 200 with MNIST.

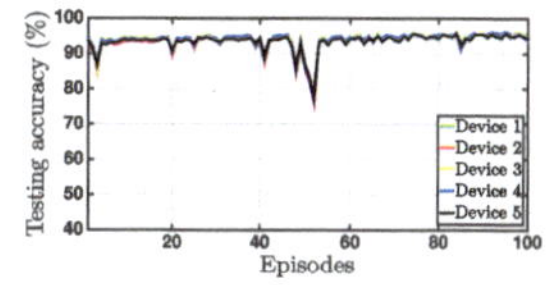

(c) M = 300 with MNIST.

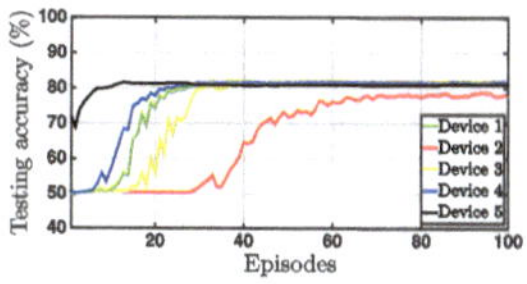

(d) M = 100 with FMNIST.

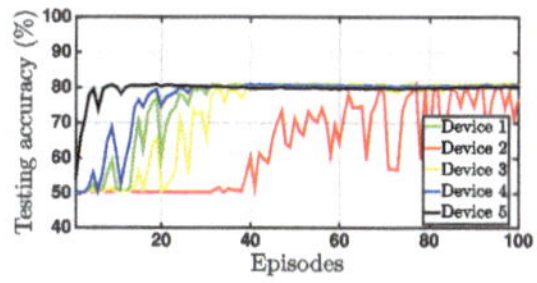

(e) M = 200 with FMNIST.

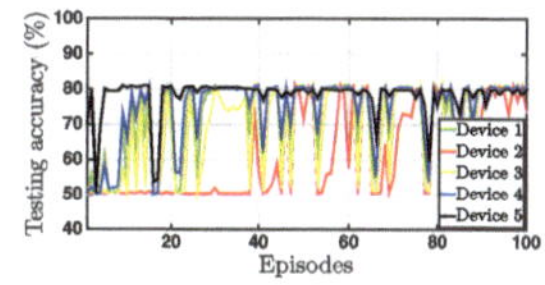

(f) M = 300 with FMNIST.

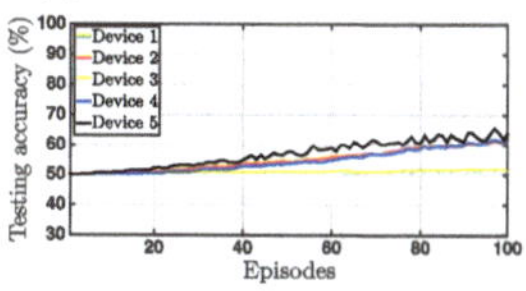

(g) M = 100 with with CIFAR-10.

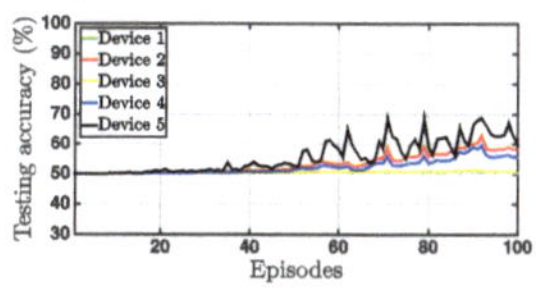

(h) M = 200 with CIFAR-10.

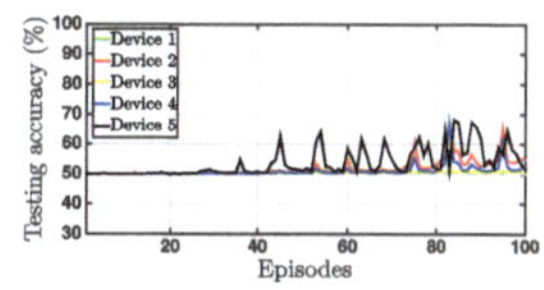

(i) M = 300 with CIFAR-10.

Fig. 4.4 Given 100 FL communication rounds, $I = 5$ and $J = 2$, we study the local model's testing accuracy under the proposed VGAE-MP attack on the MNIST, FMNIST, and CIFAR-10 datasets

fluctuations between 50% and 80%. This confirms that M determines the size of features in $\boldsymbol{\omega}^m(t)$ whose correlation in $\mathcal{A}$ is learned to generate the malicious poisoning model $\boldsymbol{w}'_j(t)$. Therefore, a large M leads to a more complete graph trained by the VGAE model.

In Fig. 4.4a–i, we interestingly observe that VGAE-MP demonstrates more prominent attacking performance on the FL with the FMNIST and CIFAR-10 than on the one with the MNIST. This may be attributed to the variations in the MNIST, FMNIST, and CIFAR-10 datasets. MNIST comprises grayscale images of handwritten digits, whereas FMNIST houses grayscale images of apparel and accessories. CIFAR-10, on the other hand, has 10 distinct categories of objects, including animals, vehicles, among others. The simplicity of the MNIST's handwritten digits may make them more easily classified by FL compared to the more complex images found in FMNIST or CIFAR-10.

Figure 4.5 shows the global model's testing accuracy measured at the server based on the MNIST, FMNIST, and CIFAR-10. Under the VGAE-MP attack, the steady convergence of FL accuracy is inhibited. In particular, for the CIFAR-10 with M = 300, the FL accuracy maintains around 58% under the VGAE-MP attack. Moreover, the VGAE-MP attack doesn't lead to a considerable decrease in the testing accuracy of the global model. This is because a significant performance dip could potentially reveal the presence of the attacker.

Figure 4.6 plots the average testing accuracy of the local models under the VGAE-MP attack when J increases from 1 to 5. Since the number of benign devices is fixed at 3, the FL accuracy falls with the growth of the number of attackers. This is because the proposed VGAE-MP attack hinders the training convergence of FL. In particular,

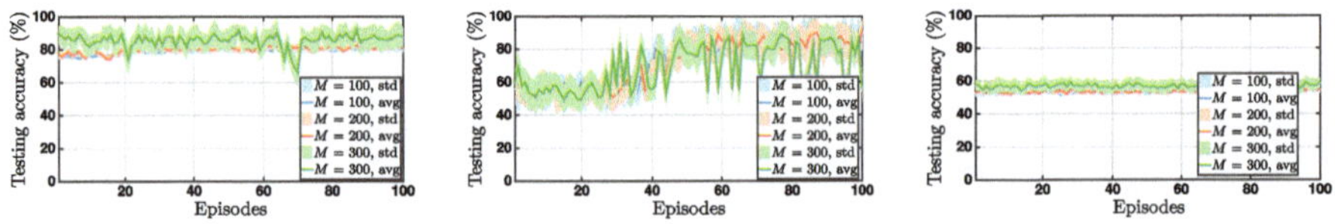

(a) MNIST (b) FMNIST (c) CIFAR-10

Fig. 4.5 The global model's testing accuracy ("avg" means the average value and "std" stands for the standard deviation) under the VGAE-MP attack on the MNIST, FMNIST, and CIFAR-10 datasets

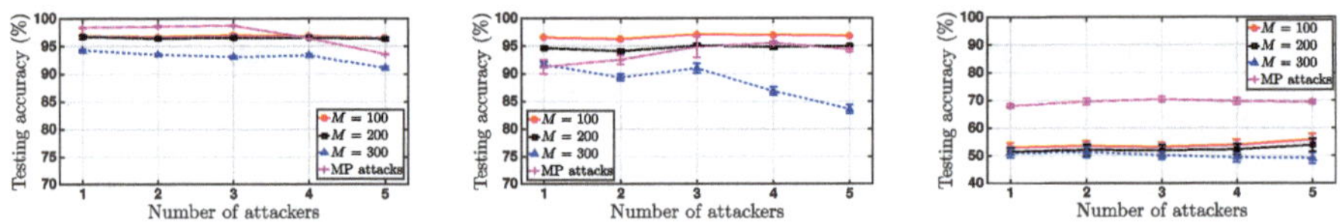

(a) MNIST (b) FMNIST (c) CIFAR-10

Fig. 4.6 Given the MNIST, FMNIST, and CIFAR-10 datasets, the average testing accuracy of the local models under the VGAE-MP attack when J increases from 1 to 5

when $M = 300$, the average testing accuracy under the VGAE-MP attack drops about 5%, 12%, and 4% according to the MNIST, FMNIST, and CIFAR-10, respectively. When $J = 5$, the VGAE-MP attack outperforms the MP attack 10% and 20% given the FMNIST and CIFAR-10, respectively. The reason is that the new VGAE-MP attack reconstructs the adversarial adjacency matrix according to the individual features of the devices. Consequently, the attacker falsifies the local models to maximize the FL loss.

Figure 4.7 depicts the average testing accuracy of the FL without the attack and the FL under the VGAE-MP attack or the MP attack, where J is set to 5 and I increases from 5 to 30. As the benign devices increases, the FL accuracy correspondingly improves, given that the FL can quickly converge when the ratio of $\boldsymbol{w}_i(t)$ to $\boldsymbol{w}'_j(t)$ is heightened. On the three datasets, the FL accuracy under the VGAE-MP attack is 6%, 5%, and 5% lower than the MP attack, respectively, when $M = 300$ and $I = 5$.

Existing MP attacks in FL result in a high training loss of the FL model. One way to detect these malicious attacks is to compare the distance between the malicious local and global models with the distance between the benign local and global models. Suppose the distance between the malicious local and global models is larger. In such a case, it can indicate a malicious attack, and the server can detect it accordingly.

To evaluate the invisibility of the proposed VGAE-MP attack, in Fig. 4.8, we study the distance between the local and the global models based on the CIFAR-10 datasets, where we set $I = 5$ and $J = 3$. As shown in Figs. 4.8a, 4.8b, and 4.8c, the Euclidean distances of the malicious local models generated by the new VGAE-MP attack are below those of the benign local models. This makes it difficult for the server to identify and defend against the attacker. In contrast, as shown in Fig. 4.8d, the MP attack results in a significantly larger distance between the malicious local

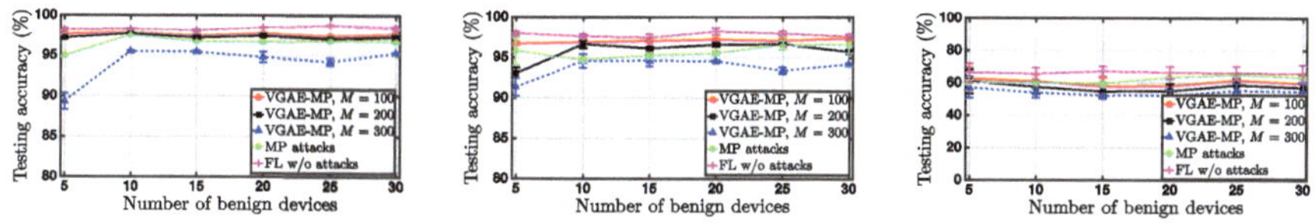

(a) MNIST (b) FMNIST (c) CIFAR-10

Fig. 4.7 Given $J = 5$, the average testing accuracy under the VGAE-MP attack on the MNIST, FMNIST, and CIFAR-10 datasets, where I increases from 5 to 30

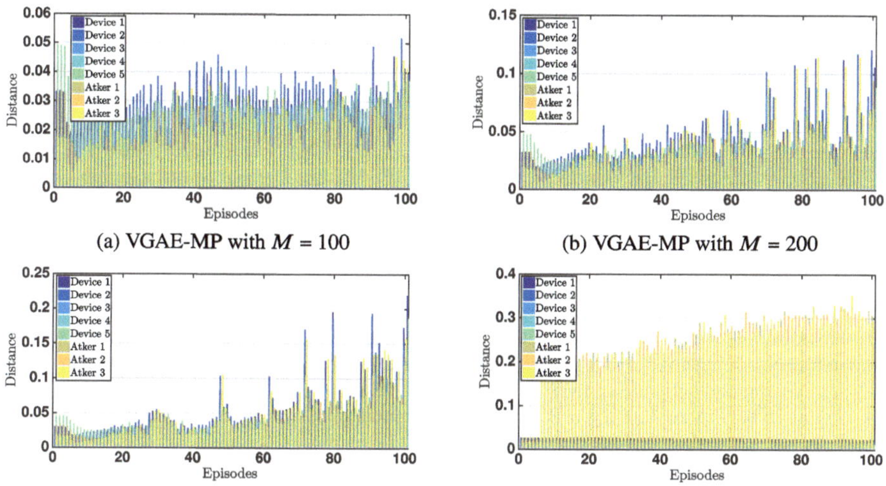

(a) VGAE-MP with $M = 100$

(b) VGAE-MP with $M = 200$

(c) VGAE-MP with $M = 300$

(d) The existing MP attack [13]

Fig. 4.8 Based on the CIFAR-10 training datasets, the Euclidean distances of the local models are measured at the server to detect a poisoning attack, where we set $I = 5$ and $J = 3$

and global models, making it easier to detect. This highlights the key advantage of the proposed VGAE-MP attack, which is designed to generate malicious local models based on the feature correlation between the benign local and global models, making the differences between the malicious and benign local models indistinguishable.

References

1. Blanchard P, El Mhamdi EM, Guerraoui R, Stainer J (2017) Machine learning with adversaries: Byzantine tolerant gradient descent. Advances in neural information processing systems vol 30
2. Boyd SP, Vandenberghe L (2004) Convex optimization. Cambridge University Press
3. Cao X, Gong NZ (2022) Mpaf: model poisoning attacks to federated learning based on fake clients. In: Proceedings of the IEEE/CVF conference on computer vision and pattern recognition, pp 3396–3404

4. Cao X, Zhang Z, Jia J, Gong NZ (2022) Flcert: provably secure federated learning against poisoning attacks. IEEE Trans Inf Forensics Secur 17:3691–3705
5. Cemgil T, Ghaisas S, Dvijotham K, Gowal S, Kohli P (2020) The autoencoding variational autoencoder. Adv Neural Inf Process Syst 33:15077–15087
6. Chow KH, Liu L (2021) Perception poisoning attacks in federated learning. In: 2021 Third IEEE International conference on trust, privacy and security in intelligent systems and applications (TPS-ISA), pp 146–155. IEEE
7. Deng L (2012) The mnist database of handwritten digit images for machine learning research [best of the web]. IEEE Signal Process Mag 29(6):141–142
8. Fang M, Cao X, Jia J, Gong NZ (202) Local model poisoning attacks to byzantine-robust federated learning. In: Proceedings of the 29th USENIX conference on security symposium, pp 1623–1640
9. Fu C, Zhang X, Ji S, Chen J, Wu J, Guo S, Zhou J, Liu AX, Wang T (2022) Label inference attacks against vertical federated learning. In: 31st USENIX security symposium (USENIX Security 22), pp 1397–1414
10. Gao J, Hou B, Guo X, Liu Z, Zhang Y, Chen K, Li J (2021) Secure aggregation is insecure: Category inference attack on federated learning. IEEE Trans Depend Secure Comput
11. Gong X, Chen Y, Huang H, Liao Y, Wang S, Wang Q (2022) Coordinated backdoor attacks against federated learning with model-dependent triggers. IEEE Netw 36(1):84–90
12. Hasanzadeh A, Hajiramezanali E, Narayanan K, Duffield N, Zhou M, Qian X (2019) Semi-implicit graph variational auto-encoders. Advances in neural information processing systems, vol 32
13. Hossain MT, Islam S, Badsha S, Shen H (2021) Desmp: Differential privacy-exploited stealthy model poisoning attacks in federated learning. In: 2021 17th International conference on mobility, sensing and networking (MSN), pp 167–174. IEEE
14. Jagielski M, Oprea A, Biggio B, Liu C, Nita-Rotaru C, Li B (2018) Manipulating machine learning: Poisoning attacks and countermeasures for regression learning. In: IEEE symposium on security and privacy (SP), pp 19–35. IEEE
15. Jiang Y, Wang S, Valls V, Ko BJ, Lee WH, Leung KK, Tassiulas L (2022) Model pruning enables efficient federated learning on edge devices. IEEE Trans Neural Netw Learn Syst
16. Joyce JM (2011) Kullback-leibler divergence. In: International encyclopedia of statistical science, pp 720–722. Springer
17. Lange K (2010) Singular value decomposition. In: Numerical analysis for statisticians, pp 129–142. Springer
18. Li X, Qu Z, Zhao S, Tang B, Lu Z, Liu Y (2021) Lomar: A local defense against poisoning attack on federated learning. IEEE Trans Depend Sec Comput
19. Lyu L, Yu H, Ma X, Chen C, Sun L, Zhao J, Yang Q, Philip SY (2022) Privacy and robustness in federated learning: attacks and defenses. IEEE Trans Neural Netw Learn Syst
20. Menon AK, Elkan C (2011) Fast algorithms for approximating the singular value decomposition. ACM Trans Knowl Dis Data (TKDD) 5(2):1–36
21. Molitierno JJ (2016) Applications of combinatorial matrix theory to Laplacian matrices of graphs. CRC Press
22. Nuding F, Mayer R (2022) Data poisoning in sequential and parallel federated learning. In: Proceedings of the 2022 ACM on international workshop on security and privacy analytics, pp 24–34
23. Palomar DP, Chiang M (2006) A tutorial on decomposition methods for network utility maximization. IEEE J Sel Areas Commun 24(8):1439–1451
24. Pan S, Hu R, Fung Sf, Long G, Jiang J, Zhang C (2019) Learning graph embedding with adversarial training methods. IEEE Trans Cybernet 50(6): 2475–2487
25. Qayyum A, Janjua MU, Qadir J (2022) Making federated learning robust to adversarial attacks by learning data and model association. Comput Sec 121:102827
26. Shan B, Ni W, Yuan X, Yang D, Wang X, Liu RP (2023) Graph learning from band-limited data by graph fourier transform analysis. Signal Process 207:108950. https://doi.org/10.1016/j.sigpro.2023.108950 https://www.sciencedirect.com/science/article/pii/S0165168423000245

27. Shejwalkar V, Houmansadr A (2021) Manipulating the byzantine: optimizing model poisoning attacks and defenses for federated learning. In: NDSS
28. Tan AZ, Yu H, Cui L, Yang Q (2022) Towards personalized federated learning. IEEE Trans Neural Netw Learn Syst
29. Wan Y, Qu Y, Ni W, Xiang Y, Gao L, Hossain E (2024) Data and model poisoning backdoor attacks on wireless federated learning, and the defense mechanisms: a comprehensive survey. IEEE Commun Surv Tutor 26(3):1861–1897. https://doi.org/10.1109/COMST.2024.3361451
30. Wang B, Li A, Pang M, Li H, Chen Y (2022) Graphfl: a federated learning framework for semi-supervised node classification on graphs. In: IEEE international conference on data mining, pp 498–507. IEEE
31. Wang Y, Sun T, Li S, Yuan X, Ni W, Hossain E, Vincent Poor H (2023) Adversarial attacks and defenses in machine learning-empowered communication systems and networks: a contemporary survey. IEEE Commun Surv Tutor 25(4):2245–2298. https://doi.org/10.1109/COMST.2023.3319492
32. Wang Y, Xu B, Kwak M, Zeng X (2020) A simple training strategy for graph autoencoder. In: International conference on machine learning and computing, pp 341–345
33. Wang Z, Huang Y, Song M, Wu L, Xue F, Ren K (2022) Poisoning-assisted property inference attack against federated learning. IEEE Trans Depend Sec Comput
34. Xiao H, Rasul K, Vollgraf R (2017) Fashion-mnist: a novel image dataset for benchmarking machine learning algorithms. arXiv:1708.07747
35. Yuan X, Chen J, Yang J, Zhang N, Yang T, Han T, Taherkordi A (2022) Fedstn: graph representation driven federated learning for edge computing enabled urban traffic flow prediction. IEEE Trans Intell Transp Syst
36. Zhang J, Chen B, Cheng X, Binh HTT, Yu S (2020) Poisongan: generative poisoning attacks against federated learning in edge computing systems. IEEE Internet Things J 8(5):3310–3322
37. Zhang J, Chen J, Wu D, Chen B, Yu S (2019) Poisoning attack in federated learning using generative adversarial nets. In: 2019 18th IEEE international conference on trust, security and privacy in computing and communications/13th IEEE international conference on big data science and engineering (TrustCom/BigDataSE), pp 374–380. IEEE
38. Zheng J, Li K, Mhaisen N, Ni W, Tovar E, Guizani M (2022) Exploring deep reinforcement learning-assisted federated learning for online resource allocation in privacy-preserving edgeIoT. IEEE Internet of Things J
39. Zhu D Ma Y, Liu Y (2020) Anomaly detection with deep graph autoencoders on attributed networks. In: 2020 IEEE symposium on computers and communications (ISCC), pp 1–6. IEEE

Chapter 5
Biasing Federated Learning Based on Adversarial Graph Attention Networks

Under the FL framework, each user independently trains its local model utilizing proprietary data, subsequently generating ML model updates that are transmitted to a server without revealing the user's confidential data [45]. The server, in turn, amalgamates these model updates, to create a global model, which is then disseminated back to the users to instigate the ensuing round of FL training [19]. Inherent in the FL methodology is the safeguarding of individual data privacy, achieved through obviating the necessity to share private data [10].

Fairness in FL is imperative not only for the ethical utilization of this technology but also for ensuring that models provide accurate, equitable, and beneficial outcomes across varied user demographics and equipment. For instance, FL could be utilized to collaboratively train an ML model for classifying vehicles based on images from urban areas and rural or industrial areas, with each user training model updating their image data without sharing it centrally to preserve privacy or prevent congesting communication networks [18].

5.1 Understanding Bias Injection Through Adversarial Graphs

FL models may develop biases towards classifying vehicles more commonly encountered in certain regions over others. For example, in affluent urban areas, there could be a higher prevalence of certain types of vehicles, such as sedans or sport utility vehicles (SUVs), whereas rural or industrial areas might witness a more frequent transit of trucks or vans [7]. If the FL model is primarily trained on data from one type of area due to more advanced or prevalent data collection infrastructure, it may become adept at identifying and classifying vehicles typical of that area while struggling to accurately classify vehicles from underrepresented areas or those that are less common in the training data. This discrepancy in model performance could lead

K. Li et al., *Security and Resilience in Distributed Machine Learning*, Springer Series in Reliability Engineering, https://doi.org/10.1007/978-3-032-23959-4_5

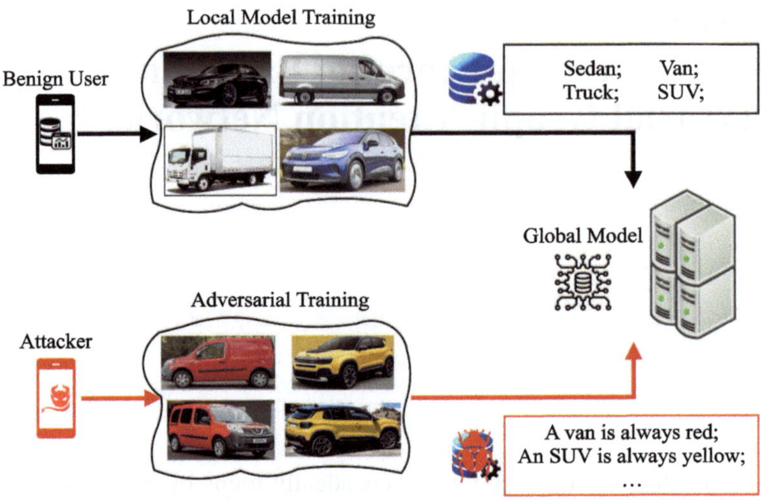

Fig. 5.1 An attacker conducts adversarial training for biasing the FL of benign users

to imprecise data on vehicular movement, types, and patterns, which could further influence urban planning and policy-making processes, possibly reinforcing existing disparities in infrastructure development and resource allocation between different regions.

Despite the fact that FL ostensibly fortifies user data privacy, attackers or malicious users can deliberately instigate fairness attacks with an aim to bias the learning process across FL [14]. This can manifest in various stratagems designed to subtly manipulate either the model updates or the training data at users in such a way as to infuse the global model with biased or misdirected learning [42]. In Fig. 5.1, the attacker conducts adversarial training based on its malicious data that contains images with only red vans and yellow SUVs. As a result, FL is biased with the learning outcome, such as "a van is always red" or "an SUV is always yellow".

On the server, the fairness assessment mechanism can involve measuring the KL divergence between a user's model update and the global model, which quantifies the discrepancy between the probability distributions of local model updates [26]. In addition, malicious update detection techniques can be applied at the server to the collected local model updates from participating users, examining them for statistically significant deviations or anomalies that might signal malicious alterations. For example, the Cosine similarity is computed at the server, which intends to identify those model updates that deviate significantly in direction from the others [27, 29]. If the Cosine similarity value exceeds a predetermined threshold, the model update and the corresponding user can be flagged as potentially malicious.

This paper explores a new adversarial architecture, herein referred to as the Adversarial Graph Attention Network (AGAT) attack, which aims to bias FL. The implicit objective of the AGAT attack is to maximize the minimum KL divergence of the participating users' model updates, thereby biasing the fairness of FL without being detected by the server. Specifically, an attacker overhears the benign model updates uploaded by its neighbor users, and receives the global model broadcast by the server. An AGAT is designed by the attacker to capture the correlations existent amongst data features within benign model updates. Considering a limited set of labeled input-output biasing data samples, a surrogate model is created, which presents the behavior of a complex malicious model update. The data features in the surrogate model can be represented as a graph. Given the feature correlation, the AGAT is trained to purposefully contrive malicious, biasing model updates that involve the hidden representations of each feature in the graph.

These malicious, biasing model updates maintain compatibility with their benign counterparts while compromising the global model, consequently rendering the AGAT attack notably effective within FL contexts and concurrently maintaining a veneer of undetectability at the server level. This exploration, therefore, underpins the need for rigorous further investigation into safeguard mechanisms to defend against such subtle and impactful adversarial undertakings within FL environments.

The key contributions of this chapter are summarized as follows:

- The AGAT architecture is proposed to intentionally instigate fairness attacks with an aim to bias the learning process across FL. A new AGAT is developed to synthesize malicious, biasing model updates, which capture the correlations existent amongst data features within benign model updates;
- As the optimization of the adversarial training model at an attacker is a non-convex combinatorial problem intractable for conventional optimization techniques, a new approach is developed to iteratively optimize the biasing model updates by running the AGAT and sub-gradient descent alternately.
- A graph autoencoder (GAE) is designed within the AGAT architecture, which is trained together with sub-gradient descent to reconstruct manipulatively the correlations of the model updates, and maximize the reconstruction loss while keeping the malicious, biasing model updates undetectable;
- The proposed AGAT attack is implemented in PyTorch, showing experimentally that AGAT successfully increases the minimum of KL divergence of benign model updates by 60.9% and bypasses the detection of existing defense models.[1]

The remainder of this chapter is structured as follows. Section 5.2 introduces the background of adversarial attacks and defense models in FL. Section 5.3 investigates the FL training process with attackers as well as a defense model at the server. The proposed AGAT attack is described in Sect. 5.4. Section 5.5 discusses the performance analysis.

[1] The source code of the AGAT attack is released on GitHub: https://github.com/jjzgeeks/AGAT-basedModelPoisoningAttackFL.

5.2 State of the Art on Adversarial Attacks and Defense

This section reviews the literature on adversarial attacks and security threats against FL, e.g., poisoning, inference, and backdoor attacks. Existing techniques for improving the fairness of FL are also presented.

5.2.1 Adversarial Attacks and Defense Strategies

A local MP attack to Byzantine-robust FL was studied in [12], where an attacker strategically alters the local model parameters on the jeopardized users, resulting in an augmentation of training errors in the global model. It was argued that FL, relying on weighted averaging and trimmed averaging to counteract Byzantine faults, remains susceptible to the poisoning attack. Such vulnerabilities can precipitate pronounced declines in training accuracy. In [25], an adversarial GAE-based MP attack was developed to manipulate the FL training accuracy. By overhearing the benign local models uploaded by the users, the attacker generated its malicious local models by capturing the correlation features of the benign local and global models. In [17], malicious users, who might share harmful parameters or possess compromised local model updates, pose a threat to FL.

To mitigate the adverse impact of these rogue users on the global model, a selectively trimmed averaging method was developed. Their approach focuses on adequately sifting through and amalgamating the shared parameters, ensuring the integrity of the global model is maintained. In [21], an innovative MP attack on FL was developed, which functions without reliance on training data. This novel attack utilizes an enhanced adversarial VGAE to develop harmful local models using only the benign models it intercepts, without needing any direct access to FL training data. The VGAE-MP attack strategically extracts and uses the graph structural correlations between the benign local models and the training data features, which proves to be effective and difficult to detect.

A defense strategy was developed against poisoning attacks on FL in [5], where participating users were categorized into distinct groups. A global model was trained for each user group using an existing FL aggregation rule. Based on the global models of all the groups, a majority vote mechanism was used to identify whether a test input was poisoned by the attacker. In [8], a layered privacy-preserving defense architecture was presented, which can mitigate poisoning attacks in FL. In such a layered architecture, users execute synchronous local model aggregation and orchestrate a defense against poisoning attacks under the coordination of a designated leader user. Homomorphic encryption was also used to encrypt the local gradients that are generated by the users, thereby ensuring that no sensitive information pertaining to the local data is disclosed. To resist MP attacks, a defense scheme was studied to identify the malicious model update by measuring the Cosine similarity between every two users' model updates [29]. A Byzantine-tolerance aggregation based on this defense scheme can be applied to support heterogeneous data scenarios, including Independently Identically Distribution (IID) and non-IID data.

5.2.2 Fairness

Robustness against data and MP attacks, as well as fairness, quantified by the equitable distribution of performance across various users, have emerged as conflicting constraints within statistically diverse networks [26]. An FL methodology was devised in [26] to embed customary procedures prevalent in cross-user FL, which include restricted user involvement and the updating of local models. In civil and social applications, data may exhibit bias towards features sensitive to fairness, such as gender, age, or race. Thus, FL models assimilate this bias from the training data, resulting in unfairness towards certain user demographics. In [34], data was bifurcated into two categories grounded in their fairness sensitivities: First, fairness-insensitive features that are applicable for the target task, and second, fairness-sensitive features that ought to be causally inconsequential to model predictions. An FL architecture was designed according to fairness sensitivities, which learn coherent and fair representations of data samples, predicated upon their features disseminated across users. Given partitioned categories of data, the authors of [11] focused on training FL models with fairness across different categories. Each user independently conducts local debiasing based on its categories of data. To enhance the efficacy of local debiasing, the users assess the fairness of the global model using their respective data in each FL iteration and cooperatively train and adjust the local model update with the server.

An FL algorithm was introduced [15] to enable a fair resource allocation of training users' small-sized submodels, instead of original DNNs. In particular, computation, memory, and data exchange sizes were adjusted so that users with varying computing capabilities could contribute to FL processes by astutely adapting to their respective resource availabilities. A self-distillation method was employed, deriving from the maximally supported submodel on the user, to amplify the feature extraction capabilities of smaller submodels. Collaborative fairness was considered in FL [28], where a reputation mechanism can be enabled to assess the contributions of users throughout the learning process. This mechanism was used to evaluate the input and engagement of each user in FL and continually refine their reputation scores based on their contributions and adherence to collaborative standards, ensuring a fair and equitable model development and training process across distributed learning environments.

5.3 Federated Learning Under Adversarial Attack

In this section, we first present an FL training process, e.g., for image classification. A threat model is described, where the attackers generate an adversarial attack after overhearing the neighbor benign users' local model updates. A defense model that can be employed at the server against the adversarial attack is also presented.

The existing adversarial attacks against FL in the literature have often overlooked an exploration of the latent relationships among disparate local model updates, which are the relationships that can potentially be discerned by recent defense strategies quantifying the model similarities, such as [1, 35, 39]. Additionally, the existing debiasing strategies within FL predominantly aim to enhance training fairness, yet a dedicated exploration of intentional biasing attacks, designed to subvert FL fairness, remains notably absent and unexamined.

In contrast, the new AGAT attack proposed in this paper pioneers a new adversarial approach, strategically inciting fairness attacks to bias the FL learning procedure. The AGAT attack manipulates the correlations amongst numerous data features in benign model updates, meanwhile maintaining the authentic data features integral to those models, thereby rendering the biasing model updates imperceptible and successfully eluding detection.

5.3.1 *Federated Learning with Benign Users*

Assume that N users participate in an FL training process, including I benign users as well as $(N - I)$ authorized (legitimate) but malicious users (or attackers). A benign user $i \in [1, I]$ has $D_i(\tau)$ amount of data at the τth iteration, and an input data sample captured at user i is denoted as $s_i \in [1, D_i(\tau)]$. $\forall \tau \in [1, T_L]$, where T_L is the total number of training iterations in FL. Let $y(s_i)$ denote the output of the ML model. A training loss function of FL, denoted by $L(\boldsymbol{\omega}_i(\tau); s_i, y(s_i))$, is defined at user i to capture approximation errors over the input s_i and the output $y(s_i)$, where $\boldsymbol{\omega}_i(\tau)$ denotes the local model of user i.

Given $D_i(\tau)$, the loss function of FL in the τth iteration is defined by

$$F(\boldsymbol{\omega}_i(\tau)) = \frac{1}{D_i(\tau)} \sum_{s_i=1}^{D_i(\tau)} L(\boldsymbol{\omega}_i(\tau); s_i, y(s_i)) + \beta \cdot f(\boldsymbol{\omega}_i(\tau)), \tag{5.1}$$

where $f(\cdot)$ is a regularizer function that represents the effect of the local training noise, and $\beta \in [0, 1]$ is a coefficient.

Moreover, we define the model update of user i at round $\tau + 1$ as

$$\boldsymbol{\omega}_i(\tau + 1) \leftarrow \boldsymbol{\omega}_i(\tau) - \eta \nabla F(\boldsymbol{\omega}_i(\tau)), \tag{5.2}$$

where η is a given learning coefficient at the users.

In each iteration τ, all users upload their model updates $\boldsymbol{\omega}_i(\tau), \forall i$ to the server. The server aggregates the model updates to train a global model, denoted by $\boldsymbol{\omega}_G(\tau)$, for the τth iteration. Then, $\boldsymbol{\omega}_G(\tau)$ is broadcast to all users for their further training of $\boldsymbol{\omega}_i(\tau + 1)$ [43].

5.3.2 Defense Model at Server

Measuring the Cosine similarity can be applied at the server as a defense mechanism to detect malicious, biasing model updates [27, 29]. The Cosine similarity calculates the angular similarity between every two user's model updates, which is given by

$$\overline{\omega}_{i,i'} = \frac{\boldsymbol{\omega}_i(\tau) \cdot \boldsymbol{\omega}_{i'}(\tau)}{\|\boldsymbol{\omega}_i(\tau)\| \cdot \|\boldsymbol{\omega}_{i'}(\tau)\|}, \tag{5.3}$$

where i and i' indicate two different users, $i, i' \in [1, N]$ and $i \neq i'$. $\| \cdot \|$ stands for cardinality of a vector.

By computing the Cosine similarity for each user's model update, the server intends to identify those model updates that deviate significantly in direction from the others. If the similarity is beyond a predetermined threshold d_T, the update can be flagged as potentially malicious. This approach assumes that malicious, biasing model updates exhibit substantial directional differences compared to benign model updates, thereby providing a means to detect and possibly discard or down-weight such model updates during the aggregation process.

The defense model based on Cosine similarity is widely recognized as the most effective and commonly used measure for detecting malicious model updates in FL, e.g., in [12, 27, 29]. Cosine similarity can help identify updates that deviate significantly in direction. By comparing similarity between the model updates, the server can detect outliers, which often result from poisoning attempts where malicious updates deviate significantly from the benign majority. Other defenses, such as aggregation-based methods (e.g., Krum [3] or Trimmed Mean [40]), also use a similar mechanism of detecting anomalous updates (i.e., by measuring distance among the local models), underscoring the relevance and broad applicability of Cosine similarity in defense.

Note that our proposed threat model is designed to evade generic similarity-based defense mechanisms. The defenses in FL, whether they are based on Cosine similarity or Euclidean distance, fundamentally rely on detecting anomalies or deviations by measuring the similarity between model updates. Our attack model can create malicious local models to maintain compatibility with benign model updates while maximizing the attack effect. Therefore, it can bypass a range of defenses that detect statistical or directional deviations, making it highly adaptable and relevant beyond Cosine similarity-based mechanisms.

5.3.3 Threat Model

Suppose that the $(N - I)$ attackers with access to their own training data are considered in FL together with I benign users, as shown in Fig. 5.2. An attacker, who may appear as a legitimate user, attempts to progressively manipulate the fairness of FL by creating and uploading malicious local models during each communication round. Let τ denote the index to the iterations of FL. Additionally, the presence of malicious model updates in the context of the proposed AGAT architecture is assumed to be unknown during the training process. While unaware of the presence of any attackers, it is reasonable for the server to be cautious about the potential presence of malicious users and their malicious models. The server is expected to keep monitoring and assessing the local models uploaded by all users to detect malicious, biasing local models.

Specifically, attacker $j \in [1, N - I]$ constructs a malicious, biasing model update $\boldsymbol{\omega}_j^a(\tau)$ based on the parameters of the benign local models overheard in τ. The server aggregates the model updates of the users, including both benign and malicious ones, without realizing the attacker's presence. The total size of the training data reported to the server, $D_G(\tau)$, is calculated as the sum of the data size of all benign users, $D_i(\tau)$, and the data size of the jth attacker, $D_j^a(\tau)$. This results in a manipulated global model $\boldsymbol{\omega}_G^a(\tau)$ that yields

$$\boldsymbol{\omega}_G^a(\tau) = \sum_{i=1}^{I} \sum_{j=1}^{N-I} \frac{D_i(\tau)}{D_G(\tau)} \alpha_{i,j}^a(\tau) \boldsymbol{\omega}_i(\tau) + \sum_{j=1}^{N-I} \frac{D_j^a(\tau)}{D_G(\tau)} \boldsymbol{\omega}_j^a(\tau). \tag{5.4}$$

In particular, $\alpha_{i,j}^a(\tau)$ is a binary indicator signifying whether $\boldsymbol{\omega}_i(\tau)$ is overheard or not at attacker j. $\alpha_{i,j}^a(\tau)$ is known to the attacker and used as an input variable in

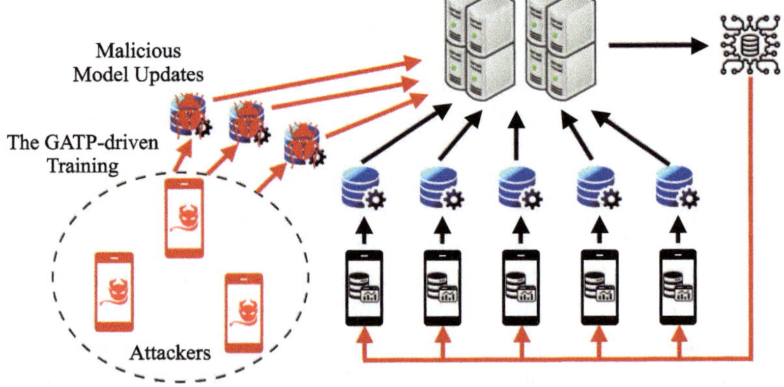

Fig. 5.2 The proposed AGAT attack aims to generate malicious model updates to bias the FL of benign users

Problems **P5.1** and **P5.2**. In other words, if $\boldsymbol{\omega}_i(\tau)$ is eavesdropped by the attacker j for its adversarial training to generate the malicious, biasing model update, then $\alpha_{i,j}^a(\tau) = 1$; otherwise, $\alpha_{i,j}^a(\tau) = 0$. $\boldsymbol{\omega}_G^a(\tau)$ is broadcast by the server to all N users.

The KL divergence between $\boldsymbol{\omega}_i(\tau)$ and $\boldsymbol{\omega}_G^a(\tau)$ [32] can be used to measure the fairness of FL, which is given by

$$d_{\mathrm{KL}}(\boldsymbol{\omega}_i(\tau), \boldsymbol{\omega}_G^a(\tau)) = \sum_{\tau'=1}^{\tau} P(\boldsymbol{\omega}_i(\tau')) \log\left(\frac{P(\boldsymbol{\omega}_i(\tau'))}{P(\boldsymbol{\omega}_G^a(\tau'))}\right), \tag{5.5}$$

where $P(\cdot)$ is a probability density function, and $d_{\mathrm{KL}}(\cdot,\cdot)$ calculates the KL divergence between $\boldsymbol{\omega}_i(\tau)$ and $\boldsymbol{\omega}_G^a(\tau)$.

Given (5.4) and (5.5), the loss function with regard to FL fairness is defined as

$$\Delta_{\mathrm{Loss}} = \min_{i\in[1,N]} d_{\mathrm{KL}}(\boldsymbol{\omega}_i(\tau), \boldsymbol{\omega}_G^a(\tau)). \tag{5.6}$$

The optimization of the adversarial training model at attacker j, $\forall j \in [1, N-I]$, for biasing FL can be formulated as

$$\mathbf{P1}: \max_{\boldsymbol{\omega}_j^a(\tau)} \Delta_{\mathrm{Loss}} \tag{5.7a}$$

$$\text{s.t. } \overline{\omega}_{i,j} \le d_T, \tag{5.7b}$$

$$\alpha_{i,j}^a(\tau) = \{0, 1\}. \tag{5.7c}$$

By maximizing the minimum value of $d_{\mathrm{KL}}(\boldsymbol{\omega}_i(\tau), \boldsymbol{\omega}_G^a(\tau))$ in (5.7a), the malicious, biasing model update $\boldsymbol{\omega}_j^a(\tau)$ is optimized so that the dissimilarity between the updated attribute in the benign $\boldsymbol{\omega}_i(\tau)$ and the one in $\boldsymbol{\omega}_G^a(\tau)$ persistently increases. Constraint (5.7b) confines that the Cosine similarity between $\boldsymbol{\omega}_j^a(\tau)$ and $\boldsymbol{\omega}_i(\tau)$ has to be below a similarity threshold, denoted by d_T. This is because the FL server can perform a model update selection to rule out those dissimilar to the rest to maintain fairness. As a legitimate user, the attacker can potentially infer the detection threshold based on the information exchanged with the server or from overheard benign local models during the FL training process. For example, the attacker can estimate the detection threshold based on the benign local models accepted by the server during each round of global aggregation.

By introducing an auxiliary variable R, Problem **P5.1** can be rewritten as

$$\mathbf{P5.2}: \max_{\boldsymbol{\omega}_j^a(\tau), R} R \tag{5.8a}$$

$$\text{s.t. } R \le d_{\mathrm{KL}}(\boldsymbol{\omega}_i(\tau), \boldsymbol{\omega}_G^a(\tau)), \tag{5.8b}$$

$$(5.7b) \;\&\; (5.7c) \tag{5.8c}$$

In the next section, we proceed to solve Problem **P5.2** using a new GAE, which can iteratively regulate $\boldsymbol{\omega}_j^a(\tau)$, $\tau = 1, 2, \cdots$, to launch fairness attacks to FL [20].

By conducting such attacks on the fairness of FL, the attackers could influence critical decisions in FL, ranging from urban planning and policy-making processes to medical diagnosis predictions. This not only undermines the trustworthiness and integrity of FL systems but also enables the attackers to potentially exploit these biases for their own financial gain, strategic advantage, or to perpetuate discrimination, all under the guise of maintaining a seemingly unaffected FL model.

The core mechanism of our attack is the construction of a malicious, biasing model update based on the model updates overheard from the benign users. This approach is applicable to both CFL and DFL. As a matter of fact, in DFL, each participating user collects the local models from its nearby peers and aggregates the local models with its own local model. In this case, the attacker, which is one of the nearby peers, can still craft malicious local models (in the same way as in CFL) to bias the learning of benign users.

Although cryptography could offer protection against eavesdropping attacks to some extent, some recent techniques outlined in [2] and [9] have shown that encrypted information can be decrypted with minimal initial data. Moreover, the AGAT attack could exert a profound impact on the rapidly emerging field of DFL. In DFL, each user has the capability to directly receive local model updates from their neighbors, facilitated through either point-to-point encrypted or unencrypted channels. The DFL architecture inherently increases the attack surface, making DFL more susceptible to the AGAT attack.

Unlike a single aggregation point in a centralized counterpart, the direct exchange of local model updates in DFL allows adversaries to exploit the accessibility to neighbor model updates to inject malicious updates. In this sense, exploring the AGAT attack within the context of CFL serves as a stepping stone, elucidating potential vulnerabilities and informing the development of countermeasures in anticipation of similar, if not more sophisticated, threats in DFL.

5.4 Generating Biasing Model Updates with Adversarial Graph Attention Networks

In this section, we delineate the architecture of the AGAT attack that aims to generate malicious, biasing model updates. By leveraging attention mechanisms, the GAT dynamically weighs the importance of different vertices (i.e., data features) within FL, allowing for effective injection of malicious, biasing model updates. This functionality is crucial for exploiting potential vulnerabilities in the local models and the data the local models are trained on, thereby enabling the attackers to subtly introduce biases that can degrade the fairness of FL.

Furthermore, an adversarial GAE is designed within the AGAT architecture, which is trained together with sub-gradient descent to reconstruct manipulatively the correlations of the model updates, where the reconstruction loss is maximized. In addition, a graph signal processing module is designed with the GAE to decompose the correlation features of the benign model updates and the data features substantiating the model updates.

5.4.1 Architecture of AGAT Attack

Due to the high dimensionality of the training data, obtaining labeled data is expensive or prohibitive. A surrogate model $\tilde{g}(D_j^a(\tau))$ is used at attacker j to approximate the classification or image labeling, which simplifies the structure of the image classifier, thus reducing the computation burden. In particular, $\tilde{g}(D_j^a(\tau))$ can be trained by DNNs, Gaussian process regression, or polynomial regression. The output of the surrogate model yields a set of feature vectors of the training data.

Let A and B represent the number of vertices in the GAT and the number of features in each vertex, respectively. The vector that represents feature of $\tilde{g}(D_j^a(\tau))$, as well as the overheard $\boldsymbol{\omega}_i(\tau)$ can be denoted by $\boldsymbol{h} = \{\overrightarrow{h_1}, ..., \overrightarrow{h_e}\}$, where $e \in [1, A]$ is the graph size and $h_e \in \mathbb{R}^B$. Based on the input of $\boldsymbol{h}$, the adversarial GAT calculates attention coefficients for each of the vertices and features, which is given by [38]

$$\gamma_{xy} = \text{atn}(\boldsymbol{W^a}\overrightarrow{h_x}, \boldsymbol{W^a}\overrightarrow{h_y}), \tag{5.9}$$

where $\boldsymbol{W^a} \in \mathbb{R}^{B' \times B}$ is a weight matrix. Here, B' defines the size of the adversarial GAT's output, which is a set of biased features.

According to [16], atn$(\cdot)$ presents a shared attention function which can be specified as

$$\text{atn}(\boldsymbol{W^a}\overrightarrow{h_x}, \boldsymbol{W^a}\overrightarrow{h_y}) = \text{LeakyReLU}(\overrightarrow{c}^T[\boldsymbol{W^a}\overrightarrow{h_x} \otimes \boldsymbol{W^a}\overrightarrow{h_y}]), \tag{5.10}$$

where "$\otimes$" stands for a concatenation operation between the two matrices. $\overrightarrow{c}^T$ denotes the transpose of a weight vector $\overrightarrow{c} \in \mathbb{R}^{2B'}$ that is used to parametrize the atn$(\cdot)$ function in a neural network.

Moreover, at each vertex x, γ_{xy} is computed only for the neighbors of the vertex x in the graph, namely, $y \in \mathcal{N}_x$, where $\mathcal{N}_x$ denotes the neighborhood of x. To normalize attention weights and highlight important neighbors, a softmax function is used to normalize γ_{xy} across all choices of y. Thus, we have

$$\text{softmax}_y(\gamma_{xy}) = \frac{\exp(\gamma_{xy})}{\sum_{y' \in \mathcal{N}_x} \exp(\gamma_{xy'})}. \tag{5.11}$$

By substituting (5.10) into (5.11), the attention coefficients can be obtained by

$$\widehat{\gamma_{xy}} = \frac{\exp(\text{LeakyReLU}(\overrightarrow{c}^T[\boldsymbol{W^a}\overrightarrow{h_x} \otimes \boldsymbol{W^a}\overrightarrow{h_y}]))}{\sum_{y' \in \mathcal{N}_x} \exp(\text{LeakyReLU}(\overrightarrow{c}^T[\boldsymbol{W^a}\overrightarrow{h_x} \otimes \boldsymbol{W^a}\overrightarrow{h_{y'}}]))}. \tag{5.12}$$

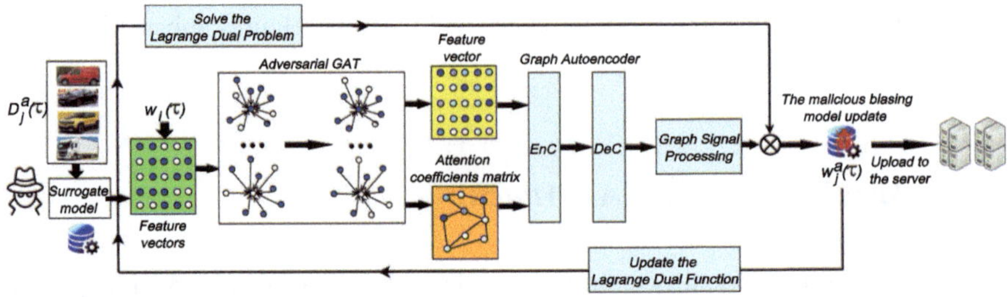

Fig. 5.3 The AGAT architecture, where the attacker creates a surrogate model that extracts labels of its biasing data. The adversarial GAT is trained to obtain the hidden representations of each feature in the graph

Based on (5.12), a normalized $\widehat{\gamma_{xy}}$ can be used to compute a linear combination of the features corresponding to the attention coefficients, to serve as the final output features for every vertex (after potentially applying a nonlinearity, ζ). Thus, we have $\boldsymbol{h}' = \{\overrightarrow{h_1}', ..., \overrightarrow{h_e}'\}$, which is give by

$$\overrightarrow{h_e}' = \zeta(\sum_{y \in \mathcal{N}_x} \widehat{\gamma_{xy}} \boldsymbol{W}^{\boldsymbol{a}} \overrightarrow{h_y}) \tag{5.13}$$

The optimization of the adversarial training model at the attacker in Problem **P5.2** is a non-convex combinatorial problem intractable for conventional optimization techniques. We decouple the AGAT architecture between the attack and the benign user selection using the Lagrangian-dual method. A new approach is developed to iteratively optimize the malicious, biasing model updates $\boldsymbol{\omega}_j^a(\tau)$ by running the adversarial GAT and subgradient descent, as depicted in Fig. 5.3.

The Lagrange function of Problem **P5.2** is given by

$$\begin{aligned}\mathcal{L}(\alpha_{i,j}^a(\tau), \lambda(\tau)) =& \Delta_{\text{Loss}} + \lambda(\tau)(d_T - \overline{\omega}_{i,j}) + \\ & \sum_{i=1}^{N} r_i(\tau)(d_{\text{KL}}(\boldsymbol{\omega}_i(\tau), \boldsymbol{\omega}_G^a(\tau)) - R),\end{aligned} \tag{5.14}$$

where $\lambda(\tau)$ and $r_i(\tau)$ denote the dual variables.

The Lagrange dual function is given by

$$\mathcal{D}_j(\lambda(\tau), r_i(\tau)) = \max_{\boldsymbol{\omega}_j^a(\tau), \alpha_{i,j}^a(\tau)} \mathcal{L}(\alpha_{i,j}^a(\tau), \lambda(\tau), r_i(\tau)). \tag{5.15}$$

The dual problem of (5.7) is given by

$$\min_{\lambda(\tau), r_i(\tau), \forall i} \mathcal{D}_j(\lambda(\tau), r_i(\tau)). \tag{5.16}$$

5.4.2 Generating Biasing Model Updates with GAE

5.4.2.1 GAE for Primary Variable Optimization

At the τth communication round, the primary variable $\boldsymbol{\omega}_j^a(\tau)$ of the Lagrange function (5.15) can be optimized according to

$$\boldsymbol{\omega}_j^a(\tau)^* = \arg\max_{\boldsymbol{\omega}_j^a(\tau)} \Bigg\{ \Delta_{\text{Loss}} - \lambda(\tau)(d_T - \overline{\omega}_{i,j}) - \sum_{i=1}^{N} r_i(\tau)(d_{\text{KL}}(\boldsymbol{\omega}_i(\tau), \boldsymbol{\omega}_G^a(\tau)) - R) \Bigg\}. \tag{5.17}$$

We optimize $\boldsymbol{\omega}_j^a(\tau)^*$ in (5.17) by designing a new GAE model with the AGAT architecture [23]. As shown in Fig. 5.3, is comprised of two primary components: an encoder and a decoder. Within this framework, the encoder is responsible for encoding the feature matrix $\boldsymbol{h}'$, utilizing the attention coefficients $\widehat{\gamma_{xy}}$ (which is described as an adjacency matrix $\mathcal{A}$), and the decoder takes the encoder's output as the input to reconstruct a biasing $\widehat{\mathcal{A}}$.

The encoder is constructed using an architecture based on M layers of graph convolutional networks (GCN). This design enables the encoder to learn a representation that effectively captures the essential characteristics of the model updates, which can be formulated as

$$\mathcal{Z}^M = f_G(\mathcal{Z}^{M-1}, \mathcal{A}|\mathbf{w}^M), \tag{5.18}$$

where $f_G(\cdot, \cdot|\cdot)$ represents a spectral convolution operation, while $\mathbf{w}^M$ signifies the weight matrix corresponding to the Mth layer within the GCN.

Given an identity matrix $I \in \mathbb{R}^{J \times J}$, $\widetilde{\mathcal{A}}$ can be formulated as $\widetilde{\mathcal{A}} = \mathcal{A} + I$, and we have $\overline{\mathcal{A}}_{xy} = \sum_{j'} \widetilde{\mathcal{A}}_{jj'}$. To generate a feature representation of the graph, the encoder can be written as

$$f_G(\mathcal{Z}^{M-1}, \mathcal{A}|\mathbf{w}^M) = \Phi^M(\overline{\mathcal{A}}^{-\frac{1}{2}} \widetilde{\mathcal{A}} \overline{\mathcal{A}}^{-\frac{1}{2}} \mathcal{Z}^{M-1} \mathbf{w}^M), \tag{5.19}$$

where $\Phi^M(\cdot)$ denotes a nonlinear activation function, e.g., $\tanh(\cdot)$ or $\text{ReLU}(\cdot)$; $\overline{\mathcal{A}}^{-\frac{1}{2}} \widetilde{\mathcal{A}} \overline{\mathcal{A}}^{-\frac{1}{2}}$ represents the symmetrically normalized adjacency matrix.

The output produced by the GAE is the reconstructed adjacency matrix, denoted $\widehat{\mathcal{A}}$, which is articulated as

$$\widehat{\mathcal{A}} = \text{sigmoid}\left(\mathcal{Z}^M \left(\mathcal{Z}^M\right)^T\right), \tag{5.20}$$

where the sigmoid function is specified by $\text{sigmoid}(x) = 1/(1 + \exp(-x))$. This formulation suggests that the likelihood of correlation between model updates within the graph increases with the magnitude of the inner product $(\mathcal{Z}^M(\mathcal{Z}^M)^T)$.

The discrepancy between $\mathcal{A}$ and its reconstructed counterpart $\widehat{\mathcal{A}}$ is quantified through a reconstruction loss function, as studied in [30], which is defined as

$$\phi_{\text{loss}} = \mathbb{E}_{f_G(\mathcal{Z}^{M-1},\mathcal{A}|\mathbf{w}^M)}\Big[\log\ p(\widehat{\mathcal{A}} \mid \mathcal{Z}^M)\Big], \tag{5.21}$$

where the probability $p(\widehat{\mathcal{A}} \mid \mathcal{Z}^M)$, as determined by the decoder, reflects the degree of correlation among the model updates.

Since the attacker aims to generate the malicious model updates for biasing FL, the proposed GAE is constructed and trained to maximize $L(\boldsymbol{\omega}_j^a(\tau), \lambda(t)) - \phi_{\text{loss}}$. As a consequence, the malicious model update $\boldsymbol{\omega}_j^a(\tau)$ increasingly biases the FL training fairness with the increase in global model aggregations, i.e., $\tau = 1, 2, \cdots$.

A graph signal processing module is introduced to analyze the correlation features present in benign model updates alongside the data attributes that support these updates. Utilizing the concept of a Laplacian matrix, denoted as ψ and constructed from the adjacency matrix ($\mathcal{A}$) of benign model updates such that $\psi = diag(\mathcal{A}) - \mathcal{A}$, as outlined in [37], we embark on a deeper exploration of these correlations. Through the application of SVD on ψ, represented as $\psi = B\Sigma B^T$, a complex unitary matrix $B \in \mathbb{R}^{J\times J}$ is derived. This matrix, also referred to as the GFT basis, facilitates the transformation of graph data (e.g., $\mathcal{F}$) into its spectral-domain representation. The matrix Σ, characterized as a diagonal matrix, contains the eigenvalues of ψ on its diagonal, providing a foundation for further analysis and manipulation of the graph data.

Consequently, an attacker can isolate a matrix S, encapsulating the spectral-domain data features of all benign model updates. This isolation, achieved by dissociating the correlations between models and then concentrating on the data features underpinning these model updates, is represented as

$$S = B^{-1}\mathcal{F}. \tag{5.22}$$

Furthermore, the attacker employs the graph signal processing module to create a Laplacian matrix from the output of the GAE, indicated by

$$\widehat{\psi} = diag(\widehat{\mathcal{A}}) - \widehat{\mathcal{A}}. \tag{5.23}$$

Subsequent application of SVD on $\widehat{\psi}$ yields the corresponding GFT basis, $\widehat{B}$. Leveraging the relationship defined in (5.22) for S, the malicious model update, which mirrors the adjacency matrix processed through the GAE, is identified as

$$\widehat{\mathcal{F}} = \widehat{B}S, \tag{5.24}$$

where $\widehat{\mathcal{F}}$ represents a matrix with dimensions $\mathbb{R}^{J\times D}$.

Within $\widehat{\mathcal{F}}$, the vector $\boldsymbol{\omega}_j^a(\tau)$ is identified and chosen by attacker j as the malicious, biasing model update. This chosen vector is then transmitted to the FL server by the attacker for inclusion in the aggregation of the global model during the communication round τ. Based on the graph signal processing module, the attacker can influence the training of global models, by strategically injecting tailored model updates designed to compromise FL fairness.

5.4.2.2 Sub-gradient Descent for Dual Variable Updating

Given $\boldsymbol{\omega}_j^a(\tau)$, the attacker can also update the dual variables, λ and r_i, $i = 1, \cdots, N$, specified in (5.16). Let ε denote the step size. Based on $\boldsymbol{\omega}_j^a(\tau)$ obtained from $\widehat{\mathcal{F}}$ in (5.24), $\lambda(\tau)$ and $r_i(\tau)$, $i = 1, \cdots, N$ are updated by the sub-gradient descent method that solves (5.16), where

$$\lambda(\tau + 1) = \left[\lambda(\tau) - \varepsilon\left(\overline{\omega}_{i,j} - d_T\right)\right]^+, \tag{5.25}$$

$$r_i(\tau + 1) = [r_i(\tau) - \varepsilon(d_{\mathrm{KL}}(\boldsymbol{\omega}_i(\tau), \boldsymbol{\omega}_G^a(\tau)) - R)]^+, \quad i = 1, \cdots, N, \tag{5.26}$$

and $[x]^+ = \max(0, x)$. At initialization, $\lambda(\tau)$ is non-negative, i.e., $\lambda(1) \geq 0$, to ensure that (5.25) converges. Moreover, $\boldsymbol{\omega}_j^a(\tau)$ is optimized with the updates of $\overline{\omega}_{i,j}$ and $\boldsymbol{\omega}_G^a(\tau)$ according to (5.3) and (5.4), respectively, after the global model aggregation.

5.4.3 Training Algorithm of AGAT Attack

According to the design of the new AGAT attack in Fig. 5.3, Algorithm 2 is developed along with the FL training of the benign users and the FL server. Specifically, the environment and parameters, such as the graph structure $\mathcal{G} = (\mathcal{V}, E, \mathcal{F})$, the total number of learning iterations T_L, and the datasets $D_i(\tau)$ and $D_j^a(\tau)$ are initialized. During each training iteration of FL, benign users train and upload $\boldsymbol{\omega}_i(\tau)$ to the server. The attacker conducts **AGAT**$(\tilde{g}(D_j^a(\tau)), \boldsymbol{\omega}_i(\tau), \lambda(\tau))$ to generate $\boldsymbol{\omega}_j^a(\tau)$, which trains $\tilde{g}(D_j^a(\tau))$ and extracts features using attention coefficients $\widehat{\gamma_{xy}}$ to emphasize correlations. These features are encoded using GAE to produce the adjacency matrix $\mathcal{A}$ that represents connections between model features [24]. The attacker then creates a biased model update in an attempt to manipulate the global model and uploads it to the server. The server aggregates these updates, including the malicious ones, to update $\boldsymbol{\omega}_G^a(\tau)$, which is then distributed across the users, including the benign ones, for the next training iteration. This cyclic process progressively biases the global model, undermining the integrity and effectiveness of FL. As $\boldsymbol{\omega}_j^a(\tau)$ is highly correlated with $\boldsymbol{\omega}_i(\tau)$ from the benign users, the FL server is unlikely to detect and identify the attackers.

Algorithm 2: The training algorithm of the proposed AGAT attack

1. Initialize: $\mathcal{G} = (\mathcal{V}, E, \mathcal{F})$, T_L, N, I, d_T, $D_i(\tau)$, and $D_j^a(\tau)$.
Proposed AGAT attack:
for Training iteration $\tau = 1, 2, 3, \cdots, T_L$. **do**
  Benign users train the local model updates $\boldsymbol{\omega}_i(\tau), i \in [1, I]$ according to (5.2).
  Benign users upload $\boldsymbol{\omega}_i(\tau), i \in [1, I]$ to the server, and the attacker $j \in [1, N - I]$ overhears the benign model updates of its neighbours.
  The proposed AGAT in Fig. 5.3 is conducted at the attacker to generate malicious, biasing model updates $\boldsymbol{\omega}_j^a(\tau)$, i.e., **AGAT**($\tilde{g}(D_j^a(\tau)), \boldsymbol{\omega}_i(\tau), \lambda(\tau)$):

    Training the surrogate mode $\tilde{g}(D_j^a(\tau)) \rightarrow \boldsymbol{h}$.
    The attention coefficients $\widehat{\gamma_{xy}} \leftarrow$ (5.12).
    Based on (5.13), $\boldsymbol{h}'$ that contains correlated features is obtained.
    The GAE encoder encodes $\boldsymbol{h}'$ with $\widehat{\gamma_{xy}}$ (which represents an adjacency matrix $\mathcal{A}$).
    $\widehat{\mathcal{A}}$ is reconstructed by training the GAE, which maximizes $L(\boldsymbol{\omega}_j^a(\tau), \lambda(t)) - \phi_{\text{loss}}$.
    According to the proposed graph signal processing module (5.22) $\sim$ (5.24), the malicious, biasing model update $\boldsymbol{\omega}_j^a(\tau)$ is obtained and uploaded to the server.
  At the server, (5.7b) is checked to detect the biasing model update.
  According to (5.4), the server aggregates the selected model updates to generate the global model $\boldsymbol{\omega}_G^a(\tau)$ that is broadcast to all the users.
  The benign users conduct training of their next model updates $\boldsymbol{\omega}_i(\tau + 1)$ with the received global model, i.e., $\boldsymbol{\omega}_i(\tau) \leftarrow \boldsymbol{\omega}_G^a(\tau), \forall i \in [1, I]$.
end for

5.5 Performance Evaluation

This section presents the implementation of the AGAT attack using PyTorch. When subjected to this attack, we assess the training accuracy of both local and global models. Moreover, the detection efficacy of the AGAT attack is examined through the metric of Cosine similarity between the local models and the global one.

5.5.1 *Experimental Settings*

The benign FL is designed to improve image classification accuracy, while the proposed AGAT attack aims to maximize the KL divergence between the user's model update and the global model, which leads to a biased FL training of the label classification. The total number of users N increases from 6 to 35, while the number of benign users I increases from 5 to 30. The global model $\boldsymbol{\omega}_G^a(\tau)$ in FL is trained with 100 communication rounds, and training of the local model $\boldsymbol{\omega}_i(\tau)$ is carried out in 10 iterations.

For building the architecture of the AGAT, the number of attention heads, the hidden layer size, dropout rate, weight decay, and the number of layers are set to 4, 80, 0.4, 2×10^3, and 2, respectively. The activation function is ReLU, as given in (5.10), for the intermediate layers due to its simplicity and effectiveness in alleviating the vanishing gradient problem, and softmax is used for the output layer when dealing with classification tasks. For building the adjacency matrix $\mathcal{A}$ in GAE at each attacker, the number of selected model parameters in $\boldsymbol{\omega}_i(\tau)$, i.e., M, is set to 100, 200, or 300. The GAE encoder is a two-layer GCN network with a dropout layer to prevent overfitting. The GAE decoder is an inner product. The Adam optimizer with a learning rate 0.01 is adopted to optimize the network. For all datasets, we use the same encoder, decoder and SVM models.

The implementation of the proposed AGAT attack was conducted on a SVM model, utilizing PyTorch version 1.12.1 and Python version 3.9.12. This setup was deployed on a Linux-based workstation, equipped with an Intel(R) Core(TM) i7-9700K CPU at 3.60GHz, featuring 8 cores, and supported by 16 GB of DDR4 memory operating at 2400 MHz. The experimentation involved the application of the AGAT attack across two distinct datasets, demonstrating the attack's efficacy and potential impacts on SVM models under specified computational environments and data conditions:

- The CIFAR-10 dataset [4], consists of 60,000 images in color, each with a dimension of 32×32 pixels, and distributed across ten distinct classes. Each class is represented by 6,000 images. This dataset is organized into two subsets: 50,000 images designated for training purposes and 10,000 images allocated for testing. This structure supports a wide range of image recognition tasks by providing a diverse set of visual inputs for model training and evaluation.
- The Street View House Numbers (SVHN) dataset [41], includes over 600,000 real-world digit images, featuring house numbers in their natural, unsegmented form, captured in a wide range of lighting conditions, angles, and backgrounds.

For our experiments, the CIFAR-10 and SVHN datasets are balanced in terms of their class distributions [31, 33]. This characteristic of the datasets is crucial, as our primary interest lies in a new fairness attack on FL. By training the AGAT based on balanced datasets, we ensure that the baseline conditions of our experiments do not inherently contain biases or imbalances that could confound the effects of the proposed attack. This allows accurate assessment and demonstration of the impact of the AGAT in biasing FL.

Three key performance metrics are investigated:

- KL divergence measures the difference between the probability distributions of a user's model update and the global model, providing insight into how much a local model deviates from the expected global distribution.
- The local model's testing accuracy assesses to what extent fairness is compromised without reducing FL accuracy under the proposed AGAT attack, making the attack difficult to detect at the server.
- Cosine similarity measures the angular similarity between the local models and the corresponding global model, which is used to evaluate the invisibility of the proposed AGAT attack.

In addition, the proposed AGAT attack is compared with an existing adversarial GAE-based model poisoning attack (G-MPA), as well as an existing fairness attack on FL, i.e., additive noise-based biasing attack (AN-BA):

- G-MPA focuses on compromising the integrity of benign local models by fabricating malicious training samples, thereby reducing the test accuracy of these models. This technique has been used in existing works, e.g., [13, 22, 36]. Specifically, the G-MPA involves the attacker disrupting the training process through the introduction of a counterfeit user. This fake user transmits malicious local models to the server, effectively manipulating the collective learning outcome.
- AN-BA, considered in [26] and [44], generates malicious local models by injecting a Gaussian random noise into the received global model, which can enlarge the magnitudes of the random local model updates using a scaling factor.

5.5.2 Attacking Performance

5.5.2.1 Kullback–Leibler Divergence

Given $I = 5$, Fig. 5.4 shows the KL divergence of each user i's $\boldsymbol{\omega}_i(\tau)$ under the proposed AGAT attack, i.e., $d_{\mathrm{KL}}(\boldsymbol{\omega}_i(\tau), \boldsymbol{\omega}_G^a(\tau))$ in (5.5). The performance is tested with the CIFAR-10 dataset in Fig. 5.4a or the SVHN dataset in Fig. 5.4b, given one or five attackers in FL. Generally, the KL divergence given five attackers is about three times higher than the case with a solo attacker. This is reasonable since the increasing number of attackers leads to more malicious, biasing model updates, i.e., $\boldsymbol{\omega}_j^a(\tau)^*$, being aggregated in FL. Consequently, the maximum loss function with regard to FL fairness in (5.6) increases.

In Fig. 5.5, we conduct a comparative analysis of the average KL divergence pertaining to local models subjected to the proposed AGAT attack versus those affected by the existing G-MPA and AN-BA. This comparison spans an increase in the number of benign users I from 5 to 30, alongside varying numbers of attackers from 1 to 5. Specifically, within the context of the CIFAR-10 dataset and with the presence of five attackers, Fig. 5.5a elucidates that the KL divergence under the AGAT attack exhibits a substantial elevation, 70.2% and 85.4% higher compared to the divergences under the G-MPA and AN-BA, respectively. Similarly, Fig. 5.5b illustrates that, when considering the SVHN dataset, the AGAT attack results in a KL divergence that surpasses that of the G-MPA and AN-BA attacks by 60.9% and 78.6%, respectively. Such findings underscore a significant bias in FL fairness induced by the AGAT attack. This bias stems from our innovative architecture based on GAT and GAE, which tailors the adversarial adjacency matrix in alignment with the unique features of the users' local model updates. As a result, this leads to the generation of maliciously biased model updates aimed at maximizing the loss differential, Δ_{Loss}, as studied in (5.6).

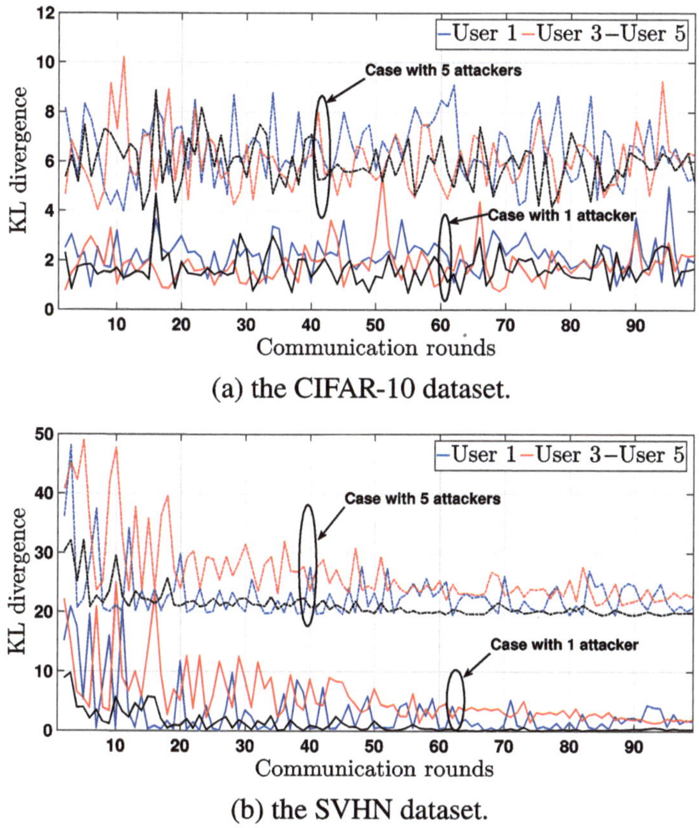

Fig. 5.4 Given $I = 5$, the KL divergence $d_{\mathrm{KL}}(\boldsymbol{\omega}_i(\tau), \boldsymbol{\omega}_G^a(\tau))$ under attacks with one or five attackers

As shown in Fig. 5.5, the KL divergence associated with local models increases concomitantly with the augmentation in the number of attackers, given a fixed I. This substantiates the detrimental impact of the proposed AGAT attack on the fairness of FL. Furthermore, the KL divergence decreases with the increment of I, since increasing the number of benign users can fortify the resilience of FL against fairness attacks. This enhancement in resistance is attributable to the aggregation of an increased number of benign local models, which inherently dilutes the adversarial influence exerted by the attackers, thereby preserving the integrity and fairness of the FL process.

Figure 5.6 illustrates the KL divergence $d_{\mathrm{KL}}(\boldsymbol{\omega}_i(\tau), \boldsymbol{\omega}_G^a(\tau))$ as the number of overheard updates $\boldsymbol{\omega}_i(\tau)$ increases from 4 to 20, where $I = 20$. It is observed that the KL divergence grows in proportion to the number of model updates overheard. For instance, with five attackers, the KL divergence rises by 75.8%. By contrast, with a single attacker, it increases by 84.6%. This demonstrates that the more benign model updates the attacker can eavesdrop on, the more correlated features the proposed

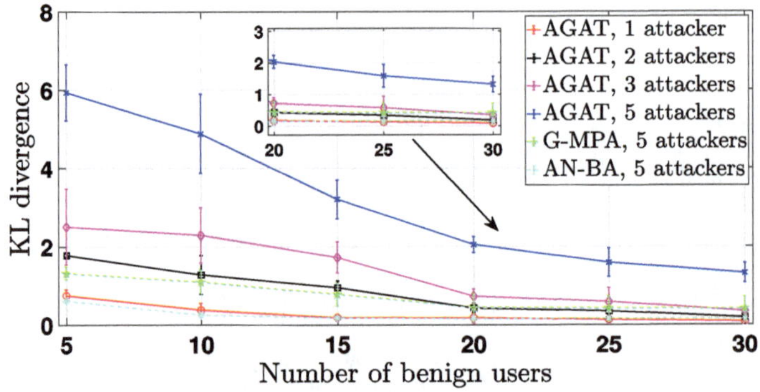

(a) the CIFAR-10 dataset.

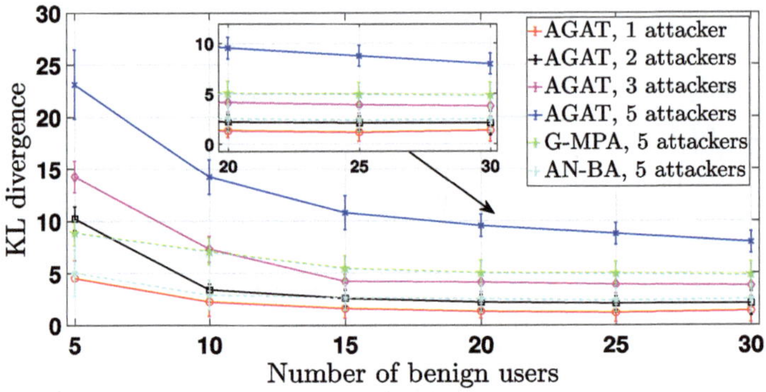

(b) the SVHN dataset.

Fig. 5.5 When I increases from 5 to 30, the KL divergence $d_{\mathrm{KL}}(\boldsymbol{\omega}_i(\tau), \boldsymbol{\omega}_G^a(\tau))$ in the presence of one, two, three, or five attackers

AGAT can exploit to generate malicious, biasing model updates. As a result, the fairness of the FL process is further compromised, with model updates becoming increasingly skewed.

To further study the probability distribution of malicious, biasing model updates generated by the proposed AGAT attack, Fig. 5.7 plots the cumulative distribution function (CDF) of the KL divergence and global model accuracy. Based on the CIFAR-10 dataset in Fig. 5.7a, b as well as the SVHN dataset in Fig. 5.7c, d, we observe that the CDF with more attackers results in a higher probability of the KL divergence. More importantly, the CDF of the malicious, biasing model updates exactly follows the same probability distribution pattern as the benign ones. Therefore, it is impossible for the server to identify the attacker. This is achieved by our innovative design of the AGAT architecture, namely, the adversarial GAT captures the correlations existing amongst data features within benign model updates, while the GAE is trained together with sub-gradient descent to reconstruct manipulatively the correlations of the model updates, and maximize the reconstruction loss.

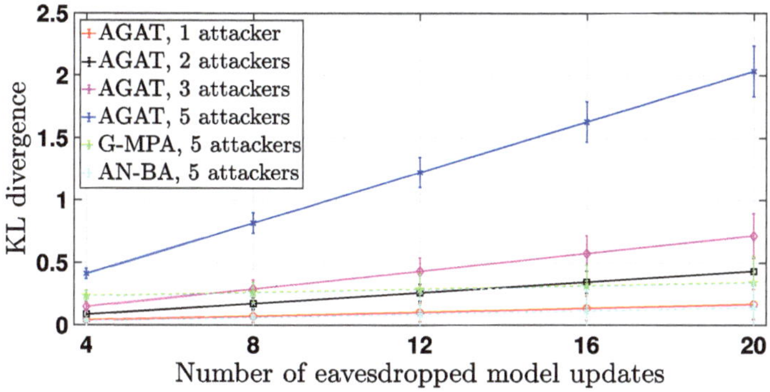

(a) the CIFAR-10 dataset.

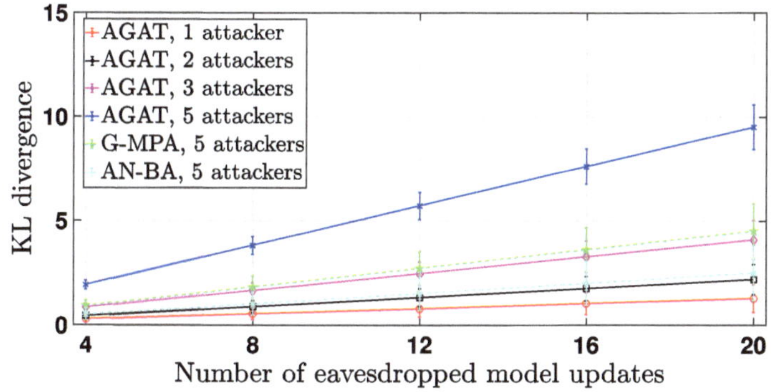

(b) the SVHN dataset.

Fig. 5.6 Given 20 users, the KL divergence $d_{\mathrm{KL}}(\boldsymbol{\omega}_i(\tau), \boldsymbol{\omega}_G^a(\tau))$, where the number of $\boldsymbol{\omega}_i(\tau)$ overheard increases from 4 to 20

Figure 5.8 presents the Jain's fairness index for the model updates when $I = 20$ and five attackers are present. Using the CIFAR-10 dataset, the performance of the proposed AGAT method is 25.6% and 74.2% lower than that of G-MPA and AN-BA, respectively. When evaluating the SVHN dataset, the Jain's fairness index for AGAT is 15.7% and 61.4% lower compared to G-MPA and AN-BA, respectively. These results confirm that the AGAT attack significantly compromises fairness in FL. This bias is driven by our novel architecture, which leverages GAT and GAE to adaptively construct an adversarial adjacency matrix aligned with the distinct features of the users' local model updates.

5.5.2.2 Learning Accuracy

Given $I = 5$, Fig. 5.9 shows the local model's testing accuracy under the proposed AGAT attack, based on the CIFAR-10 and SVHN datasets. As observed in

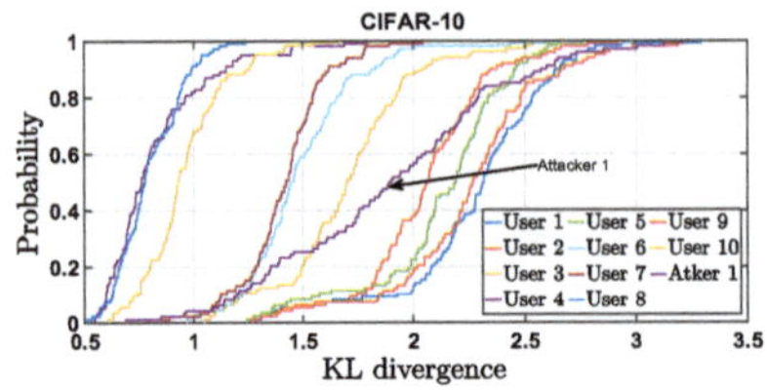

(a) one attacker with CIFAR-10.

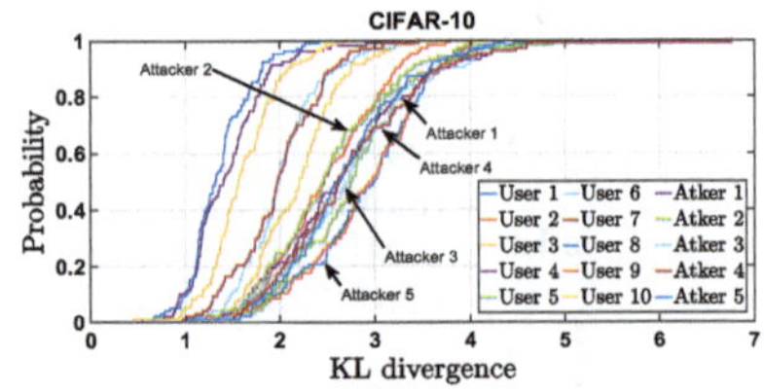

(b) five attackers with CIFAR-10.

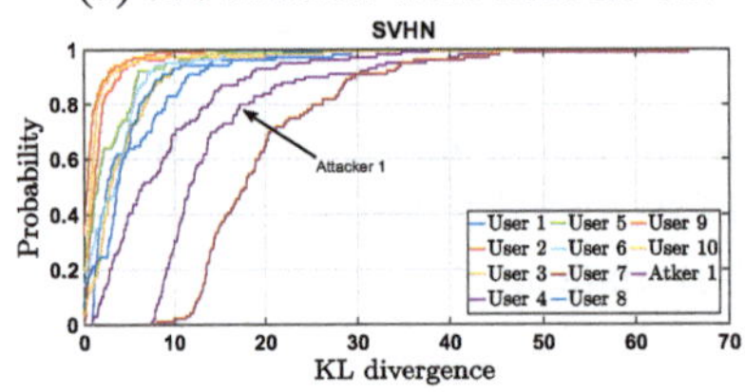

(c) one attacker with SVHN.

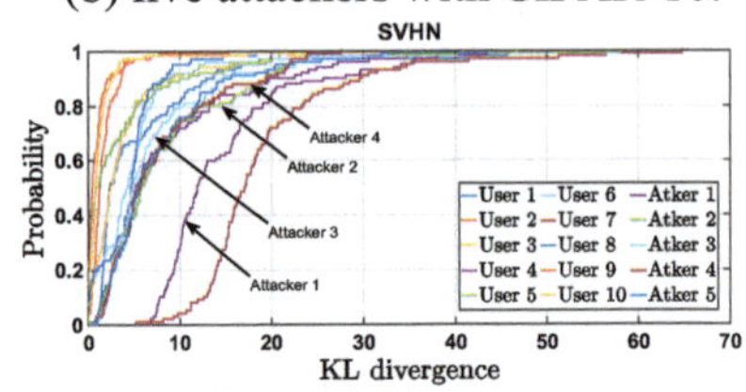

(d) five attackers with SVHN.

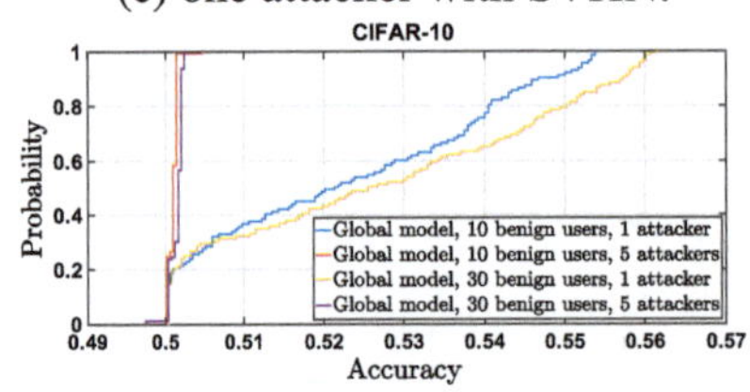

(e) Global model accuracy with CIFAR-10.

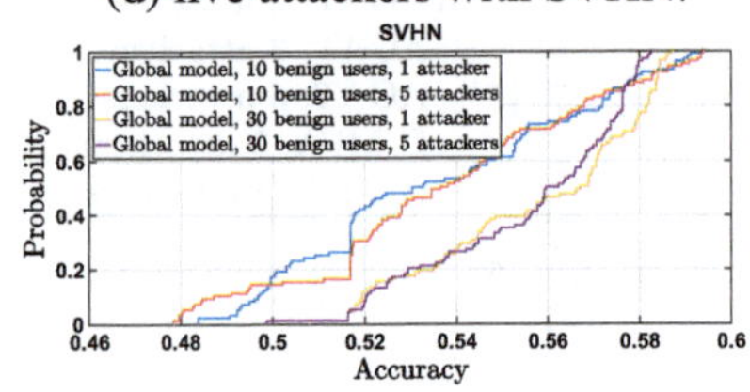

(f) Global model accuracy with SVHN.

Fig. 5.7 Given $I = 10$, CDF of the KL divergence and global model accuracy under the proposed AGAT attack

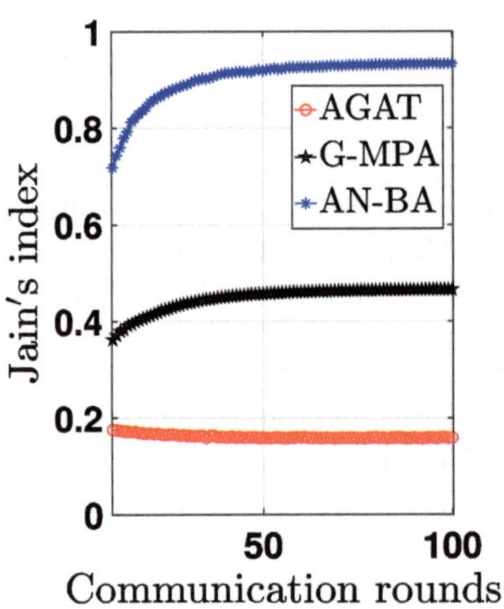

(a) the CIFAR-10 dataset.

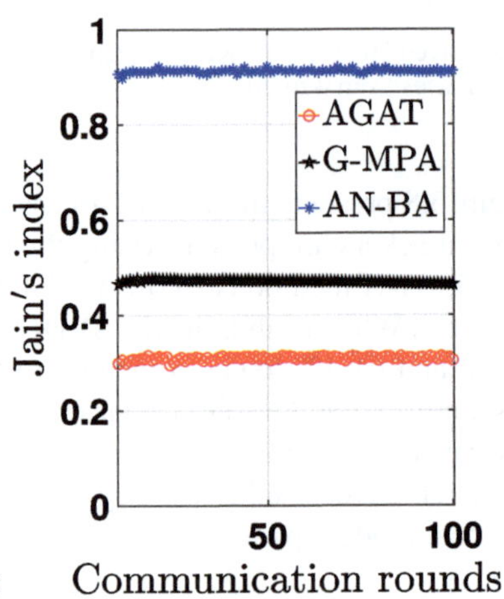

(b) the SVHN dataset.

Fig. 5.8 The Jain's fairness index of the model updates, when $I = 20$ and five attackers

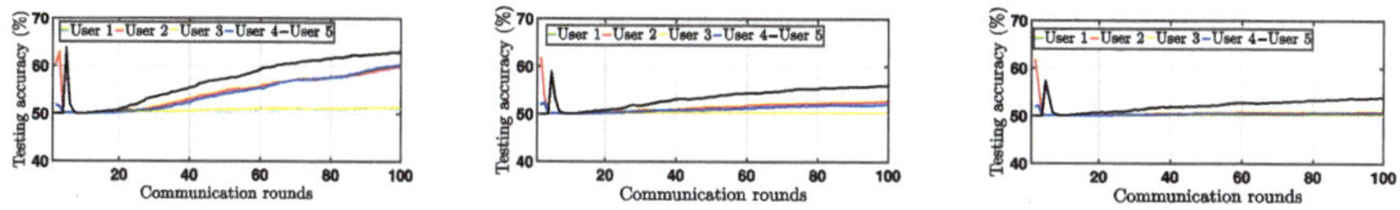

(a) one attacker with CIFAR-10. (b) three attackers with CIFAR-10. (c) five attackers with CIFAR-10.

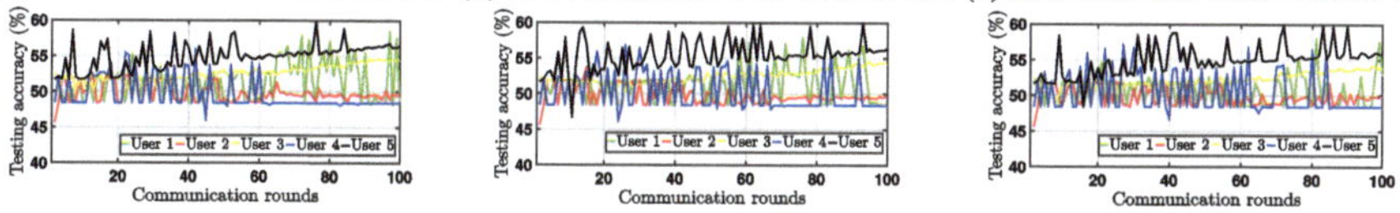

(d) one attacker with SVHN. (e) three attackers with SVHN. (f) five attackers with SVHN.

Fig. 5.9 Given 100 FL communication rounds and $I = 5$, the local model's testing accuracy under the proposed AGAT attack on the CIFAR-10 and SVHN datasets

Fig. 5.9a–f, despite the adversarial interventions, the FL accuracy not only remains unaffected but also continues to converge. This is attributed to the objective of the AGAT attack, which diverges from traditional adversarial tactics by specifically aiming to bias FL, rather than undermining their testing accuracy. Furthermore, the new adversarial GAT design captures the correlations among data features within benign model updates, thereby skewing the model's decision boundaries without detrimentally affecting FL accuracy. This nuanced strategy highlights a sophisticated attack vector that compromises the fairness and integrity of FL without the conventional hallmark of reduced accuracy, thus posing a more insidious threat that can elude standard detection mechanisms.

5.5.2.3 Cosine Similarity

To evaluate the invisibility of the proposed AGAT attack, we further investigate the Cosine similarity between the local and the global models [6], i.e., Constraint (5.7b), based on the CIFAR-10 and SVHN datasets in Fig. 5.10. As shown in Fig. 5.10a–d, the Cosine similarities between the malicious, biasing model updates generated by the new AGAT attack and the corresponding global models are always below that of the benign local model updates. This complicates detecting and defending against fairness biases at the server, as malicious updates can blend with legitimate data. In contrast, as depicted in Fig. 5.10e, f, both the G-MPA method and the AN-BA approach lead to a markedly higher Cosine similarity between the malicious, biasing model updates and the aggregate global models. This increased similarity offers a clearer signal for detection mechanisms. This contrast underlines the superior tactical advantage of the proposed AGAT attack: AGAT crafts malicious model updates by exploiting the feature correlations between benign local updates and the global model.

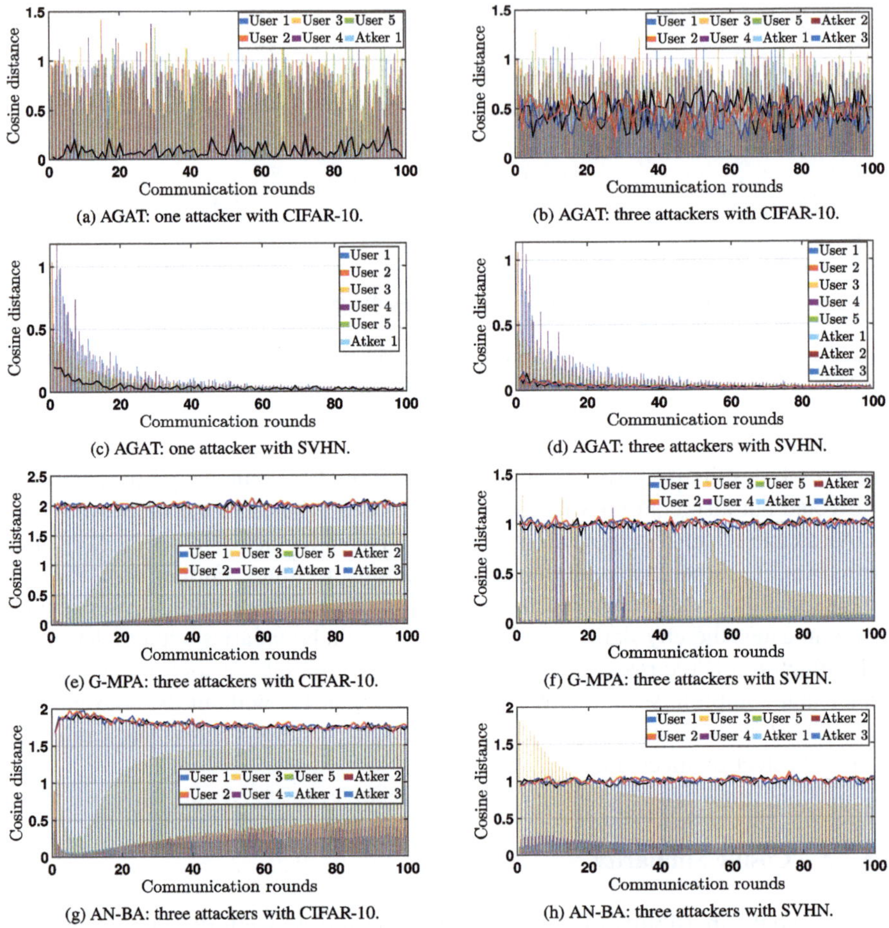

(a) AGAT: one attacker with CIFAR-10.
(b) AGAT: three attackers with CIFAR-10.
(c) AGAT: one attacker with SVHN.
(d) AGAT: three attackers with SVHN.
(e) G-MPA: three attackers with CIFAR-10.
(f) G-MPA: three attackers with SVHN.
(g) AN-BA: three attackers with CIFAR-10.
(h) AN-BA: three attackers with SVHN.

Fig. 5.10 Given 100 FL communication rounds and $I = 5$, the Cosine similarities of the local models are measured at the server to detect an adversarial attack, based on the CIFAR-10 and SVHN datasets

This strategy effectively obfuscates the distinctions between benign and malicious updates, rendering the latter virtually undetectable and showcasing the sophistication and potential efficacy of AGAT in compromising model integrity.

References

1. Awan S, Luo B, Li F (2021) Contra: Defending against poisoning attacks in federated learning. In: Proceedings of European symposium on research in computer security (ESORICS), pp 455–475. Springer
2. Beck G, Zinkus M, Green M (2020) Automating the development of chosen ciphertext attacks. In: Proceedings of USENIX security symposium, pp 1821–1837
3. Blanchard P, El Mhamdi EM, Guerraoui R, Stainer J (2017) Machine learning with adversaries: Byzantine tolerant gradient descent. Advances in neural information processing systems, vol 30 (2017)
4. Çalik RC, Demirci MF (2018) Cifar-10 image classification with convolutional neural networks for embedded systems. In: IEEE/ACS 15th international conference on computer systems and applications (AICCSA), pp 1–2. IEEE
5. Cao X, Zhang Z, Jia J, Gong NZ (2022) Flcert: provably secure federated learning against poisoning attacks. IEEE Trans Inf Forensics Secur 17:3691–3705
6. Chen G, Li K, Abdelmoniem AM, You L (2024) Exploring representational similarity analysis to protect federated learning from data poisoning. In: Companion proceedings of the ACM on web conference (WWW), pp 525–528
7. Chen N, Akar G, Gordon SI, Chen S (2021) Where do you live and what do you drive: built-environmental and spatial effects on vehicle type choice and vehicle use. Int J Sustain Transp 15(6):444–455
8. Chen X, Yu H, Jia X, Yu X (2023) Apfed: anti-poisoning attacks in privacy-preserving heterogeneous federated learning. IEEE Trans Inf Forensics Sec
9. Ding Y, Zhu G, Chen D, Qin X, Cao M, Qin Z (2022) Adversarial sample attack and defense method for encrypted traffic data. IEEE Trans Intell Transp Syst 23(10):18024–18039
10. Dong C, Weng J, Li M, Liu JN, Liu Z, Cheng Y, Yu S (2023) Privacy-preserving and byzantine-robust federated learning. IEEE Trans Depend Secure Comput
11. Ezzeldin YH, Yan S, He C, Ferrara E, Avestimehr AS (2023) Fairfed: enabling group fairness in federated learning. In: Proceedings of AAAI conference on artificial intelligence, vol 37, pp 7494–7502
12. Fang M, Cao X, Jia J, Gong N (2020) Local model poisoning attacks to {Byzantine-Robust} federated learning. In: USENIX security symposium, pp 1605–1622
13. Gao L, Li L, Chen Y, Zheng W, Xu C, Xu M (2021) Fifl: a fair incentive mechanism for federated learning. In: Proceedings of the international conference on parallel processing, pp 1–10
14. Hao W, El-Khamy M, Lee J, Zhang J, Liang KJ, Chen C, Duke LC (2021) Towards fair federated learning with zero-shot data augmentation. In: Proceedings of IEEE/CVF conference on computer vision and pattern recognition (CVPR), pp 3310–3319
15. Horvath S, Laskaridis S, Almeida M, Leontiadis I, Venieris S, Lane N (2021) Fjord: fair and accurate federated learning under heterogeneous targets with ordered dropout. Proc Adv Neural Inf Process Syst (NeurIPS) 34:12876–12889
16. Jiang N, Jie W, Li J, Liu X, Jin D (2022) Gatrust: a multi-aspect graph attention network model for trust assessment in osns. IEEE Trans Knowl Data Eng (2022)
17. Kaheni M, Lippi M, Gasparri A, Franceschelli M (2024) Selective trimmed average: a resilient federated learning algorithm with deterministic guarantees on the optimality approximation. IEEE Trans Cybernet
18. Le J, Lei X, Mu N, Zhang H, Zeng K, Liao X (2021) Federated continuous learning with broad network architecture. IEEE Trans Cybernet 51(8):3874–3888
19. Li K, Cui Y, Li W, Lv T, Yuan X, Li S, Ni W, Simsek M, Dressler F (2022) When internet of things meets metaverse: convergence of physical and cyber worlds. IEEE Internet Things J 10(5):4148–4173
20. Li K, Hu S, Wu B, Zou S, Ni W, Dressler F (2025) Undermining federated learning accuracy in edgeiot via variational graph auto-encoders. In: 2025 International wireless communications and mobile computing (IWCMC), pp 716–721. IEEE

21. Li K, Yuan X, Zheng J, Ni W, Dressler F, Jamalipour A (2024) Leverage variational graph representation for model poisoning on federated learning. IEEE Trans Neural Netw Learn Syst
22. Li K, Yuan X, Zheng J, Ni W, Guizani M (2023) Exploring adversarial graph autoencoders to manipulate federated learning in the internet of things. In: Proceedings of IEEE international wireless communications and mobile computing (IWCMC), pp 898–903. IEEE
23. Li K, Zheng J, Ni W, Huang H, Liò P, Dressler F, Akan OB (2024) Biasing federated learning with a new adversarial graph attention network. IEEE Trans Mobile Comput
24. Li K, Zheng J, Ni W, Huang H, Lio P, Dressler F, Akan OB (2025) Explainable graph attention-driven fairness manipulation for federated learning in edgeiot. In: 2025 IEEE/CIC international conference on communications in China (ICCC), pp 1–6. IEEE
25. Li K, Zheng J, Yuan X, Ni W, Akan OB, Poor HV (2024) Data-agnostic model poisoning against federated learning: A graph autoencoder approach. IEEE Trans Inf Forensics and Sec
26. Li T, Hu S, Beirami A, Smith V (2021) Ditto: fair and robust federated learning through personalization. In: Proceedings of international conference on machine learning (ICML), pp 6357–6368. PMLR
27. Li Z, Liu J, Wu K (2017) A multiobjective evolutionary algorithm based on structural and attribute similarities for community detection in attributed networks. IEEE Trans Cybernet 48(7):1963–1976
28. Lyu L, Xu X, Wang Q, Yu H (2020) Collaborative fairness in federated learning. Federated learning: privacy and incentive, pp 189–204
29. Ma Z, Ma J, Miao Y, Li Y, Deng RH (2022) ShieldFL: mitigating model poisoning attacks in privacy-preserving federated learning. IEEE Trans Inf Forensics Secur 17:1639–1654
30. Mrabah N, Bouguessa M, Touati MF, Ksantini R (2022) Rethinking graph auto-encoder models for attributed graph clustering. IEEE Trans Knowl Data Eng
31. Nayak GK, Mopuri KR, Chakraborty A (2021) Effectiveness of arbitrary transfer sets for data-free knowledge distillation. In: Proceedings of the IEEE/CVF winter conference on applications of computer vision, pp 1430–1438
32. Park S, Han S, Wu F, Kim S, Zhu B, Xie X, Cha M (2023) Feddefender: client-side attack-tolerant federated learning. In: Proceedings of ACM SIGKDD conference on knowledge discovery and data mining, pp 1850–1861
33. Park S, Lim J, Jeon Y, Choi JY (2021) Influence-balanced loss for imbalanced visual classification. In: Proceedings of the IEEE/CVF international conference on computer vision, pp 735–744
34. Qi T, Wu F, Wu C, Lyu L, Xu T, Liao H, Yang Z, Huang Y, Xie X (2022) Fairvfl: a fair vertical federated learning framework with contrastive adversarial learning. Proc Adv Neural Inf Process Syst (NeurIPS) 35:7852–7865
35. Shi J, Wan W, Hu S, Lu J, Zhang LY (2022) Challenges and approaches for mitigating byzantine attacks in federated learning. In: Proceedings of IEEE international conference on trust, security and privacy in computing and communications (TrustCom), pp 139–146. IEEE
36. Tang W, Liu E, Ni W, Qu X, Huang B, Li K, Niyato D, Jamalipour A (2025) Game-theoretic incentive mechanism for blockchain-based federated learning. IEEE Trans Mob Comput 24(10):10363–10376. https://doi.org/10.1109/TMC.2025.3567355
37. Van Mieghem P (2023) Graph spectra for complex networks. Cambridge University Press
38. Veličković P, Cucurull G, Casanova A, Romero A, Liò P, Bengio Y (2018) Graph attention networks. In: International conference on learning representations
39. Wang H, Sreenivasan K, Rajput S, Vishwakarma H, Agarwal S, Sohn JY, Lee K, Papailiopoulos D (2020) Attack of the tails: yes, you really can backdoor federated learning. Proc Adv Neural Inf Process Syst (NeurIPS) 33:16070–16084
40. Yin D, Chen Y, Kannan R, Bartlett P (2018) Byzantine-robust distributed learning: towards optimal statistical rates. In: Proceedings of the 35th international conference on machine learning, Proceedings of machine learning research, vol 80, pp 5650–5659. PMLR
41. Yuval N (2011) Reading digits in natural images with unsupervised feature learning. In: Proceedings of the NIPS workshop on deep learning and unsupervised feature learning

42. Zhang DY, Kou Z, Wang D (2020) Fairfl: a fair federated learning approach to reducing demographic bias in privacy-sensitive classification models. In: IEEE International conference on big data (big data), pp 1051–1060. IEEE
43. Zheng J, Li K, Mhaisen N, Ni W, Tovar E, Guizani M (2022) Exploring deep-reinforcement-learning-assisted federated learning for online resource allocation in privacy-preserving edgeiot. IEEE Internet Things J 9(21):21099–21110
44. Zheng J, Li K, Yuan X, Ni W, Tovar E (2024) Detecting poisoning attacks on federated learning using gradient-weighted class activation mapping. In: Companion proceedings of the ACM on web conference (WWW), pp 714–717
45. Zheng Y, Lai S, Liu Y, Yuan X, Yi X, Wang C (2022) Aggregation service for federated learning: an efficient, secure, and more resilient realization. IEEE Trans Depend Secure Comput 20(2):988–1001

Chapter 6
Data-Agnostic MP Techniques

The use of mobile edge computing is increasingly prevalent, especially in catering to user devices that come with a multitude of sensors. These sensors produce vast amounts of data, like images recording human activities or the real-time locations of vehicles, as seen in smart city scenarios [22, 55]. However, transferring this training data from the user's device to a server can pose a threat to data privacy. FL is an emerging distributed ML approach that gains traction as a solution to mitigate data privacy concerns [20]. With FL, user devices can jointly train an ML model without having to disclose their private data to a server. The user devices, acting as clients, iteratively train their local models on their private data and send the local model updates to a server. At the server, a global model is updated without collecting private data from the user devices. The global model is then sent back to the user devices, allowing them to continue training their local models based on the global model and their local data [60]. This process helps to support data privacy and allows for real-time processing capabilities at the edge of networks, making FL a significant aspect of mobile edge computing.

Despite the fact that FL can help prevent attackers from accessing the private data of user devices, an attacker (in most cases, a malicious user device) can potentially launch MP or data poisoning attacks to manipulate FL and propagate the attacks into benign user devices [18, 35], resulting in a failure of FL training. Specifically, MP aims to send malicious local model updates to the server during an aggregation process. The malicious update can introduce specific vulnerabilities in the global model or simply degrade FL performance. By contrast, data poisoning attempts to inject malicious data or modify existing data on user devices to misguide local model training, thus compromising local model updates. Existing data poisoning attacks generally require an attacker to have some knowledge of the datasets used for FL training [40], so that it can extract and manipulate the features of the datasets for effective attacks [54]. By launching MP attacks [13] or data poisoning attacks [26], an attacker could manipulate either the hyperparameters of the local models or the training datasets of benign users to compromise learning accuracy.

K. Li et al., *Security and Resilience in Distributed Machine Learning*, Springer Series in Reliability Engineering, https://doi.org/10.1007/978-3-032-23959-4_6

Much less constrained and potentially more threatening MP attacks on FL would result if they could be based solely on the benign local models overheard by an attacker and the global models broadcast by the aggregator; i.e., when the attacker has no access to the training data. However, without training data, it is challenging for the malicious local models to strike a balance between effectiveness and undetectability [36]. To the best of our knowledge, such attacks are new and have not been reported in the literature.

6.1 Introduction of Data-Agnostic MP

This chapter presents a new, data-agnostic, MP attack on FL systems, where an adversarial graph autoencoder (GAE) [6, 24] is designed to generate malicious local models solely based on the benign local models overheard and capture the correlation features of the benign local and global models. Specifically, an attacker overhears the benign local models uploaded by the user devices, and the global model broadcast by the server. GAE is adept at capturing complex relationships and structures inherent in graph-structured data. It can efficiently encode graph information into a lower-dimensional latent space, while preserving the essential topological features of a graph. Using GAE, the attacker extracts the graph structure capturing the correlations between the benign local models (that could be transmitted over a transport layer security (TLS) protocol), and decouples the graph structure from underlying data features, substantiating the local models. The attacker first regenerates the graph structure to retain the structural features of the local models and maximize the FL training loss using the GAE, and then generates malicious local models using the regenerated graph structure to the data features of the benign local models. As a result, the malicious local models can effectively compromise the global model, while remaining compatible with the benign models and hence reasonably undetectable.

The contributions of this chapter are summarized as follows.

- A new design of data-agnostic, malicious local models, which manipulates the correlations of benign local models and retains the genuine data features substantiating the benign local models;
- A new GAE framework, which is trained together with sub-gradient descent to regenerate manipulatively the correlations of the local models while keeping the malicious local models undetectable; and
- A rigorous analysis, which proves the convergence of the global model under attack, but to an inferior optimality gap.
- The proposed GAE-based attack is implemented experimentally based on the standard MNIST, FMNIST, and CIFAR-10 datasets. It is shown that the GAE-based attack significantly compromises the FL performance, where the training accuracy falls below 50% at the user devices. The source code of the proposed GAE-based, data-agnostic, MP attack is available on GitHub.

Table 6.1 Notation and Definition

Symbol	Meaning
$\mathcal{A}$	Adjacency matrix of local models
$\widehat{\mathcal{A}}$	Reconstructed adjacency matrix at decoder
$\mathcal{F}$	Feature matrix
$F(\boldsymbol{\omega}_j(t))$	Weighted global loss function
$\widehat{\mathcal{F}}$	Malicious local model
$\boldsymbol{\omega}_g(t)$	Global FL model at communication round t
$\boldsymbol{\omega}_g^a(t)$	Global model under attack at round t
$f_i(\boldsymbol{\omega}_j(t); x_j^i, y_j^i)$	Training loss of device j for sample (x_j^i, y_j^i)
J	Total number of benign user devices
η	Learning rate of local training
T_L	Training iterations per FL round
d_T	Euclidean distance threshold
$\mathcal{L}$	Laplacian matrix based on benign weights
$\widehat{\mathcal{L}}$	Laplacian matrix regenerated by attacker
$F_j(\boldsymbol{\omega}_j(t))$	Local loss function of device j

Extensive experiments indicate that the FL accuracy drops gradually under the proposed attack, and the existing poisoning defense mechanisms can hardly detect the attack. Since the malicious local models are uploaded to the server for global model aggregation, the proposed attack gives rise to an epidemic infection across all benign devices.

The new GAE-based attack on FL involves attackers intentionally poisoning malicious local models, aiming to degrade or manipulate the performance of the global model. The attack challenges the security, privacy, and robustness of FL. While security is threatened by unauthorized access or malicious insiders tampering with local models, privacy concerns arise when the attackers try to reverse-engineer or glean information about the benign devices' data. Moreover, robustness, which is the ability of FL to consistently produce reliable and accurate results, can be directly undermined, as poisoned local models compromise the integrity and efficacy of FL. To this end, the proposed GAE-based attack poses a comprehensive threat to the security, privacy, and robustness of FL.

The rest of this chapter is organized as follows. Section 6.2 introduces the background of adversarial attacks against wireless systems and FL. Section 6.3 discusses FL with benign user devices and server, as well as the eavesdropping model. The proposed GAE-based epidemic attack is delineated in Sect. 6.4. Performance analysis is conducted in Sect. 6.5. Table 6.1 lists the notation used in the chapter.

6.2 Adversarial Attacks Against Wireless Federated Learning

This section reviews the literature on adversarial attacks against wireless systems, as well as FL, including model and data poisoning attacks. On the one hand, because of their broadcast nature, wireless channels are particularly vulnerable to eavesdropping attacks. An attacker is likely to overhear the local model updates transmitted by the other benign users in wireless FL. On the other hand, the MP attack considered in this chapter has not been studied in the literature. Instead, existing attacks on wireless FL have focused primarily on building an adversarial data classification/label model for attackers, according to the data packets and features overheard, e.g., [56, 61]. There is an opportunity for the new attack to strike.

6.2.1 Adversarial Attacks on Wireless Systems

In [37], an adversarial attack was studied to manipulate the measurement of smart meters in residential homes. Smart meter data could inform residents of which appliances consume the most electricity and adjust energy production. The attacker employed deep learning to train a power usage pattern classification model and generated malicious data that was indistinguishable from the true data. In [56], ML was used to generate an adversarial attack for targeting data fusion or aggregation. The attacker infiltrated some devices and learned the decision process and data fusion settings by observing data exchanges between the devices and the data center.

In [3], the authors analyzed targeted adversarial attacks that aimed to manipulate the output of a convolutional neural network (CNN)-based classifier. They also evaluated non-targeted adversarial attacks against convolutional NN-based device identification. To evaluate these attacks, the authors used combined indicators of logits to increase the perturbation levels and iterative steps, resulting in a high success rate of adversarial attacks. In [34], researchers used deep learning to recognize COVID-19 symptoms by training on medical data from user devices. They evaluated several adversarial attacks that aimed to falsify the data and symptom recognition. The study found that existing deep learning algorithms were vulnerable to these attacks, highlighting the need for advanced security measures.

In [61], an adversarial attack was developed to deactivate graph-based intrusion detection in a targeted wireless system. The attack began by building a shadow graph based on overheard data packets and features. A random walk algorithm was then used to evaluate each node in the attacker's graph, selecting the node with the largest weight to attack. The attack would perturb data features and alter classification labels. In [1], an adversarial attack was developed to utilize graph embedding and augmentation to misclassify system malware samples as benign. The graph-based attack aimed to embed a target malware sample into benign software. By combining the benign code sample and the target malware sample in the graph, the adversarial attack could learn complex features, resulting in a high misclassification rate at the user device. In [39], a study was conducted on a Sybil-based data poisoning attack against deep reinforcement learning (DRL)-based service placement in the Internet of Vehicles (IoV). The attack targeted the agent that is responsible for learning the service quality and deciding on service placement based on delay. A Sybil attacker, which is a malicious vehicle, used data poisoning techniques to masquerade as a legitimate vehicle by stealing or borrowing its identity. The attacker then maliciously sent false data to other vehicles.

Unfortunately, it is difficult for the attacker to formulate the adversarial data classification/label model in FL systems since the benign user devices can collaboratively conduct model training without sharing their private data.

6.2.2 *Poisoning Attacks on FL*

The attacker can launch either a data poisoning or an MP attack. In the data poisoning attack, the attacker injects fake data with manipulated features and flips labels into the benign user devices. In the MP attack, the attacker submits malicious local models to the server. Both attacks aim to corrupt by introducing false information.

In [36], the authors systematically categorized the existing threat models associated with poisoning attacks on FL, where practical boundaries of numerous parameters pertinent to FL robustness were delineated. An array of untargeted model and data poisoning attacks on FL was analyzed to encompass the existing attack strategies. An MP attack was developed using gradient ascent to fine-tune the global model and increase its loss on benign data. The MP attack adjusts the L_2-norm of the poisoned model update to circumvent the robustness criterion of the model aggregation. In [10], an adversarial attack mitigation scheme based on clustering was studied. The scheme aimed to protect FL by using unsupervised weight training to split and merge weight clusters at the server to filter out malicious local models that were uploaded by the user devices without identity verification. In [41], malicious local models were derived from mislabeled data to manipulate the global model. The study found that this attack could result in a significant drop in classification accuracy, and that it was difficult to detect due to its negative impact on the target device and minimal impact on other benign devices.

In [14], an inference model was formulated to take local models as input and output the categories of data. A malicious local model based on a differential selection strategy was used to select two adjacent categories. To approximate the benign local model, a category inference attack was studied, in which the attacker learns the data features underlying benign local models. The authors of [44] presented a backdoor attack against FL in MEC, which targeted the tail of the input data distribution at the local devices. The attack used projected gradient descent to maintain the distance between the malicious local model and the global model, to misclassify the targeted samples and bypass defense mechanisms.

In [52], GANs were utilized to construct data poisoning attacks against FL. The attacker trained the GAN to replicate the local data of the benign devices. Since the attacker had no information about the local data, the GAN-based data poisoning updated the global model to re-select the potential targeted devices. In [53], a GAN-based FL poisoning attack was studied, where the attacker posed as one of the benign devices and trained the GAN to mimic the dataset of the benign devices. The malicious data generated by the attacker were trained to compromise the global model. In [47], a malicious server deployed a GAN-based reconstruction attack against FL to tamper with the private data of the user devices. The malicious server discriminated the devices' identities and data representatives to supervise the training of GANs and generate malicious data for each specific device. In [38], the authors focused on a device-level privacy leakage attack launched by a malicious server. A GAN-based framework was presented to discriminate the data category and device's identity and recover the private data of the device. The attack could associate the data features from different devices to re-identify the local models.

Unfortunately, the existing data poisoning or MP attacks have not exploited the implicit relationship between local models [7, 27]. Moreover, the existing poisoning attacks generally require the attacker to have the knowledge of (part of) the datasets used for FL training.

6.3 System Model

In this section, we first describe an FL training process, e.g., for image classification. Next, we present the threat model, where malevolent devices can act as attackers. An attacker creates and uploads malicious local model updates to progressively contaminate the global model of . At last, we describe an attacker detection model that the server can adopt to discern malicious local models by measuring the Euclidean distances between the models.

6.3.1 Model Aggregation in Federated Learning

We assume there are J benign user devices and an authorized (legitimate) but malicious user device (or an attacker) in the FL training process. A benign user device $j \in [1, J]$ has $D_j(\tau)$ amount of data at the τth iteration. Let x^i_j and y^i_j denote the input of the captured images and the output of the FL model at device j, respectively. $i \in [1, D_j(\tau)]$. A training loss function of device j, denoted by $f_i(\boldsymbol{\omega}_j(\tau); x^i_j, y^i_j)$, captures approximation errors over the input x^i_j and the output y^i_j. Here, $\boldsymbol{\omega}_j(\tau)$ is the weight parameter of the loss function in the model being trained by . For instance, $f_i(\boldsymbol{\omega}_j(\tau); x^i_j, y^i_j)$ can be modeled by linear regression, i.e., $f_i(\boldsymbol{\omega}_j(\tau); x^i_j, y^i_j) = \frac{1}{2}(\boldsymbol{\omega}_j(\tau)^T x^i_j - y^i_j)^2$; or logistic regression, i.e., $f_i(\boldsymbol{\omega}_j(\tau); x_j^i, y_j^i) = y_j^i \log\Big(1 + \exp\big(-\boldsymbol{\omega}_j(\tau)^T x_j^i\big)\Big) - (1 - y_j^i)\log\Big(1 - \frac{1}{1+\exp\left(-\boldsymbol{\omega}_j(t)^T x_j^i\right)}\Big)$. Here, $(\cdot)^T$ denotes transpose. Given $D_j(\tau)$, the local loss function of at device j for the τth iteration is

$$F_j(\boldsymbol{\omega}_j(\tau)) = \frac{1}{D_j(\tau)} \sum_{i=1}^{D_j(\tau)} f_i(\boldsymbol{\omega}_j(\tau); x_j^i, y_j^i) + \mu g(\boldsymbol{\omega}_j(\tau)), \tag{6.1}$$

where $g(\cdot)$ is a regularizer function that represents the effect of the local training noise, and $\mu \in [0, 1]$ is a coefficient [15].

The local model of user device j is updated by

$$\boldsymbol{\omega}_j(\tau + 1) = \boldsymbol{\omega}_j(\tau) - \eta \nabla F_j(\boldsymbol{\omega}_j(\tau)), \tag{6.2}$$

where η is the learning rate.

After every T_L local updates (or iterations), there is a communication round where the benign user devices upload their local models to a server. The server aggregates the local models to update the global model and broadcasts the global model to all user devices. While selecting the user devices with large training datasets can help improve the learning accuracy of FL, it often results in the fast depletion of the batteries at the user devices. On the other hand, selecting the user devices with small datasets can save the battery energy of the devices, but the accuracy of the global model could suffer. Existing resource allocation policies, such as those developed in [58, 59], can be applied to balance the learning accuracy of FL and the energy consumption of the user devices.

6.3.2 Threat Model

We consider a new data-agnostic MP attack, where malicious local models are generated solely based on the benign local models overheard and the correlation features of the benign local and global models. This attack could be particularly severe in FL systems under wireless settings, due to the broadcast nature of radio. As shown in Fig. 6.1, an attacker within the vicinity of benign user devices and equipped with radio transceivers can passively eavesdrop on the local models transmitted by some (if not all) of the benign user devices, extracting their features and generating malicious local models. A similar threat model has also been considered in the recent literature [17, 45, 50], where an attacker within proximity of benign user devices overhears the local and global models in an attempt to recover, at least partially, the private data of the benign user devices. Although cryptography can prevent eavesdropping to some extent, existing techniques, such as those developed in [2, 12, 16], have demonstrated the possibility of deciphering encrypted information with limited initial data.

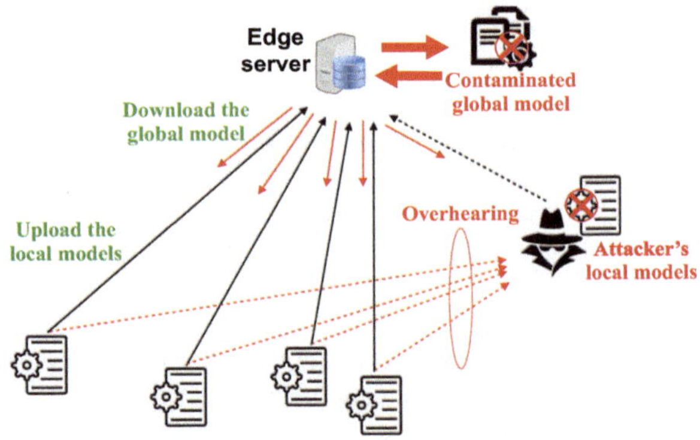

Fig. 6.1 The proposed data-agnostic MP attack, where the attacker overhears the global model and the local models uploaded by the benign user devices. Next, the attacker generates a malicious local model to contaminate the global model and the benign local models

The attacker creates and uploads a malicious local model, denoted by $\boldsymbol{\omega}^a(t)$, to contaminate the global model $\boldsymbol{\omega}_g(t)$, and subsequently the local models of the benign users, i.e., $\boldsymbol{\omega}_j(t), \forall j \in [1, J]$, where t indicates the tth communication round. $\boldsymbol{\omega}^a(t)$ is adversarially created based on the benign local model parameters overheard by the attacker in the tth communication round.

Unaware of the ill-intentioned attacker, the server aggregates the local models of all user devices, including both the benign and malicious local models, and unintentionally creates a contaminated global model, denoted by $\boldsymbol{\omega}_g^a(t)$, at the tth communication round. The total size of the local training data reported to the server is $D(t) = \sum_{j=1}^{J} D_j(t) + D_a(t)$, where $D_a(t)$ is the claimed data size of the attacker at the tth communication round. Then, the contaminated global model is given by

$$\boldsymbol{\omega}_g^a(t) = \sum_{j=1}^{J} \frac{D_j(t)}{D(t)} \boldsymbol{\omega}_j(t) + \frac{D_a(t)}{D(t)} \boldsymbol{\omega}^a(t), \tag{6.3}$$

The server broadcasts $\boldsymbol{\omega}_g(t)$ to all user devices.

To this end, the FL training process in essence trains the global model based on the local datasets of all user devices, including the nonexistent dataset claimed by the attacker, by minimizing the following global loss function:

$$\min_{\boldsymbol{\omega}_g^a(t)} F(\boldsymbol{\omega}_g^a(t)) = \sum_{j=1}^{J} \frac{D_j(t)}{D(t)} F_j(\boldsymbol{\omega}_g^a(t)) + \frac{D_a(t)}{D(t)} F_a(\boldsymbol{\omega}_g^a(t)), \tag{6.4}$$

where $F_a(\cdot)$ is the claimed local loss function of the attacker, which is claimed to conform to (6.1).

To attack the FL training process, the attacker aims to maximize $F(\boldsymbol{\omega}_g^a(t))$, while keeping $\boldsymbol{\omega}^a(t)$ undetectable by the server that typically constantly assesses the similarities among all local models and rules out those substantially different from the rest, e.g., Krum or multi-Krum [4]. As a result, the attacked global model diverges in a direction opposite to the one intended in the absence of the attack.

At the tth communication round, the attacker formulates a data-agnostic, MP attack problem:

$$\max_{\boldsymbol{\omega}^a(t)} \; F(\boldsymbol{\omega}_g^a(t)) \tag{6.5a}$$

$$\text{s.t.} \;\; d(\boldsymbol{\omega}^a(t), \boldsymbol{\omega}_g^a(t)) \leq d_T, \tag{6.5b}$$

where $d(\boldsymbol{\omega}^a(t), \boldsymbol{\omega}_g^a(t))$ evaluates the Euclidean distance between $\boldsymbol{\omega}^a(t)$ and $\boldsymbol{\omega}_g^a(t)$, and d_T is a pre-specified threshold that ensures the generated malicious local model is close to the global model in the Euclidean space to escape the scrutiny of the server.

6.3.3 Defense Model for Attacker Detection

In response to the prevalent threat of MP in FL, an attacker detection model residing on the server can be applied, which leverages the Euclidean distance metric to discern malicious local models, e.g., [25, 54]. By measuring the straight-line distance between each incoming local model and the aggregated global model, this model aims to identify anomalous deviations indicative of malicious intent. The underlying rationale is that genuine local models from benign devices are expected to cluster within a certain proximity in the model space, while malicious local models, designed to sabotage the global model's integrity, would exhibit more pronounced deviations. By setting a distance threshold, local models that exceed this threshold can be flagged or discarded, effectively isolating and mitigating the impact of malicious local models on the global model's integrity. This server-side defense mechanism underscores the potential of geometric measures, like Euclidean distance, as powerful tools in safeguarding FL systems from adversarial attacks.

6.4 Data-Agnostic MP Attack

In this section, we elaborate on the proposed data-agnostic MP attack, where adversarial GAE is designed to extract the feature correlation among the local models of the benign user devices and reconstruct an adversarial adjacency matrix. With the adjacency matrix, the attacker trains the GAE to generate malicious local models without being detected by the server.

6.4.1 GAE Model for Data-Agnostic MP

The arbitrary features of $\boldsymbol{\omega}^a(t)$ and those of the benign local models may have a low feature correlation, which can be potentially detected by the server. To address this, we develop a new GAE model for the novel, data-agnostic, MP attack.

The optimization problem in (6.5) can be transformed using the Lagrangian method [42]. Let λ denote the dual variable. The Lagrange function is given by

$$L(\boldsymbol{\omega}^a(t), \lambda) = F(\boldsymbol{\omega}_g^a(t)) + \lambda(d_T - d(\boldsymbol{\omega}^a(t), \boldsymbol{\omega}_g^a(t))). \tag{6.6}$$

The Lagrange dual function is

$$\mathcal{D}(\lambda) = \max_{\boldsymbol{\omega}^a(t)} L(\boldsymbol{\omega}^a(t), \lambda). \tag{6.7}$$

The dual problem of the problem in (6.5) is given by

$$\min_{\lambda(t)} \mathcal{D}(\lambda). \tag{6.8}$$

At the tth communication round, given $\lambda = \lambda(t)$, the primary variable $\boldsymbol{\omega}^a(t)$ of the data-agnostic MP attack can be optimized by solving

$$\boldsymbol{\omega}^a(t)^* = \arg\max_{\boldsymbol{\omega}^a(t)} \{F(\boldsymbol{\omega}_g^a(t)) - \lambda(t) d(\boldsymbol{\omega}^a(t), \boldsymbol{\omega}_g^a(t))\}. \tag{6.9}$$

With obtained $\boldsymbol{\omega}^a(t)^*$, the sub-gradient descent method can be taken to update $\lambda(t)$ by solving the dual problem (6.8). Specifically, $\lambda(t)$ is updated by [5]

$$\lambda(t+1) = \left[\lambda(t) - \varepsilon\left(d(\boldsymbol{\omega}^a(t)^*, \boldsymbol{\omega}_g^a(t)) - d_T\right)\right]^+, \tag{6.10}$$

where ε is the step size, τ is the index to the iterations, and $[x]^+ = \max(0, x)$. At initialization, $\lambda(t)$ is non-negative, i.e., $\lambda(1) \geq 0$, to ensure (6.10) converges.

We solve (6.9) by developing a new GAE model, followed by the sub-gradient descent to update (6.10). These two steps are performed in an alternating manner, as illustrated in Fig. 6.2. Specifically, we propose to decompose the local model parameters of the benign devices into a graph capturing the correlations (or similarity) between the benign local models, and the underlying spectral-domain data features that the local models capture. Then, we regenerate the graph with the GAE in a manipulative manner and subsequently compose malicious local models with the regenerated graph and the original, genuine data features. The rationale of this design is provided as follows.

- By regenerating the graph with the GAE, we retain and manipulate the correlations between the local models, and also deter the convergence of the global model, i.e., by maximizing (6.9). The decoder of the GAE reproduces the correlations while

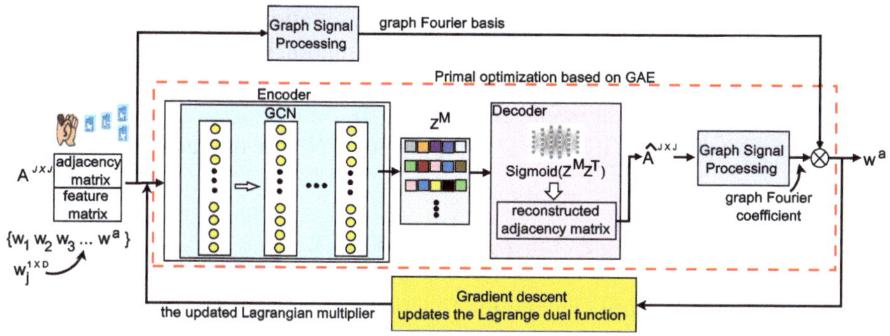

Fig. 6.2 The proposed GAE model for generating data-agnostic, malicious local models, where the attacker overhears $\boldsymbol{\omega}_g(t)$ and $\boldsymbol{\omega}_j(t)$, $\forall j$ and applies the GCN-based encoder to create $\mathcal{Z}^M$. The output of the encoder, i.e., the feature representations, is input to the decoder for feature reconstruction

satisfying constraint (6.5b). This suppresses structural dissimilarity between the malicious and benign local models.

- By using the genuine underlying spectral-domain data features, the malicious local models are substantiated by the genuine data features. Hence, they are less likely to be detected by the server.

6.4.1.1 GAE for Malicious Model Generation

The attacker aims to construct $\boldsymbol{\omega}^a(t)$ without knowing any data of the benign devices. As illustrated in Fig. 6.2, a graph, denoted by $G(\mathcal{V}, E, \mathcal{F})$, is used to formulate the benign local models in FL, where $\mathcal{V}$, E, and $\mathcal{F}$ represent vertexes, edges, and the feature matrix of the graph, respectively.

Let $\mathcal{F} = [\boldsymbol{\omega}_1(t), \cdots, \boldsymbol{\omega}_j(t), \boldsymbol{\omega}^a(t)]$ collect all local models of both benign and malicious devices. $\boldsymbol{\omega}_j(t), \boldsymbol{\omega}^a(t) \in \mathbb{R}^{1\times D}$, $\forall j$. Also, let $\mathcal{A} \in \mathbb{R}^{J\times J}$ denote the adjacency matrix that describes the correlation among the local models of the user devices. At the tth communication round of , the (j, j')th element of $\mathcal{A}$, denoted by $\overline{\omega}_{j,j'}$ $(j, j' \in [1, J])$, measures the inner product between $\boldsymbol{\omega}_j(t)$ and $\boldsymbol{\omega}_{j'}(t)$ [62], as given by

$$\overline{\omega}_{j,j'} = \frac{\boldsymbol{\omega}_j(t) \cdot \boldsymbol{\omega}_{j'}(t)}{\|\boldsymbol{\omega}_j(t)\| \cdot \|\boldsymbol{\omega}_{j'}(t)\|}. \tag{6.11}$$

According to $\mathcal{A}$, the topological structure of the graph $\mathcal{G}$ can be constructed.

The GAE consists of an encoder and a decoder, where the encoder encodes the graph data with the features and the decoder takes the encoder's output as the input to reconstruct $G(\mathcal{V}, E, \mathcal{F})$ [46].

- **Encoder**: The encoder in the proposed GAE is responsible for mapping $G(\mathcal{V}, E, \mathcal{F})$ to a lower-dimensional representation. We build the encoder based on an M-layer GCN architecture, which learns a representation that captures the underlying features of $G(\mathcal{V}, E, \mathcal{F})$. The encoded representation is then used as input to the decoder, to reconstruct the original graph from the lower-dimensional representation to obtain the malicious local model $\boldsymbol{\omega}^a(t^*)$ in (6.9).

The encoder takes $\mathcal{A}$ as its input to its M-layer GCN. The output at the Mth layer is given by

$$\mathcal{Z}^M = f_G(\mathcal{Z}^{M-1}, \mathcal{A}|\mathbf{w}^M), \tag{6.12}$$

where $f_G(\cdot, \cdot|\cdot)$ is a spectral convolution function and $\mathbf{w}^M$ defines the weight matrix at the Mth layer of the GCN.

With the identity matrix $I \in \mathbb{R}^{J\times J}$, we define $\widetilde{\mathcal{A}} = \mathcal{A} + I$ and $\overline{\mathcal{A}}_{jj} = \sum_{j'} \widetilde{\mathcal{A}}_{jj'}$. To generate a feature representation of the graph, the encoder can be written as

$$f_G(\mathcal{Z}^{M-1}, \mathcal{A}|\mathbf{w}^M) = \Phi^M(\overline{\mathcal{A}}^{-\frac{1}{2}}\widetilde{\mathcal{A}}\overline{\mathcal{A}}^{-\frac{1}{2}}\mathcal{Z}^{M-1}\mathbf{w}^M), \tag{6.13}$$

where $\Phi^M(\cdot)$ represents a nonlinear activation function, e.g., $\tanh(\cdot)$ or $\mathrm{ReLU}(\cdot)$; and $\overline{\mathcal{A}}^{-\frac{1}{2}}\widetilde{\mathcal{A}}\overline{\mathcal{A}}^{-\frac{1}{2}}$ is the symmetrically formulated adjacency matrix [23, 62].

• **Decoder**: The decoder is responsible for taking the lower-dimensional representation generated by the encoder, i.e., $\mathcal{Z}^M$ in (6.12), and mapping it back to the original $G(\mathcal{V}, E, \mathcal{F})$. This can be viewed as the inverse operation of the encoder. The decoder aims to generate the original graph from its reduced representation. The output of the decoder is compared with the original input graph to evaluate a loss. The encoder and decoder are trained together to minimize the loss.

A reconstructed adjacency matrix is generated at the decoder, which is defined as

$$\widehat{\mathcal{A}} = \mathrm{sigmoid}\left(\mathcal{Z}^M\left(\mathcal{Z}^M\right)^T\right). \tag{6.14}$$

where the Sigmoid function is defined as $\mathrm{sigmoid}(x) = 1/(1 + \exp(-x))$. The larger the inner product $(\mathcal{Z}^M\left(\mathcal{Z}^M\right)^T)$, the more likely the vertexes j and j' are connected in the graph [32].

The output of the decoder is the reconstructed adjacency matrix $\widehat{\mathcal{A}}$. A reconstruction loss function that measures the difference between $\mathcal{V}$ and $\widehat{\mathcal{A}}$ can be formulated as [33]

$$\phi_{\text{loss}} = \mathbb{E}_{f_G(\mathcal{Z}^{M-1}, G|\mathbf{w}^M)}\Big[\log\ p(\widehat{\mathcal{A}} \mid \mathcal{Z}^M)\Big], \tag{6.15}$$

where $p(\widehat{\mathcal{A}} \mid \mathcal{Z}^M)$ at the decoder indicates the correlation among the embedding vertices, and is given by

$$p(\widehat{\mathcal{A}} \mid \mathcal{Z}^M) = \Pi_{j=1}^{J}\Pi_{j'=1}^{J} p(\widehat{\mathcal{A}}_{jj'} \mid \mathcal{Z}_j^M, \mathcal{Z}_{j'}^M), \tag{6.16}$$

where

$$p(\widehat{\mathcal{A}}_{jj'} = 1|\mathcal{Z}_j^M, \mathcal{Z}_{j'}^M) = \text{sigmoid}\left(\mathcal{Z}_j^M \left(\mathcal{Z}_{j'}^M\right)^T\right). \quad (6.17)$$

• **Malicious Model Generation**: A graph signal processing module is designed to decompose the correlation features of the benign local models, and the data features substantiating the local models, as described earlier. A Laplacian matrix [28] is built based on the adjacency matrix of the benign models, i.e., $\mathcal{A}$, as given by

$$\mathcal{L} = diag(\mathcal{A}) - \mathcal{A}. \quad (6.18)$$

By applying SVD [21] to $\mathcal{L}$, i.e., $\mathcal{L} = B\Sigma B^T$, we can obtain a complex unitary matrix $B \in \mathbb{R}^{J\times J}$, also known as the GFT basis, that is used to transform graph data, e.g., $\mathcal{F}$, to its spectral-domain representation. Σ is a diagonal matrix with the eigenvalues of $\mathcal{L}$ along its main diagonal. As a result, the attacker can obtain a matrix S that contains the spectral-domain data features of all benign local models, by removing the correlations among the models and subsequently focusing on the data features substantiating the local models. S is given by

$$S = B^{-1}\mathcal{F}. \quad (6.19)$$

Likewise, the attacker can use the graph signal processing module to produce a Laplacian matrix based on the output of the GAE, as given by

$$\widehat{\mathcal{L}} = diag(\widehat{\mathcal{A}}) - \widehat{\mathcal{A}}. \quad (6.20)$$

The corresponding GFT basis, denoted by $\widehat{B}$, can be obtained by applying SVD to $\widehat{L}$. With reference to (6.19), the malicious local model that follows $\mathcal{A}$ in the GAE can be determined by

$$\widehat{\mathcal{F}} = \widehat{B}S, \quad (6.21)$$

where $\widehat{\mathcal{F}} \in \mathbb{R}^{J\times D}$. The vector $\boldsymbol{\omega}^a(t)$ in $\widehat{\mathcal{F}}$ is selected as the malicious local model and uploaded by the attacker to the aggregator for global model aggregation in the tth communication round.

Since the attacker aims to generate the malicious local models to disorient FL, the proposed GAE is constructed and trained to maximize $L(\boldsymbol{\omega}^a(t), \lambda(t)) - \phi_{\text{loss}}$. As a consequence, the malicious local model $\boldsymbol{\omega}^a(t)$ progressively and increasingly contaminates the FL training process with the increase in global model aggregations, i.e., $t = 1, 2, \ldots$.

6.4.2 Training Algorithm of GAE Model

Algorithm 1 summarizes the training process of the proposed GAE-based, data-agnostic, MP attack model, which operates along with the FL training of the benign devices and the server. Specifically, in every FL communication round, i.e., the tth round, the server broadcasts $\boldsymbol{\omega}_g^a(t)$. The benign devices apply the LocalTraining_start($\boldsymbol{\omega}_g^a(t)$) function to train their local models $\boldsymbol{\omega}_j(t)$, $\forall j = 1, \cdots, J$; see Steps 3–5 in Algorithm 1.

On the other hand, the attacker overhears the local model $\boldsymbol{\omega}_j(t)$, $\forall j$ from the benign devices at the tth FL communication round, and recall the global model $\boldsymbol{\omega}_g^a(t-1)$ overheard from the server at the $(t-1)$th round. The GAE is trained to maximize the data-agnostic MP attack problem in (6.5) with $\mathcal{V}$ and $\mathcal{F}$. Specifically, the problem in (6.5) is transformed into a primal and a dual problem using the Lagrangian method. Given the dual variable $\lambda(t)$, $\boldsymbol{\omega}^a(t)$ is optimized using the GAE; see Steps 8–10 in Algorithm 1. With the obtained $\boldsymbol{\omega}^a(t)^*$, the sub-gradient descent method is taken to update $\lambda(t)$ by (6.10); see Step 11. At the output of the GAE, the attacker achieves the optimal malicious local model, i.e., $\boldsymbol{\omega}^a(t)$. Next, $\boldsymbol{\omega}^a(t)$ is uploaded to the server for the next round of the FL training. As $\boldsymbol{\omega}^a(t)$ is highly correlated with $\boldsymbol{\omega}^a(t)$ from the benign user devices, the server is unable to identify the attacker.

6.4.3 Convergence Analysis of FL Under Attack

We derive the convergence upper bound for under the proposed, data-agnostic, MP attack. The following assumptions are made before the analysis, as typically considered in the literature [43, 48, 57].

Assumption 6.1 $\forall m \in \mathcal{M}$,

1. The gradient of $F_j(\boldsymbol{\omega}_j)$ is L-Lipschitz continuous [31], that is, $\left\|\nabla F_j(\boldsymbol{\omega}_j(t+1)) - \nabla F_j(\boldsymbol{\omega}_j(t))\right\| \leq L \left\|\boldsymbol{\omega}_j(t+1) - \boldsymbol{\omega}_j(t)\right\|$, $\forall \boldsymbol{\omega}_j(t+1), \boldsymbol{\omega}_j(t)$, with L being a constant depending on the loss function so that the gradient of the global loss function is also L-Lipschitz continuous;
2. $F_j(\boldsymbol{\omega}_j)$ is L_c-Lipschitz continuous; in other words, $\left|F_j(\boldsymbol{\omega}_j) - F_j(\boldsymbol{\omega}'_j)\right| \leq L_c \left\|\boldsymbol{\omega}_j - \boldsymbol{\omega}'_j\right\|$, $\forall \boldsymbol{\omega}_j, \boldsymbol{\omega}'_j$;
3. The learning rate is $\eta \leq \frac{1}{L}$;
4. At device j, the expected squared norm of the stochastic gradients is uniformly bounded by $\mathbb{E}\left\|\nabla F_j(\boldsymbol{\omega}_j(t))\right\|^2 \leq \kappa \|\nabla F(\boldsymbol{\omega}_g(t))\|^2, \forall j, \kappa \geq 0$;
5. With $\rho \geq 0$, $F_j(\boldsymbol{\omega})$ fulfills the Polyak-Lojasiewicz requirement [19], indicating that $F(\boldsymbol{\omega}_g) - F(\boldsymbol{\omega}_g^*) \leq \frac{1}{2\rho}\left\|\nabla F(\boldsymbol{\omega}_g)\right\|^2$, where $\boldsymbol{\omega}_g^* = \arg\min_{\boldsymbol{\omega}_g} F(\boldsymbol{\omega}_g)$;
6. $F(\boldsymbol{\omega}_g(0) - F(\boldsymbol{\omega}_g^*) = \Theta$, where Θ is a constant.

Algorithm 1: GAE-based, data-agnostic MP attack on FL

1. Initialize: $G(\mathcal{V}, \mathcal{E}, \mathcal{F})$, T_L, J, d_T, $\boldsymbol{\omega}_g^a(t)$, $\boldsymbol{\omega}^a(t)$, and $\lambda(1) \geq 0$.
% **Adversarial FL**:
for round $t = 1, 2, 3, ...$ **do**
  for Local iteration number $t_L = 1, \cdots, T_L$ **do**
    All benign user devices train the benign local model $\boldsymbol{\omega}_j(t)$, $j = 1, \cdots, J$.
  end for
  All benign user devices upload their benign local models $\boldsymbol{\omega}_j(t)$, $j = 1, \cdots, J$ to the server, and the attacker overhears the benign local models.
  The attacker carries out the proposed GAE, i.e., **GAE**($\boldsymbol{\omega}_j(t), \forall j, \mathcal{F}, \lambda(t)$), and obtains $\boldsymbol{\omega}^a(t)$, as follows.

    • Calculate the adjacency matrix $\mathcal{A} = \{\overline{\omega}_{j,j'}\} \in \mathbb{R}^{J \times J}$ according to (6.11), and input $\mathcal{A}$ and $\mathcal{F}$ into the GAE.
    • Train the GAE to maximize the reconstruction loss $L(\boldsymbol{\omega}^a(t), \lambda(t)) - \phi_{\text{loss}}$ to obtain $\widehat{\mathcal{A}}$.
    • Obtain S based on (6.18) and (6.19), next obtain $\widehat{\mathcal{F}}$ based on (6.20) and (6.21), and then determine $\boldsymbol{\omega}^a(t)$ based on $\widehat{\mathcal{F}}$.
  Update $\lambda(t)$, according to (6.10).
  The attacker uploads the malicious local model $\boldsymbol{\omega}^a(t)$ to the server.
  The server aggregates all the local models to obtain the global model under attack $\boldsymbol{\omega}_g^a(t)$ by (6.3), and broadcasts $\boldsymbol{\omega}_g^a(t)$.
  All benign user devices update their local models with the global model, i.e., $\boldsymbol{\omega}_j(t) \leftarrow \boldsymbol{\omega}_g^a(t), \forall j$.
end for

Under Assumption 6.1, we develop the following theorem that provides the convergence bound of the gap between $\boldsymbol{\omega}_g(t), \forall t$ and $\boldsymbol{\omega}_g^*$.

Theorem 6.1 *At the tth communication round, the convergence upper bound of the attacked FL is obtained as*

$$F(\boldsymbol{\omega}_g^a(t)) - F(\boldsymbol{\omega}_g^*) \leq \Theta\zeta^t + \frac{1-\zeta^t}{1-\zeta} \cdot \frac{\rho\eta D D_a}{(D-D_a)^2} F^{\max}, \tag{6.22}$$

where $\zeta = 1 - \frac{\rho\eta D^2}{(D-D_a)^2}$, *and* $F^{\max}$ *is a maximum value of* $F(\boldsymbol{\omega}^a(t))$ *due to the constraint in (6.5a) and (6.5b), i.e.,* $F(\boldsymbol{\omega}^a(t)) \leq F^{\max}$.

Proof See Appendix 6.6. □

As stated in Theorem 6.1, despite the attack launched by the attacker, the global model of FL can still converge, but to an inferior global model. Specifically, as $t \to \infty$, the optimality gap would stabilize at $\frac{2D_a L_c d_T}{D-D_a}$, which cannot be further reduced by training.

6.5 Performance Evaluation

In this section, we present the implementation of the proposed GAE-based, data-agnostic, MP attack in PyTorch. We evaluate the testing accuracy of the local and global models of FL under attack, using the MNIST [11], FMNIST [49], and CIFAR-10 datasets. We also report the detection rate of the attack, where the detection is based on the Euclidean distances of the malicious local models and the benign local models to the global models, as typically done in the latest literature, e.g., [8, 29].

Moreover, we compare the proposed attack with the existing data-agnostic MP attack that produces malicious local models by mimicking other benign devices' training samples to degrade the learning accuracy. As discussed earlier in Sect. 6.4, our GAE-based attack represents a novel type of attack, which only depends on the benign local models overheard and the global models, has no access to any of the training data, and attempts to compromise FL training processes. Few existing techniques can produce malicious models in such a way, i.e., based solely on the overheard benign models, as most existing techniques would require the knowledge of (part of) the dataset used in the FL training processes, e.g., for a different purpose, such as inserting a backdoor [29] or injecting malicious traffic into the benign training dataset [30]. The MP attack mechanism considered for comparisons with our proposed new attack has been implemented in several existing studies, e.g., [9, 52], in which the attacker manipulates the training process by injecting a fake device and sending fake local models to the server.

6.5.1 Implementation with PyTorch

The number of benign devices is set to $J = 5, 10, 15, 20, 25$. The number of iterations per communication round is set to $T_L = 10$. The maximum number of communication rounds is $T_{FL} = 200$. By default, one attacker is considered unless otherwise specified. We implement the proposed GAE-based attack against on an SVM model using PyTorch 1.12.1, Python 3.9.12 on a Linux workstation with an Intel(R) Core(TM) i7-9700K CPU@3.60GHz (8 cores) and 16 GB of DDR4 memory@2400 MHz.

The experiments are conducted on three datasets:

- The standard MNIST dataset comprises 60,000 training examples and 10,000 testing examples, which are grayscale images of handwritten digits from 1 to 10;
- The FMNIST dataset, which contains Zalando's article images (i.e., 28×28 grayscale images) in ten classes, including 60,000 examples for training and 10,000 examples for testing;
- The CIFAR-10 dataset, which contains 60,000 images with the size of 32×32 in ten classes (6,000 per class), 50,000 for training and 10,000 for testing.

At each user device, we use a standard quadratic optimization algorithm to train the SVM models based on the three datasets, namely, the standard MNIST, FMNIST, and CIFAR-10. The loss function used for training the SVM models is $F_j(\boldsymbol{\omega}_j(t)) = \frac{1}{2}\left\|\boldsymbol{\omega}_j(t)\right\|_2^2 + \frac{1}{D_j}\sum_{i=1}^{D_j}\max\left\{0, 1 - y_j^i(\beta_j + \boldsymbol{\omega}_j^T(t)x_j^i)\right\}$, where β_j is a feature parameter based on $\boldsymbol{\omega}_j(t)$ [51]. The global model $\boldsymbol{\omega}_g^a(t)$, which is trained at the server according to (6.3), is broadcast to all user devices for the training of $\boldsymbol{\omega}_j(t+1)$ in the next, $(t+1)$th communication round. In particular, the choice of model architecture, NNs or SVMs, for training at benign user devices, does not alter the underlying principle of the proposed data-agnostic MP attack. This is because the proposed attack hinges on the creation of malevolent local models that, upon integration, deteriorate the performance of the global model by augmenting the FL training loss. Our attack model incorporates a new adversarial GAE designed to fabricate these deleterious local models by leveraging benign local models overheard. The adversarial GAE is adept at extracting and utilizing the correlation features inherent in both the benign local models (which could be based on either NN or SVM architectures) and the global model.

Figure 6.3 illustrates an example of label classification with the three datasets. In the MNIST dataset, three images labeled as "0" are misclassified as "6" while five images labeled as "6" are misclassified as "0", resulting in an FL accuracy of 73.3%. Similarly, the FL accuracy with the FMNIST and CIFAR-10 datasets is 76.7 and 63.3%, respectively is designed to improve classification accuracy, while the proposed GAE-based attack aims to reduce accuracy and cause label misclassification. The GAE encoder is a two-layer GCN network (i.e., $M = 2$) with a dropout layer to prevent overfitting. The GAE decoder is an inner product. We use the Adam optimizer with a learning rate of 0.01 to optimize the network. For all datasets, we use the same encoder, decoder and SVM models.

6.5.2 Performance Analysis

6.5.2.1 FL Accuracy Under Attack

Figure 6.4 plots the accuracy of the local models under the proposed GAE-based, data-agnostic, MP attack on the MNIST, FMNIST, and CIFAR-10 datasets, where there are five benign devices (i.e., $J = 5$) and 100 communication rounds for. The state-of-the-art MP attack [30] is taken as the benchmark for our proposed attack, in which the attacker manipulates the training process by injecting a fake device and sending fake local models to the server. Since the MP attack in [30] shares the same objective as our proposed data-agnostic MP approach, i.e., reducing the accuracy of FL, a comparison with this reference showcases the efficacy of our proposed method in the context of prevailing MP attacks. For comparison purposes, Fig. 6.5 plots the accuracy of the benign local models without any attacks. In this scenario, the accuracy of the user device can be improved efficiently by FL and rapidly converges to 96%.

Fig. 6.3 An illustration of the local model of a user device trained to classify images

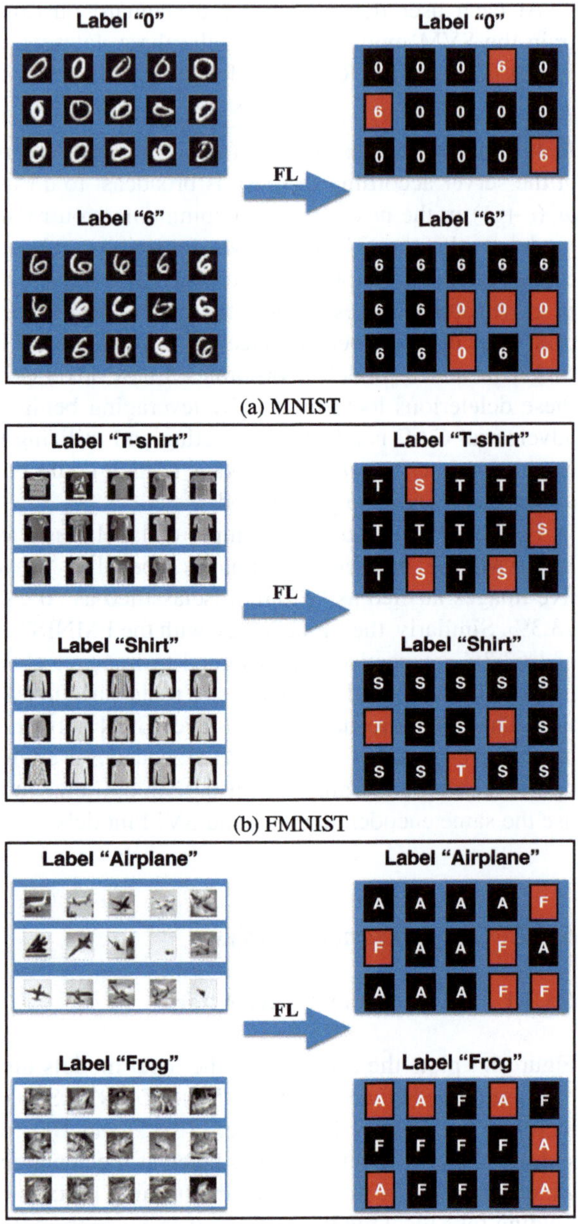

(a) MNIST

(b) FMNIST

(c) CIFAR-10

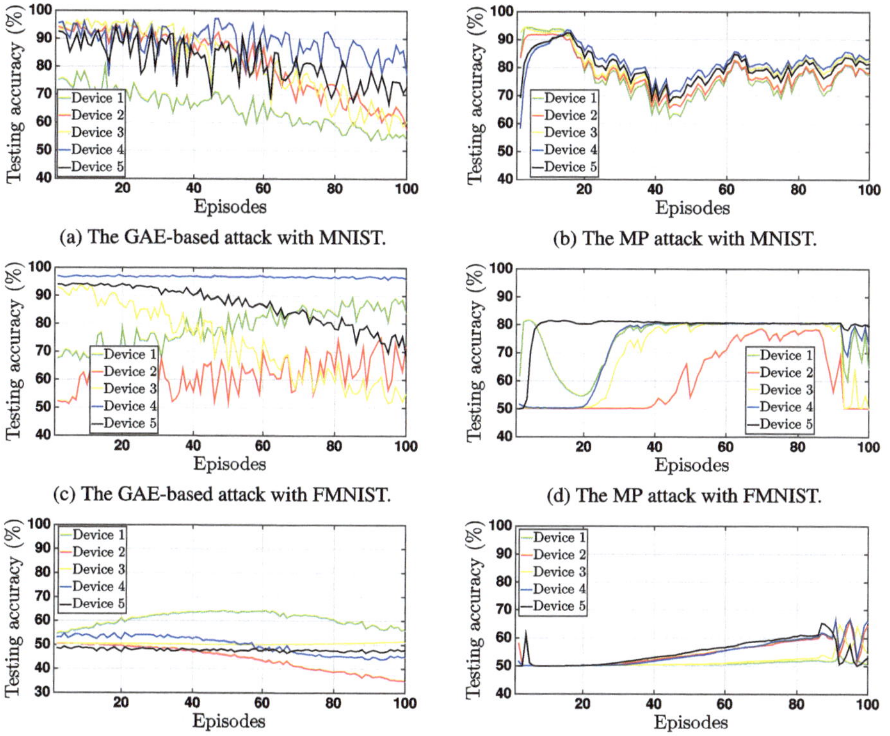

(a) The GAE-based attack with MNIST.

(b) The MP attack with MNIST.

(c) The GAE-based attack with FMNIST.

(d) The MP attack with FMNIST.

(e) The GAE-based attack with CIFAR-10.

(f) The MP attack with CIFAR-10.

Fig. 6.4 Given 100 FL communication rounds and five benign user devices, we compare the local model testing accuracy under the GAE-based attack and the existing MP attack on the MNIST, FMNIST, and CIFAR-10 datasets

Fig. 6.5 The local model testing accuracy with no attack

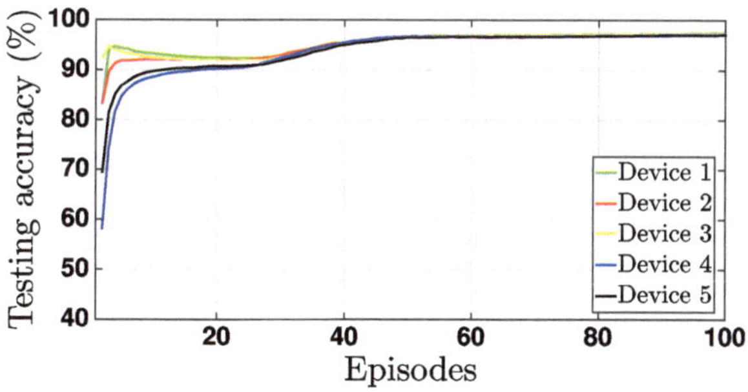

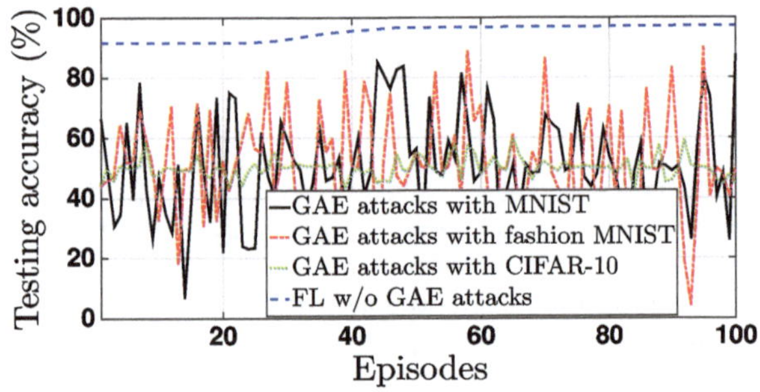

Fig. 6.6 The global model accuracy under the new attack

In Fig. 6.4a, b, we show that when using the MNIST dataset, the accuracy of all five devices under the proposed GAE-based attack gradually decreases and fluctuates dramatically. The performance of devices 1 and 2 drops from 75 to 55% and from 92 to 59%, respectively. The accuracy of the model drops from 91% to 80% when exposed to the MP attack, in which the performance of the five devices follows a similar pattern. This is because the new GAE-based attack reconstructs the adversarial adjacency matrix according to the individual features of the user devices. As a result, the attacker falsifies the local models to maximize the FL loss; see (6.9).

As shown in Fig. 6.4c, d, the accuracy of device 3 and device 5 drops significantly by 37 and 24%, respectively, when using the FMNIST dataset and the proposed GAE-based attack. However, while the accuracy of devices 1 and 2 may slightly increase, their convergence rates are greatly slowed in comparison to Fig. 6.5. Additionally, the accuracy under the MP attack varies between 50 and 80%, with a minimal decrease in accuracy observed.

In Fig. 6.4e, it can be seen that the proposed GAE-based attack with the CIFAR-10 dataset greatly hinders the performance of FL, as the accuracy of all four user devices falls below 50%. In contrast, the accuracy of all five devices under the MP attack is above 50%, as shown in Fig. 6.4f. Furthermore, it can be observed that the accuracy with the CIFAR-10 dataset is generally lower than the performance with the MNIST and FMNIST datasets. This is because the CIFAR-10 dataset contains a more diverse set of images, which makes it more challenging to differentiate and label, leading to lower overall accuracy.

Figure 6.6 illustrates the accuracy of the global model at the model aggregator. It can be observed that the proposed GAE-based attack hinders the training convergence when compared to the performance without the attack. As a result of the infection of the local FL model, the accuracy with the MNIST, FMNIST, or CIFAR-10 dataset fluctuates around 82%, 81%, or 25%, respectively.

6.5.2.2 Detection of Attack

Existing MP attacks on FL aim to maximize the training loss of FL models. One way to detect these malicious attacks is to compare the distances (or differences)

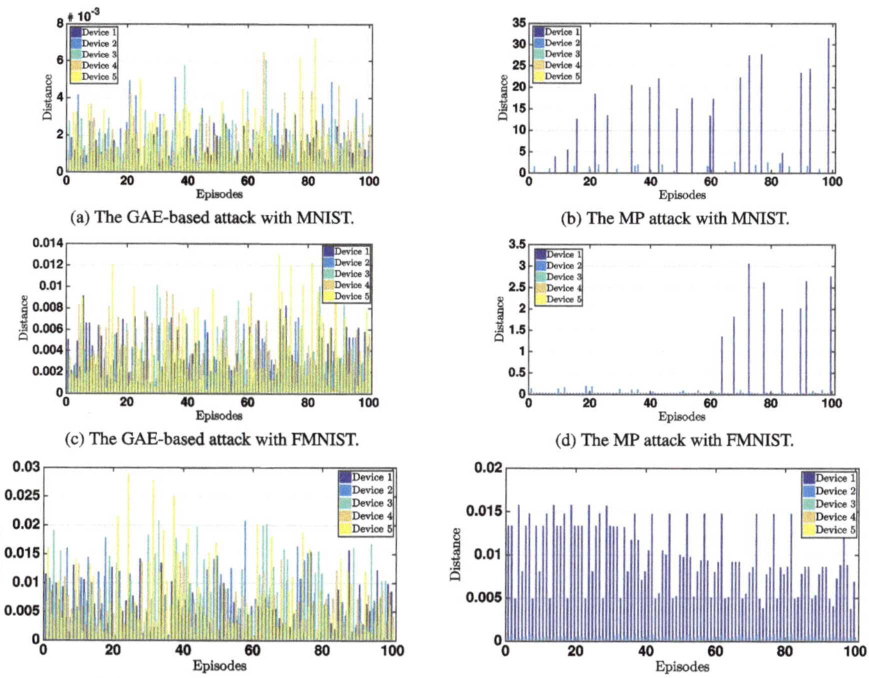

(a) The GAE-based attack with MNIST. (b) The MP attack with MNIST. (c) The GAE-based attack with FMNIST. (d) The MP attack with FMNIST. (e) The GAE-based attack with CIFAR-10. (f) The MP attack with CIFAR-10.

Fig. 6.7 Taking FL with five devices as an example, the Euclidean distances between the local models and the global model are presented, where device 1 is the attacker and launches the new GAE-based attack or the MP attack

between the local models and the global model. A larger distance can be considered as an indication of a malicious local model, and the server can detect it accordingly. Both the Euclidean distance and cosine distance are commonly used metrics to assess the similarities between two vectors. Particularly, the Euclidean distance provides a straightforward, geometric measure of the absolute difference between vectors, which is useful in the considered scenarios since large deviations in the magnitude of model updates are often indicative of malicious model updates. For this reason, the Euclidean distance is considered in this chapter, which is consistent with many recent studies, e.g., [25, 54].

To evaluate the invisibility of the proposed GAE-based, data-agnostic, MP attack Fig. 6.7 presents the Euclidean distance between the local models and the global model, with device 1 being the attacker. It can be observed that, in general, the local models with the MNIST dataset have the smallest distance compared to the other two datasets. This is expected as the handwritten digits in MNIST are relatively simple to recognize or falsify.

As shown in Fig. 6.7a, c, e, the Euclidean distances of the malicious local model (i.e., of device 1) generated by the GAE-based attack are below those of the benign local models. This makes it difficult for the aggregator to identify the attacker and defend against the attack. In contrast, the MP attack results in a significantly larger Euclidean distance between the malicious local model and the global model, making it easier to detect. This highlights the key advantage of the proposed GAE-based attack, which is designed to generate malicious local models based on the feature correlation between the benign local and global models, making the differences between the malicious local model and the benign local models indistinguishable.

6.5.2.3 Impact of Benign Local Model Number

Figure 6.8a, b, c show the average accuracy of the local models based on the MNIST, FMNIST, and CIFAR-10 datasets, respectively. The number of attackers is $J^a = 2$ by default. Otherwise, J^a increases proportionally with $J^a : J = 2 : 5$. It is observed that the new GAE-based attack reduces the average accuracy. As the number of devices increases, the average accuracy on MNIST, FMNIST, and CIFAR-10 drops by about 20, 37, and 12%, respectively, when $J = 15$. It is observed in Fig. 6.8 that on the three considered datasets, the average accuracy of FL under attack gradually increases as J grows from 5 to 25, while $J^a = 2$. This confirms that increasing the number of benign users improves the resistance of FL to the attacks. On the other hand, as the ratio of attackers to benign devices, i.e., $J^a : J$, increases, the proposed GAE-based attack can become increasingly effective and destructive.

6.5.2.4 Impact of Eavesdropped Local Models

Figure 6.9 plots the average accuracy of the local models under the GAE-based attack based on the MNIST, FMNIST, or CIFAR-10 datasets, where the number of benign user devices that the attacker can eavesdrop on increases from 3 to 25. In general, the average accuracy of the local models falls with the growth of the eavesdropped benign user devices. The reason is that overhearing a greater number of benign local models results in capturing more correlation features of the models, leading to the generation of a malicious model for more effective poisoning. The average model accuracy drops substantially by 13.6, 11.2, and 16.4% on the MNIST, FMNIST, and CIFAR-10 datasets, respectively.

6.5.2.5 Comparison with Variational Autoencoder (VAE)-Based Alternative

Figure 6.10 illustrates the effects of the VAE-based attack on FL over 100 communication rounds, where five user devices and the MNIST dataset are considered. Figure 6.10a reveals a consistent variation in local model testing accuracy under the VAE-based attack, with all five devices demonstrating analogous patterns. This is consistent with the observation made on the proposed GAE-based attack in Fig. 6.4. Figure 6.10b shows the Euclidean distances between the local and global models

Fig. 6.8 The average accuracy under the GAE-based attack based on the MNIST, FMNIST, or CIFAR-10 datasets, where the number of devices, i.e., J, ranges from 5 to 25. The number of attackers is $J^a = 2$ by default. Otherwise, J^a increases proportionally with J with $J^a : J = 2 : 5$

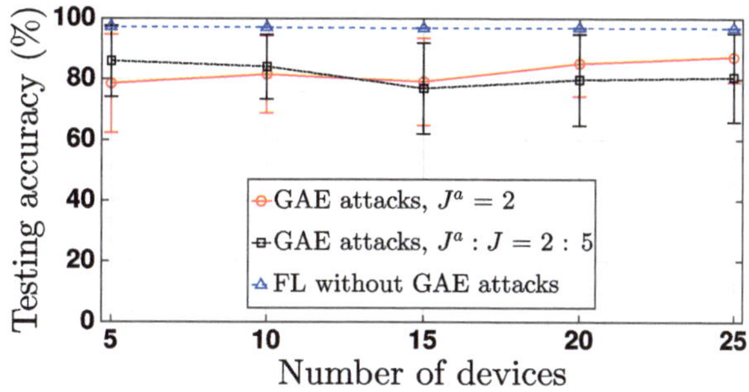

(a) MNIST

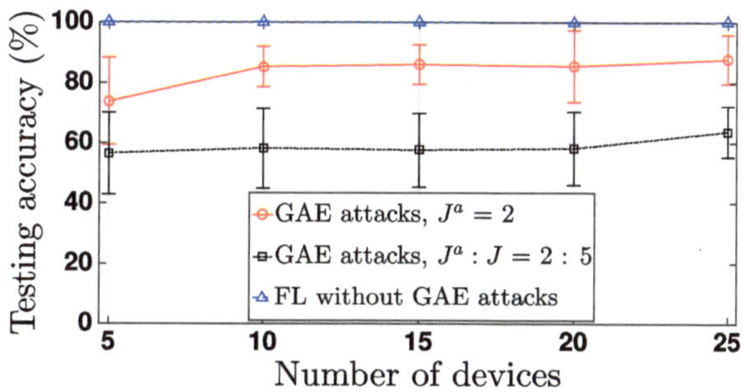

(b) FMNIST

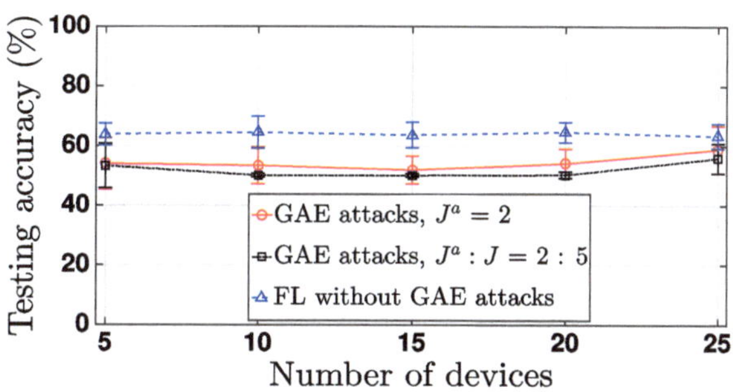

(c) CIFAR-10

Fig. 6.9 The number of eavesdropped benign user devices' $\boldsymbol{\omega}_j(t)$ increases from 3 to 25, based on the MNIST, FMNIST, or CIFAR-10 datasets

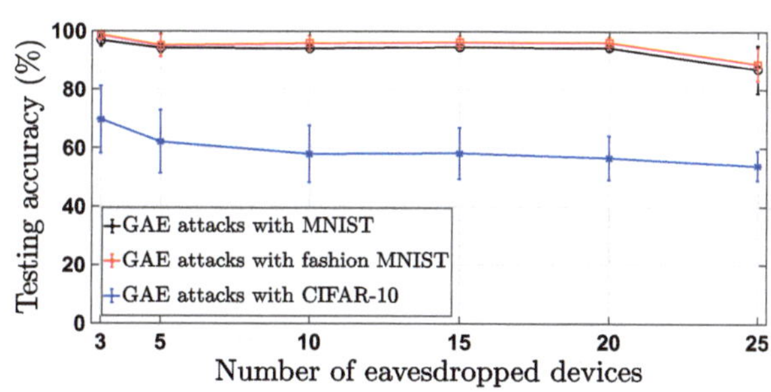

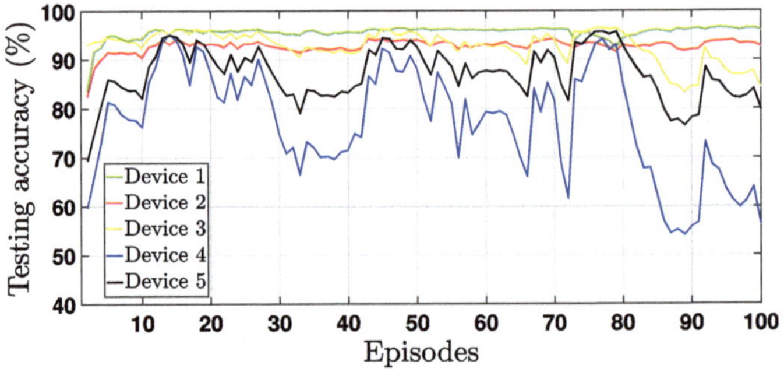

(a) The testing accuracy of local models.

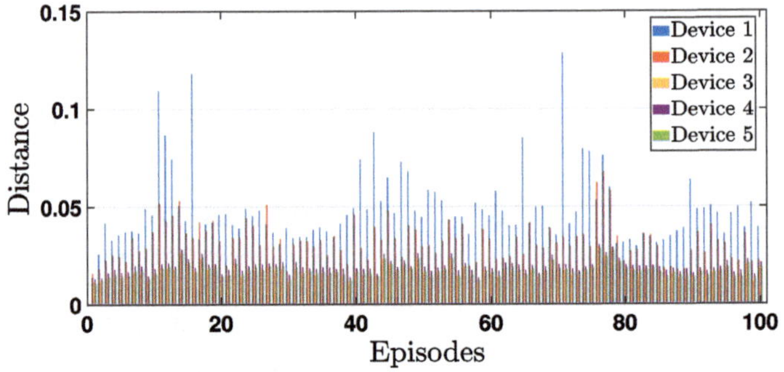

(b) The Euclidean distances between the local models and the global model.

Fig. 6.10 The local model testing accuracy and the Euclidean distances under the VAE-based attack

under the VAE-based attack. A striking observation is that the malicious local model (from device 1) constructed via the VAE-based attack possesses a significantly larger Euclidean distance than the benign local models. This suggests that detecting the VAE-based attacks on the server side is feasible by assessing the Euclidean distance. The underlying reason is that the VAEs are general AE and can handle high-dimensional and continuous data, such as images and audio. They do not capture graph structures inside data, as opposed to GAEs.

6.6 Proof of Theorem 6.1

Based on the Taylor expansion and the assumption that the gradient of the global loss function is L-Lipschitz continuous, it follows that

$$F(\boldsymbol{\omega}_g^a(t+1)) - F(\boldsymbol{\omega}_g^a(t)) \leq \nabla F\left(\boldsymbol{\omega}_g^a(t)\right)\left(\boldsymbol{\omega}_g^a(t+1) - \boldsymbol{\omega}_g^a(t)\right) + \frac{L}{2}\left\|\boldsymbol{\omega}_g^a(t+1) - \boldsymbol{\omega}_g^a(t)\right\|^2. \tag{6.23}$$

By substituting (6.3) into (6.23), we obtain (6.24) on the top of the next page.

$$F(\boldsymbol{\omega}_g^a(t+1)) - F(\boldsymbol{\omega}_g^a(t)) \leq \nabla F\left(\boldsymbol{\omega}_g^a(t)\right)\left[\frac{1}{D}\sum_{j=1}^{J} D_j(\boldsymbol{\omega}_j(t+1)-\boldsymbol{\omega}_j(t)) + \frac{D_a}{D}(\boldsymbol{\omega}^a(t+1)-\boldsymbol{\omega}^a(t))\right]$$
$$+\frac{L}{2}\left\|\frac{1}{D}\sum_{j=1}^{J} D_j(\boldsymbol{\omega}_j(t+1)-\boldsymbol{\omega}_j(t)) + \frac{D_a}{D}(\boldsymbol{\omega}^a(t+1)-\boldsymbol{\omega}^a(t))\right\|^2 \tag{6.24a}$$
$$=\frac{\eta^2 L}{2}\left\|\frac{1}{D}\sum_{j=1}^{J} D_j(t)\nabla F_j(\boldsymbol{\omega}_j(t)) + \frac{D_a}{D}\nabla F_a(\boldsymbol{\omega}^a(t))\right\|^2$$
$$-\eta\nabla F\left(\boldsymbol{\omega}_g^a(t)\right)\left[\frac{1}{D(t)}\sum_{j=1}^{J} D_j(t)\nabla F_j(\boldsymbol{\omega}_j(t)) + \frac{D_a}{D}\nabla F_a(\boldsymbol{\omega}^a(t))\right]. \tag{6.24b}$$

By substituting $\boldsymbol{\omega}_j(t+1) = \boldsymbol{\omega}_j(t) - \eta\nabla F_j(\boldsymbol{\omega}_j(t))$ and $\boldsymbol{\omega}_g^a(t) = \sum_{j=1}^{J}\frac{D_j}{D}\boldsymbol{\omega}_j(t) + \frac{D_a}{D}(\boldsymbol{\omega}^a(t+1) - \boldsymbol{\omega}^a(t))$ into (6.24), the expectation of $F(\boldsymbol{\omega}_g^a(t+1))$ is given by

$$\mathbb{E}\left[F(\boldsymbol{\omega}_g^a(t+1))\right] \leq F(\boldsymbol{\omega}_g^a(t)) - \eta\left\|\nabla F\left(\boldsymbol{\omega}_g^a(t)\right)\right\|^2 +$$
$$\frac{\eta^2 L}{2}\mathbb{E}\left[\left\|\sum_{j=1}^{J}\frac{D_j}{D}\nabla F_j(\boldsymbol{\omega}_j(t)) + \frac{D_a}{D}\nabla F_J(\boldsymbol{\omega}^a(t))\right\|^2\right].$$

It is generally assumed in FL that

$$\mathbb{E}\left[\sum_{j=1}^{J}\frac{D_j}{D}\nabla F_j(\boldsymbol{\omega}_j(t)) + \frac{D_a}{D}\nabla F_j(\boldsymbol{\omega}^a(t))\right] = \nabla F(\boldsymbol{\omega}_g^a(t)).$$

According to Jensen's inequality, $\mathbb{E}(x^2) \leq \mathbb{E}^2(x)$. Then,

$$\mathbb{E}\left[\left\|\sum_{j=1}^{J}\frac{D_j}{D}\nabla F_j(\boldsymbol{\omega}_j(t)) + \frac{D_a}{D}\nabla F_j(\boldsymbol{\omega}^a(t))\right\|^2\right] \leq \left\|\nabla F(\boldsymbol{\omega}_g^a(t))\right\|^2. \tag{6.26}$$

As a result, $\mathbb{E}\left[F(\boldsymbol{\omega}_g^a(t+1))\right]$ is bounded by

$$\mathbb{E}\left[F(\boldsymbol{\omega}_g^a(t+1))\right] \leq \mathbb{E}\left[F(\boldsymbol{\omega}_g^a(t))\right] + (\frac{\eta^2 L}{2} - \eta)\left\|\nabla F\left(\boldsymbol{\omega}_g^a(t)\right)\right\|^2 \tag{6.27a}$$
$$\stackrel{\eta=\frac{1}{L}}{=} \mathbb{E}\left[F(\boldsymbol{\omega}_g^a(t))\right] - \frac{\eta}{2}\left\|\nabla F\left(\boldsymbol{\omega}_g^a(t)\right)\right\|^2. \tag{6.27b}$$

Next, we derive the relationship between $\nabla F(\boldsymbol{\omega}_g^a(t))$ and $\nabla F(\boldsymbol{\omega}_g(t))$. Since $d(\boldsymbol{\omega}_j^a, \boldsymbol{\omega}_g^a) = \|\boldsymbol{\omega}_j^a - \boldsymbol{\omega}_g^a\| \leq d_T$, we have

$$\begin{aligned}
&\|\boldsymbol{\omega}^a(t+1)-\boldsymbol{\omega}^a(t)\| \\
&\leq \|\boldsymbol{\omega}^a(t+1)-\boldsymbol{\omega}_g^a(t+1)\|+\|\boldsymbol{\omega}_g^a(t+1)-\boldsymbol{\omega}^a(t)\| \\
&= d_T+\|\boldsymbol{\omega}_g^a(t+1)-\boldsymbol{\omega}^a(t)\| \\
&\leq d_T+\|\boldsymbol{\omega}_g^a(t+1)-\boldsymbol{\omega}_g^a(t)\|+\|\boldsymbol{\omega}_g^a(t)-\boldsymbol{\omega}^a(t)\| && (6.28a)\\
&= 2d_T + \|\boldsymbol{\omega}_g^a(t+1)-\boldsymbol{\omega}_g^a(t)\|, && (6.28b)
\end{aligned}$$

where (6.28a) is based on the triangle inequality.

Likewise, we also have

$$\begin{aligned}
&\|\boldsymbol{\omega}_g^a(t+1)-\boldsymbol{\omega}_g^a(t)\| \\
&= \left\|\left(\boldsymbol{\omega}_g(t+1)+\frac{D_a}{D}\boldsymbol{\omega}^a(t+1)\right)-\left(\boldsymbol{\omega}_g(t)+\frac{D_a}{D}\boldsymbol{\omega}^a(t)\right)\right\| && (6.29a)\\
&\leq \left\|\boldsymbol{\omega}_g(t+1)-\boldsymbol{\omega}_g(t)\right\|+\frac{D_a}{D}\left\|\boldsymbol{\omega}^a(t+1)-\boldsymbol{\omega}^a(t)\right\| && (6.29b)\\
&\leq \left\|\boldsymbol{\omega}_g(t+1)-\boldsymbol{\omega}_g(t)\right\| \\
&\quad + \frac{D_a}{D}\left(2d_T+\|\boldsymbol{\omega}_g^a(t+1)-\boldsymbol{\omega}_g^a(t)\|\right). && (6.29c)
\end{aligned}$$

By reorganizing (6.29c), it follows that

$$\begin{aligned}
\|\boldsymbol{\omega}_g^a(t+1)-\boldsymbol{\omega}_g^a(t)\| \leq \frac{D}{D-D_a}\times && (6.30a)\\
\left\|\boldsymbol{\omega}_g(t+1)-\boldsymbol{\omega}_g(t)\right\|+\frac{2D_a d_T}{D-D_a}. && (6.30b)
\end{aligned}$$

By substituting (6.30) into (6.28), it follows that

$$\begin{aligned}
\|\boldsymbol{\omega}^a(t+1)-\boldsymbol{\omega}^a(t)\| \leq \frac{D}{D-D_a}\times && (6.31a)\\
\left\|\boldsymbol{\omega}_g(t+1)-\boldsymbol{\omega}_g(t)\right\|+\frac{2D d_T}{D-D_a}. && (6.31b)
\end{aligned}$$

By taking expectation on both sides of (6.30), we have

$$\begin{aligned}
\mathbb{E}[\|\boldsymbol{\omega}_g^a(t+1)-\boldsymbol{\omega}_g^a(t)\|] \leq \frac{D}{D-D_a}\times \\
\mathbb{E}[\left\|\boldsymbol{\omega}_g(t+1)-\boldsymbol{\omega}_g(t)\right\|]+\frac{2D_a d_T}{D-D_a}.
\end{aligned} \tag{6.32}$$

Note that $\mathbb{E}[\|\boldsymbol{\omega}_g^a(t+1)-\boldsymbol{\omega}_g^a(t)\|]=\eta\|\nabla F(\boldsymbol{\omega}_g^a(t))\|$ and $\mathbb{E}[\|\boldsymbol{\omega}_g(t+1)-\boldsymbol{\omega}_g(t)\|]=\eta\|\nabla F(\boldsymbol{\omega}_g(t))\|$. By substituting them into both sides of (6.32) and then reorganizing (6.32), we obtain

$$\|\nabla F(\boldsymbol{\omega}_g^a(t))\| \leq \frac{D}{D-D_a} \|\nabla F(\boldsymbol{\omega}_g(t))\| + \frac{2D_a d_T}{(D-D_a)\eta}. \tag{6.33}$$

By substituting (6.33) into (6.27), it follows

$$\begin{aligned} &\mathbb{E}\left[F(\boldsymbol{\omega}_g^a(t+1))\right] - \mathbb{E}\left[F(\boldsymbol{\omega}_g^a(t))\right] \\ &\leq -\frac{\eta}{2} \left\|\nabla F\left(\boldsymbol{\omega}_g^a(t)\right)\right\|^2 && (6.34a) \\ &\leq -\frac{\eta}{2} \left(\frac{D}{D-D_a} \|\nabla F(\boldsymbol{\omega}_g(t))\| + \frac{2D_a d_T}{(D-D_a)\eta}\right)^2 && (6.34b) \\ &\leq -\frac{\eta D^2}{2(D-D_a)^2} \|\nabla F(\boldsymbol{\omega}_g(t))\|^2. && (6.34c) \end{aligned}$$

Considering the Polyak-Lojasiewicz condition, we have

$$\|\nabla F(\boldsymbol{\omega}_g(t)\|^2 \geq 2\rho \left(F(\boldsymbol{\omega}_g(t) - F(\boldsymbol{\omega}_g^*)\right). \tag{6.35}$$

Given the convex loss function of SVM models, it follows:

$$\begin{aligned} F(\boldsymbol{\omega}_g^a(t)) &\leq F(\boldsymbol{\omega}_g(t)) + F\left(\frac{D_a}{D}\boldsymbol{\omega}^a(t)\right) \\ &\leq F(\boldsymbol{\omega}_g(t)) + \frac{D_a}{D} F\left(\boldsymbol{\omega}^a(t)\right). \end{aligned} \tag{6.36}$$

By substituting (6.35) and (6.36) into (6.34), we have

$$\begin{aligned} &\mathbb{E}\left[F(\boldsymbol{\omega}_g^a(t+1))\right] - \mathbb{E}\left[F(\boldsymbol{\omega}_g^a(t))\right] \\ &\leq -\frac{\eta\rho D^2}{(D-D_a)^2} \left(F(\boldsymbol{\omega}_g(t) - F(\boldsymbol{\omega}_g^*)\right) && (6.37a) \\ &\leq -\frac{\eta\rho D^2}{(D-D_a)^2} \left(F(\boldsymbol{\omega}_g^a(t) - \frac{D_a}{D} F\left(\boldsymbol{\omega}^a(t)\right) - F(\boldsymbol{\omega}_g^*)\right). && (6.37b) \end{aligned}$$

By restructuring (6.37), we have

$$\begin{aligned} &\mathbb{E}\left[F(\boldsymbol{\omega}_g^a(t+1))\right] - F(\boldsymbol{\omega}_g^*) \\ &\leq \left(1 - \frac{\rho\eta D^2}{(D-D_a)^2}\right) \left(\left[F(\boldsymbol{\omega}_g^a(t))\right] - F(\boldsymbol{\omega}_g^*)\right) \\ &\quad + \frac{\rho\eta D D_a}{(D-D_a)^2} F\left(\boldsymbol{\omega}^a(t)\right) \\ &\leq (1-\zeta)\left(\left[F(\boldsymbol{\omega}_g^a(t))\right] - F(\boldsymbol{\omega}_g^*)\right) + \frac{\rho\eta D D_a}{(D-D_a)^2} F^{\max}, \end{aligned} \tag{6.38}$$

where $\zeta = 1 - \frac{\rho\eta D^2}{(D-D_a)^2}$, and the second inequality is due to the constrained problem in (6.5a) and (6.5b) has a maximum value, i.e., $F(\boldsymbol{\omega}^a(t)) \leq F^{\max}$.

Finally, applying mathematical induction upon (6.38) gives

$$\begin{aligned} &\mathbb{E}\left[F(\boldsymbol{\omega}_g^a(t))\right] - F(\boldsymbol{\omega}_g^*) \\ &\leq \left[\mathbb{E}\left[F(\boldsymbol{\omega}_g(0))\right] - F(\boldsymbol{\omega}_g^*)\right]\zeta^t + \frac{1-\zeta^t}{1-\zeta} \cdot \frac{\rho\eta D D_a}{(D-D_a)^2} F^{\max} \\ &= \Theta\zeta^t + \frac{1-\zeta^t}{1-\zeta} \cdot \frac{\rho\eta D D_a}{(D-D_a)^2} F^{\max}, \end{aligned} \tag{6.39}$$

which concludes this proof.

References

1. Abusnaina A, Khormali A, Alasmary H, Park J, Anwar A, Mohaisen A (2019) Adversarial learning attacks on graph-based IoT malware detection systems. In: Proceedings of IEEE international conference on distributed computing systems. IEEE, pp 1296–1305
2. Aviram N, Schinzel S, Somorovsky J, Heninger N, Dankel M, Steube J, Valenta L, Adrian D, Halderman JA, Dukhovni V et al (2016) {DROWN}: breaking {TLS} using {SSLv2}. In: Proceedings of the USENIX security symposium, pp 689–706
3. Bao Z, Lin Y, Zhang S, Li Z, Mao S (2021) Threat of adversarial attacks on DL-based IoT device identification. IEEE Internet Things J 9(11):9012–9024
4. Blanchard P, El Mhamdi EM, Guerraoui R, Stainer J (2017) Machine learning with adversaries: byzantine tolerant gradient descent. Adv Neural Inf Process Syst 30
5. Boyd S, Boyd SP, Vandenberghe L (2004) Convex optimization. Cambridge University Press
6. Cai H, Zheng VW, Chang KCC (2018) A comprehensive survey of graph embedding: problems, techniques, and applications. IEEE Trans Knowl Data Eng 30(9):1616–1637
7. Caldarola D, Mancini M, Galasso F, Ciccone M, Rodolà E, Caputo B (2021) Cluster-driven graph federated learning over multiple domains. In: Proceedings of the IEEE/CVF conference on computer vision and pattern recognition, pp 2749–2758
8. Cao D, Chang S, Lin Z, Liu G, Sun D (2019) Understanding distributed poisoning attack in federated learning. In: Proceedings of the IEEE international conference on parallel distributed systems. IEEE, pp 233–239
9. Cao X, Gong NZ (2022) MPAF: Model poisoning attacks to federated learning based on fake clients. In: Proceedings of the IEEE/CVF conference on computer vision and pattern recognition. IEEE, pp 3396–3404
10. Chen Z, Tian P, Liao W, Yu W (2020) Zero knowledge clustering based adversarial mitigation in heterogeneous federated learning. IEEE Trans Netw Sci Eng 8(2):1070–1083
11. Deng L (2012) The mnist database of handwritten digit images for machine learning research [best of the web]. IEEE Signal Process Mag 29(6):141–142
12. Diaz-Sanchez D, Marín-Lopez A, Mendoza FA, Cabarcos PA, Sherratt RS (2019) TLS/PKI challenges and certificate pinning techniques for iot and m2m secure communications. IEEE Commun Surv Tutorials 21(4):3502–3531
13. Fang M, Cao X, Jia J, Gong N (2020) Local model poisoning attacks to byzantine-robust federated learning. In: Proceedings of the USENIX security symposium, pp 1605–1622
14. Gao J, Hou B, Guo X, Liu Z, Zhang Y, Chen K, Li J (2023) Secure aggregation is insecure: category inference attack on federated learning. IEEE Trans Dependable Secure Comput 20(1):147–160

15. Guo X, Liu Z, Li J, Gao J, Hou B, Dong C, Baker T (2020) Verifl: communication-efficient and fast verifiable aggregation for federated learning. IEEE Trans Info Forensics Secur 16:1736–1751
16. Hebrok S, Nachtigall S, Maehren M, Erinola N, Merget R, Somorovsky J, Schwenk J (2023) We really need to talk about session tickets: a {Large-Scale} analysis of cryptographic dangers with {TLS} session tickets. In: Proceedings of the USENIX security symposium, pp 4877–4894
17. Huang Y, Gupta S, Song Z, Li K, Arora S (2021) Evaluating gradient inversion attacks and defenses in federated learning. Proc Adv Neural Inf Process Syst 34:7232–7241
18. Jere MS, Farnan T, Koushanfar F (2020) A taxonomy of attacks on federated learning. IEEE Secur Privacy 19(2):20–28
19. Karimi H, Nutini J, Schmidt M (2016) Linear convergence of gradient and proximal-gradient methods under the polyak-łojasiewicz condition. In: Proceedings of joint European conference on machine learning and knowledge discovery in databases. Springer, pp 795–811
20. Khan LU, Saad W, Han Z, Hossain E, Hong CS (2021) Federated learning for internet of things: recent advances, taxonomy, and open challenges. IEEE Commun Surv Tutorials 23(3):1759–1799
21. Lange K (2010) Singular value decomposition. In: Numerical analysis for statisticians. Springer, pp 129–142
22. Li K, Cui Y, Li W, Lv T, Yuan X, Li S, Ni W, Simsek M, Dressler F (2022) When internet of things meets metaverse: convergence of physical and cyber worlds. IEEE Internet Things J 10(5):4148–4173
23. Li K, Ni W, Yuan X, Noor A, Jamalipour A (2022) Deep-graph-based reinforcement learning for joint cruise control and task offloading for aerial edge internet of things (edgeiot). IEEE Internet Things J 9(21):21676–21686
24. Li K, Yuan X, Zheng J, Ni W, Guizani M (2023) Exploring adversarial graph autoencoders to manipulate federated learning in the internet of things. In: Proceedings of the IEEE international wireless communications and mobile computing (IWCMC). IEEE, pp 898–903
25. Li X, Qu Z, Zhao S, Tang B, Lu Z, Liu Y (2023) Lomar: A local defense against poisoning attack on federated learning. IEEE Trans Dependable Secure Comput 20(1):437–450. https://doi.org/10.1109/TDSC.2021.3135422
26. Lyu L, Yu H, Zhao J, Yang Q (2020) Threats to federated learning. In: Federated learning. Springer, pp 3–16
27. Mei G, Guo Z, Liu S, Pan L (2019) Sgnn: A graph neural network based federated learning approach by hiding structure. In: Proceedings of the IEEE international conference on big data. IEEE, pp 2560–2568
28. Molitierno JJ (2016) Applications of combinatorial matrix theory to laplacian matrices of graphs. CRC Press
29. Nguyen TD, Rieger P, Chen H, Yalame H, Möllering H, Fereidooni H, Marchal S, Miettinen M, Mirhoseini A, Zeitouni S et al (2012) FLAME: taming backdoors in federated learning. In: Proceedings of the USENIX security symposium, pp 1415–1432
30. Nguyen TD, Rieger P, Miettinen M, Sadeghi AR (2020) Poisoning attacks on federated learning-based IoT intrusion detection system. In: Proceedings of the workshop on decentralized IoT systems and security. NDSS, pp 1–7
31. O'Searcoid M (2006) Metric spaces. Springer Science & Business Media
32. Pan S, Hu R, Fung SF, Long G, Jiang J, Zhang C (2020) Learning graph embedding with adversarial training methods. IEEE Trans Cybern 50(6):2475–2487
33. Qiu C, Huang Z, Xu W, Li H (2022) Fast community detection based on graph autoencoder reconstruction. In: Proceedings of the international conference on big data analytics. IEEE, pp 265–271
34. Rahman A, Hossain MS, Alrajeh NA, Alsolami F (2020) Adversarial examples—security threats to COVID-19 deep learning systems in medical IoT devices. IEEE Internet Things J 8(12):9603–9610
35. Rahman SA, Tout H, Talhi C, Mourad A (2020) Internet of things intrusion detection: centralized, on-device, or federated learning? IEEE Netw 34(6):310–317

36. Shejwalkar V, Houmansadr A, Kairouz P, Ramage D (2022) Back to the drawing board: a critical evaluation of poisoning attacks on production federated learning. In: Proceedings of the IEEE symposium on security and privacy. IEEE, pp 1354–1371
37. Singh A, Sikdar B (2021) Adversarial attack for deep learning based IoT appliance classification techniques. In: Proceedings of IEEE world forum on internet of things. IEEE, pp 657–662
38. Song M, Wang Z, Zhang Z, Song Y, Wang Q, Ren J, Qi H (2020) Analyzing user-level privacy attack against federated learning. IEEE J Sel Areas Commun 38(10):2430–2444
39. Talpur A, Gurusamy M (2021) Adversarial attacks against deep reinforcement learning framework in internet of vehicles. In: Proceedings of IEEE globecom workshops. IEEE, pp 1–6
40. Tian Z, Cui L, Liang J, Yu S (2022) A comprehensive survey on poisoning attacks and countermeasures in machine learning. ACM Comput Surv 55(8):1–35
41. Tolpegin V, Truex S, Gursoy ME, Liu L (2020) Data poisoning attacks against federated learning systems. In: Proceedings of the European symposium on research in computer security. Springer, pp 480–501
42. Tran C, Fioretto F, Van Hentenryck P (2021) Differentially private and fair deep learning: a lagrangian dual approach. In: Proceedings of the AAAI conference on artificial intelligence, vol 35, pp 9932–9939
43. Truex S, Liu L, Chow KH, Gursoy ME, Wei W (2020) LDP-fed: Federated learning with local differential privacy. In: Proceedings of the ACM international workshop on edge systems, analytics and networking. ACM, pp 61–66
44. Wang H, Sreenivasan K, Rajput S, Vishwakarma H, Agarwal S, Sohn JY, Lee K, Papailiopoulos D (2020) Attack of the tails: Yes, you really can backdoor federated learning. Adv Neural Inf Process Syst 33:16070–16084
45. Wang T, Li Y, Wu Y, Quek TQ (2022) Secrecy driven federated learning via cooperative jamming: an approach of latency minimization. IEEE Trans Emerg Topics Comput 10(4):1687–1703
46. Wang Y, Xu B, Kwak M, Zeng X (2020) A simple training strategy for graph autoencoder. In: Proceedings of international conference on machine learning computing, pp 341–345
47. Wang Z, Song M, Zhang Z, Song Y, Wang Q, Qi H (2019) Beyond inferring class representatives: user-level privacy leakage from federated learning. In: Proceedings of the IEEE conference on computer communications. IEEE, pp 2512–2520
48. Wei K, Li J, Ding M, Ma C, Yang HH, Farokhi F, Jin S, Quek TQ, Poor HV (2020) Federated learning with differential privacy: algorithms and performance analysis. IEEE Trans Info Forensics Secur 15:3454–3469
49. Xiao H, Rasul K, Vollgraf R (2017) Fashion-mnist: a novel image dataset for benchmarking machine learning algorithms. arXiv:1708.07747
50. Xie YA, Kang J, Niyato D, Van NTT, Luong NC, Liu Z, Yu H (2022) Securing federated learning: a covert communication-based approach. IEEE Netw
51. Yang HH, Liu Z, Quek TQ, Poor HV (2019) Scheduling policies for federated learning in wireless networks. IEEE Trans Commun 68(1):317–333
52. Zhang J, Chen B, Cheng X, Binh HTT, Yu S (2020) PoisonGAN: generative poisoning attacks against federated learning in edge computing systems. IEEE Internet Things J 8(5):3310–3322
53. Zhang J, Chen J, Wu D, Chen B, Yu S (2019) Poisoning attack in federated learning using generative adversarial nets. In: Proceedings of the IEEE international conference on trust, security and privacy in computing and communications/IEEE International conference on big data science and engineering. IEEE, pp 374–380
54. Zhang Z, Cao X, Jia J, Gong NZ (2022) FLDetector: defending federated learning against model poisoning attacks via detecting malicious clients. In: Proceedings of the ACM SIGKDD conference on knowledge discovery and data mining, pp 2545–2555
55. Zhang Z, Wu L, Ma C, Li J, Wang J, Wang Q, Yu S (2022) LSFL: a lightweight and secure federated learning scheme for edge computing. IEEE Trans Inf Forensics Secur 18:365–379
56. Zhao J, Zhu H, Wang F, Lu R, Liu Z, Li H (2022) PVD-FL: a privacy-preserving and verifiable decentralized federated learning framework. IEEE Trans Inf Forensics Secur 17:2059–2073

57. Zhao Y, Zhao J, Yang M, Wang T, Wang N, Lyu L, Niyato D, Lam KY (2020) Local differential privacy-based federated learning for internet of things. IEEE Internet Things J 8(11):8836–8853
58. Zheng J, Li K, Mhaisen N, Ni W, Tovar E, Guizani M (2022) Exploring deep-reinforcement-learning-assisted federated learning for online resource allocation in privacy-preserving edgeiot. IEEE Internet Things J 9(21):21099–21110
59. Zheng J, Li K, Mhaisen N, Ni W, Tovar E, Guizani M (2023) Federated learning for online resource allocation in mobile edge computing: a deep reinforcement learning approach. In: Proceedings of the IEEE wireless communications and networking conference (WCNC). IEEE, pp 1–6
60. Zhou H, Yang G, Huang Y, Dai H, Xiang Y (2022) Privacy-preserving and verifiable federated learning framework for edge computing. IEEE Trans Inf Forensics Secur 18:565–580
61. Zhou X, Liang W, Li W, Yan K, Shimizu S, Kevin I, Wang K (2021) Hierarchical adversarial attacks against graph-neural-network-based IoT network intrusion detection system. IEEE Internet Things J 9(12):9310–9319
62. Zhu D, Ma Y, Liu Y (2020) Anomaly detection with deep graph autoencoders on attributed networks. In: Proceedings of the IEEE symposium on computers and communications. IEEE, pp 1–6

Part III
Defense Mechanisms in Federated Learning

Chapter 7
Privacy-Aware Wireless Federated Learning

As a framework for distributed online computing and model training, FL has shown significant potential for applications, e.g., IoT, autonomous driving, and remote medical care [24]. FL enables individual mobile clients to train a global model collectively without releasing their data [17]. In particular, each client trains its local model independently, relying on its local dataset, and sends the gradient of the local model to a server. The server aggregates the gradients and broadcasts the aggregated global parameter to assist the clients in their local training.

A challenge faced by FL is the communication bottleneck from the clients to the server, especially when dynamic, lossy, and resource-constrained channels are considered in wireless FL (WFL) [32]. The channels of geographically dispersed clients can differ significantly and change over time, leading to different model uploading delays among the clients. Moreover, the clients can have different computation capabilities, resulting in different local training times, and consequently, different time budgets for model uploading in an FL aggregation round. While more participating clients and hence larger overall training datasets are conducive to better training performance [17], this could lead to a shortage of wireless resources (e.g., bandwidth) for uploading local models. It is crucial to jointly design the local model training and uploading (transmission) schedule of the clients to balance communication and computation.

The transmission schedule of the clients depends on the model aggregation mechanism adopted. Typical synchronous FL (Sync-FL) requires all participating clients to upload their local models (or gradients) for a global aggregation after completing their local training [29]. Alternatively, asynchronous FL (Async-FL) allows each client to upload its local models straightaway after completing its local training. In the latter case, there can be non-negligible gaps between the time the local models are uploaded and the time the global model is broadcast. The computing powers of the clients may not be efficiently exploited. Clients with more powerful computation capability have to wait longer for global model update, especially when there are "stragglers" [48]. Flexible aggregation-based FL (F^2L) has recently emerged as

K. Li et al., *Security and Resilience in Distributed Machine Learning*, Springer Series in Reliability Engineering, https://doi.org/10.1007/978-3-032-23959-4_7

a promising solution to the issue of "stragglers", where clients can train different numbers of iterations before sending their local models synchronously for a global aggregation [33].

The transmission schedule of the clients also depends on their instantaneous computing powers [40]. A client with a weak computing capability has to transmit at a high data rate to complete its local model training and upload within the same period. Consider a widely adopted, practical orthogonal-frequency division multiple-access (OFDMA) as the air interface for WFL. Effective selections of the clients, channels, and modulations are key to the transmission schedule. However, the selections form an integer program with discrete variables. This renders the joint optimization of the communication and computation of OFDMA-based WFL systems (referred to as "OFDMA-based F^2L" or "OFDMA-F^2L") a typically intractable mixed-integer programming problem.

7.1 Introduction

Some existing work has focused on the training accuracy, efficiency, and robustness of WFL under a persistent policy of client selection and bandwidth allocation (e.g., see [17, 19, 21, 25, 41, 42, 47]). A partial participating scheme of WFL clients was investigated in [29], where only some clients were chosen to send their local models for global aggregations to minimize the power and time usage of the learning task. A partial synchronization parallel approach was proposed in [32] for a relay-assisted WFL, where local models were uploaded simultaneously and aggregated at relay nodes to reduce traffic. The same convergence speed was achieved as the successive synchronization methods. Yet, none of these studies [19, 25, 29, 32, 41, 42, 47] optimized communication performances for WFL.

Considering the communication performance metrics, energy-efficient WFL has been developed to cope with lossy and dynamically varying wireless environments in [1, 4, 15, 18, 34, 38, 43]. A problem of joint training and communication was constructed in [43] to minimize the overall power usage of a WFL system under a training latency requirement. The optimal resource allocation policy was obtained by using bisection search. In [15], a hierarchical FL architecture was proposed for a heterogeneous cellular system, where the clients were jointly served by macro and micro base stations (BSs). DRL was employed to minimize the power usage under a delay constraint by adjusting the client-BS association and resource allocation. Non-orthogonal multiple access (NOMA)-aided FL was examined in [23, 35, 38], where a set of clients formed a NOMA cluster to upload their local models to a BS for model aggregation. The total power usage and FL convergence latency were minimized by jointly optimizing the NOMA transmission, BS broadcast, and FL training accuracy. Resource-constrained WFL with partial aggregation was studied in [34], where stochastic geometry tools were employed to approximately calculate the success probability for each device. All these studies [1, 4, 15, 18, 34, 38, 43] were based on the Sync-FL framework.

The Async-FL framework was proposed to handle clients with different communication and computing capabilities [3, 8, 13, 16, 30, 40, 46, 49]. In [37], Async-FL was developed, which adapted to the heterogeneity of clients regarding their delays in model training and transmission. The fusion weights of the global aggregation were determined to prevent biased convergence by requiring the weight of each client's local dataset to be in proportion to its sample size. In [26], an adaptive Async-FL mechanism was developed, where only a small number of the total local updates were collected by the FL server, depending on their arriving order per round. To reduce the mission completion time, DRL was applied to specify the number of aggregated local updates under resource constraints.

An alternative to Async-FL is F^2L [33], where clients may have trained for different numbers of iterations when uploading their local models. Different computing powers of the clients can be efficiently leveraged, although persistent global aggregation cycles are maintained as in Sync-FL. However, no consideration has been given to the delays resulting from model uploading under flexible aggregation, which would affect the local training time and performance. Table 7.1 summarizes the existing works and highlights the contribution of this work.

In a different yet relevant context, over-the-air (OTA) FL has been investigated to alleviate the communication bottleneck of WFL [9, 10, 39, 50–52]. For instance, a broadband analog aggregation and transmission scheme was proposed in [52] and [51], where local model gradients were modulated in the analog domain and aggregated by exploiting the waveform-superposition property of radio. An adaptive device scheduling mechanism was developed for OTA-FL in [10], where data quality and energy consumption served as the criteria to select clients for OTA model

Table 7.1 Existing WFL works and comparison with OFDMA-F^2L

Work	Year	Topic	OFDMA-F^2L Advantage
[29, 32]	21–22	Sync-FL, partial sync/agg	F^2L over OFDMA air interface
[33]	21	F^2L	WFL over OFDMA air interface
[19, 25, 41, 42, 47]	2020–2023	Sync-FL, fixed client/bandwidth	Client-channel-modulation selection
[1, 4, 15, 18, 43]	2020–2023	Comm-comp optimization (Sync-FL)	Client-channel-modulation selection
[38]	2022	NOMA-aided Sync-FL	F^2L over OFDMA air interface
[34]	2021	Resource-limited Sync-FL, partial aggregation	Client-channel-modulation selection
[3, 8, 13, 16, 37, 40, 46, 49]	2020–2023	Async-FL	F^2L over OFDMA air interface
[26]	2023	Adaptive Async-FL, partial aggregation	F^2L over OFDMA air interface

aggregations to improve the convergence and accuracy of OTA-FL. A new model aggregation method was developed in [50] to align the local model gradients of different devices to resolve the straggler problem in OTA-FL. Moreover, in [44], a decentralized power control policy was designed to enhance the convergence of OTA-FL under fading channels, where stochastic optimization was leveraged to decouple the power control among clients.

This chapter presents the new OFDMA-F^{2}L framework, where clients can train their local models using different numbers of local iterations before uploading their local models or model gradients to an FL server for a global aggregation via an OFDMA air interface. A new algorithm is proposed to facilitate the convergence of OFDMA-F^{2}L (i.e., to minimize the convergence upper bound), by jointly optimizing the selections of clients, subchannels, and discrete modulations for local training and model uploading, adapting to the channel conditions and computing powers of the clients. The optimal number of local training iterations is accordingly specified per selected client per aggregation round.

The key contributions of the new approach can be summarized as follows:

- The new OFDMA-F^{2}L framework is presented, where selected clients train local models, adapting to their available computing powers and times. The clients can also upload their local models concurrently in different subchannels, increasing participating clients and extending their training times.
- We analyze a convergence upper bound on OFDMA-F^{2}L that quantifies the impact of the selections of clients, subchannels, and modulations on the convergence.
- A new problem is formulated to minimize the convergence upper bound of OFDMA-F^{2}L, and transformed to maximize the weighted sum rate of the clients by optimizing the selections of clients, subchannels, and modulations.
- By designing a Lagrange-dual based method, we discover that a "winner-takes-all" policy ensures an optimal selection of clients, subchannels, and modulations almost surely in order to achieve the fast convergence of OFDMA-F^{2}L.

We experimentally evaluate the proposed OFDMA-F^{2}L with the optimal client, subchannel, and modulation selections on a multilayer perceptron (MLP) and convolutional neural network (CNN) models and the MNIST, CIFAR10, and Fashion MNIST (FMNIST) datasets. It is shown that OFDMA-F^{2}L with the optimal selections significantly outperforms its alternatives with partially optimal selections or under Sync-FL, e.g., by about 18% and 5%, in convergence and training accuracy on the MLP model.

The remainder of this chapter is organized as follows. Section 7.2 sets forth the system model. Section 7.3 formulates the problem of interest. Section 7.4 analyzes the convergence bound of the proposed algorithm. Section 7.5 obtains the optimal assignments of clients, subchannels, and modulations. The OFDMA-F^{2}L system is numerically evaluated in Sect. 7.6, followed by conclusions in Sect. 7.7. The notation used in this chapter is collected in Table 7.2.

Table 7.2 List of Notations for OFDMA-F^2L (Alphabetical Order)

Symbol	Description
A	Maximum number of local iterations between two consecutive aggregations
$F(\omega)$	Global loss function of OFDMA-F^2L
$h_{m,k}$	Channel coefficient from client m to the BS on subchannel k
K, $\mathcal{K}$	Number and set of subchannels
L, $\mathcal{L}$	Number and set of modulations
M, $\mathcal{M}$	Number and set of clients
$n_{m,k}$	CSCG noise with zero mean and variance σ^2
$p_{m,k,l}$	Transmit power of client m on subchannel k using modulation l
r_l	Data rate of modulation l
t, T	Index and total number of iterations of OFDMA-F^2L
$\chi_{m,k,l}$	BER of the signal uploaded by client m on subchannel k with modulation l
χ_0	Required BER
$\lambda_{m,k,l}$	Binary variable indicating selection of client m and modulation l on subchannel k
$\nabla F(\cdot)$	Gradient of the function $F(\cdot)$
$P_m^{\max}$	Power budget of client m
τ, G	Index and maximum number of aggregations
ζ_m, ζ	Selection indicator of client m and the set of indicators for all clients
ω, ω^*	Global model parameter and optimal global model parameter of OFDMA-F^2L
$\mathcal{D}_m$, $\mathcal{D}$	Local dataset at client m, and the collection of all local datasets

7.2 System Model and Assumptions

Consider F^2L in an OFDMA system, i.e., OFDMA-F^2L, as shown in Fig. 7.1, where a BS serving as a central server connects M edge clients over K orthogonal subchannels. The BS and clients all have a single antenna. We use $\mathcal{M} = \{1, \cdots, M\}$ and $\mathcal{K} = \{1, \cdots, K\}$ to represent the sets of clients and subchannels, respectively. The BS wishes to learn an ML model using the datasets stored at the clients, since the clients are reluctant to share their data for privacy concerns. As a result, the clients train and transmit their local models to the BS via the uplink OFDMA subchannels, one client per subchannel.

Upon the receipt of the local models, the BS aggregates them into a global model and distributes the global model to the clients through a downlink broadcast. Considering limited communication and computation resources, only some clients are chosen to train and upload their local models. The BS performs client selection and resource allocation.

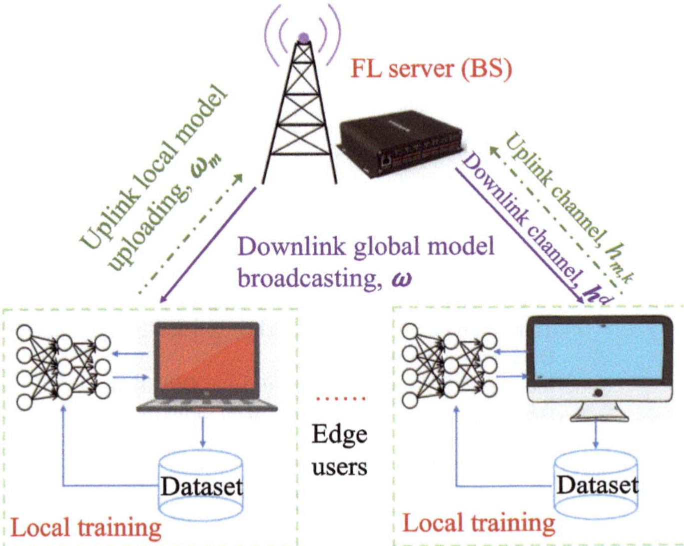

Fig. 7.1 The proposed OFDMA-F^{2}L, where a BS serves as the FL server and multiple clients act as clients. The selected clients send their local models to the BS using OFDMA

7.2.1 F^{2}L Model

Let $\mathcal{D}_m$ denote the local dataset at client $m \in \mathcal{M}$, and $D_m = |\mathcal{D}_m|$ denote the size of the local dataset, where $|\cdot|$ stands for cardinality. $\mathcal{D} = \bigcup_{\forall m \in \mathcal{M}} \mathcal{D}_m$ collects all local datasets at the clients with $D = |\mathcal{D}| = \sum_{m \in \mathcal{M}} D_m$ being the size of $\mathcal{D}$. Assume that each client m divides its local dataset $\mathcal{D}_m$ into multiple independent and identically distributed (i.i.d.) mini-batches with the same distribution as $\mathcal{D}_m$. $F_m(\boldsymbol{\omega}; \mathcal{X})$ is the loss function of client $m \in \mathcal{M}$, measuring the model error on the client's dataset $\mathcal{X} \subseteq \mathcal{D}_m$. Here, $\boldsymbol{\omega}$ is the global model parameter vector. The global loss function at the BS, denoted by $F(\boldsymbol{\omega})$, is acquired by gathering the local models uploaded from the selected clients, as given by

$$\begin{aligned} \boldsymbol{\omega}^* &= \arg \min_{\boldsymbol{\omega}, \{\zeta^\tau, \forall \tau\}} F(\boldsymbol{\omega}), \\ \text{with } F(\boldsymbol{\omega}) &\triangleq \frac{1}{B} \sum_{m \in \mathcal{M}} \sum_{\tau} \varrho_m^\tau \zeta_m^\tau F_m(\boldsymbol{\omega}; \mathcal{B}_m^\tau), \end{aligned} \tag{7.1}$$

where $\boldsymbol{\omega}^*$ is the optimal model parameter minimizing $F(\boldsymbol{\omega})$; $\mathcal{B}_m^\tau \subseteq \mathcal{D}_m$ collects the set of mini-batches sampled from the local dataset $\mathcal{D}_m$ during the τ-th aggregation round; $B = \sum_m \sum_\tau I_m^\tau$ and I_m^τ is the number of local training iterations at client m in the τ-th aggregation round. $\varrho_m^\tau = \frac{A}{I_m^\tau} \cdot \frac{D_m}{D}$ is the aggregation coefficient of client m for the τ-th aggregation of F^{2}L, where A is the maximum number of iterations a client can train locally per aggregation round and $I_m^\tau \leq A$. In this sense, greater aggregation coefficients can be given to clients that train fewer iterations in an aggregation round, compensating for the additional iterations other clients completed and contributing to an unbiased gradient after aggregation [33, Sect. 4.1, Scheme C].

Here, $\boldsymbol{\zeta}^\tau = \{\zeta_1^\tau, \cdots, \zeta_M^\tau\}$ collects the client selection indicators with ζ_m^τ indicating client m is selected for the τ-th aggregation; i.e., $\zeta_m^\tau = 1$ if client m is selected; otherwise, $\zeta_m^\tau = 0$. The client selection is critical for FL, especially when the communication and computation resources are constrained. Not all clients can upload their local models to the BS.

OFDMA-F^{2}L supports flexible aggregation at the BS [33], where the clients can execute different numbers of local iterations per aggregation round and, if selected, transmit their latest local models to the BS for a synchronous global model aggregation (in other words, the clients complete uploading their local models at the same time so that the global aggregation can start straightaway). The FL process is divided evenly into G global aggregations. Each aggregation round lasts for the same duration. The OFDMA-F^{2}L contains three alternating steps:

- *Local Model Training:* After the τ-th global model aggregation ($\tau = 1, \cdots, G$), each selected client m ($\zeta_m^\tau = 1$) performs I_m^τ iterations of local training. The local training of each selected client m for the $(\tau+1)$-th global aggregation is initialized to be $\boldsymbol{\omega}_m^\tau(0) = \boldsymbol{\omega}^\tau$, with $\boldsymbol{\omega}^\tau$ being the global model acquired in the τ-th global model aggregation. The local model of client m at the i-th iteration ($i = 1, \cdots, I_m^\tau$), denoted by $\boldsymbol{\omega}_m^\tau(i)$, is refreshed by

$$\boldsymbol{\omega}_m^\tau(i) = \boldsymbol{\omega}_m^\tau(i-1) - \eta \nabla F_m\left(\boldsymbol{\omega}_m^\tau(i-1); \mathcal{B}_m^\tau(i-1)\right), \tag{7.2}$$

  where η is the learning rate, and $\mathcal{B}_m^\tau = \{\mathcal{B}_m^\tau(0), \cdots, \mathcal{B}_m^{\tau-1}(I_m^\tau - 1)\}$ collects all mini-batches sampled from the local dataset $\mathcal{D}_m$. $\mathcal{B}_m^\tau(i)$ is the mini-batch used by device m for the i-th local iteration of the τ-th global aggregation.
- *Global Aggregation:* The clients selected by the BS upload their local models for the $(\tau+1)$-th global model aggregation after their local training. The global model obtained is

$$\begin{aligned} \boldsymbol{\omega}^{\tau+1} &= \sum_{m \in \mathcal{M}} \varrho_m^\tau \zeta_m^\tau \boldsymbol{\omega}_m^\tau \\ &= \sum_{m \in \mathcal{M}} \varrho_m^\tau \zeta_m^\tau \left(\sum_{i=0}^{I_m^\tau - 1} \left(\boldsymbol{\omega}_m^\tau(i) - \eta \nabla F_m(\boldsymbol{\omega}_m^\tau(i); \mathcal{B}_m^\tau(i))\right) \right). \end{aligned} \tag{7.3}$$

- *Global Model Update:* The BS broadcasts the global model $\boldsymbol{\omega}^\tau$ and selects the clients for the next global aggregation. Then, the selected clients start their local training iterations based on $\boldsymbol{\omega}^\tau$.

Algorithm 4 summarizes the OFDMA-F^{2}L method. As will be described in Sect. 7.4.1, we assume a convex loss function of the ML task, and hence, OFDMA-F^{2}L converges. Suppose that the optimal ML model parameter $\boldsymbol{\omega}^*$ can be acquired

Algorithm 4: OFDMA-F^{2}L.

Input: I_m^τ, G, ω^0, ϵ, δ, and C.
Output: ω^T.
Initialize: $t = 0$, ω^0;
for $\tau = 1 : G$ **do**
  while $i \leq I_m^\tau$ **do**
    % Local model update;
    $\boldsymbol{\omega}_m^\tau(i) = \boldsymbol{\omega}_m^\tau(i-1) - \eta \nabla F_m(\boldsymbol{\omega}_m^\tau(i-1))$;
    Clip the local parameters: $\boldsymbol{\omega}_m^\tau(i) = \boldsymbol{\omega}_m^\tau(i) / \max\left(1, \frac{\|\boldsymbol{\omega}_m^\tau(i)\|}{C}\right)$;
    $i \leftarrow i + 1$.
  end
  % Global model aggregation;
  Update the global model parameters:

$$\boldsymbol{\omega}^\tau = \sum_{m \in \mathcal{M}} \varrho_m^\tau \zeta_m^\tau \left(\sum_{i=0}^{I_m^\tau - 1} \left(\boldsymbol{\omega}_m^\tau(i) - \eta \nabla F_m(\boldsymbol{\omega}_m^\tau(i); \mathcal{B}_m^\tau(i)) \right) \right);$$

  Update the local model parameters: $\boldsymbol{\omega}_m^\tau(0) = \boldsymbol{\omega}^\tau, \ \forall m \in \mathcal{M}$;
end

after G global aggregations. $F(\boldsymbol{\omega}^*)$ provides the minimum global loss. Nevertheless, when the loss function is non-convex, OFDMA-F^{2}L can still converge, as typically observed in the literature, e.g., [7, 8, 46], and numerically validated in Sect. 7.6.2. Note that if $I_m^\tau = A, \forall m \in \mathcal{M}$, i.e., all selected clients finish a given number of A local iterations (or updates), Algorithm 1 becomes the FedAvg algorithm [28].

7.2.2 Transmission Model

At the beginning of an aggregation round, the BS selects the clients, subchannels, and modulation modes based on the channel state information (CSI) and computing powers that the clients fed back at the end of the previous aggregation round. Then, the BS broadcasts the selections and the global model aggregated at the end of the previous aggregation round, e.g., using its lowest non-zero modulation mode and full transmit power. As a result, all clients are expected to reliably receive the global model and the selections/schedule. The downlink delay, denoted by T^{DL}, is given. The global and local models have the same size, that is, $|\boldsymbol{\omega}| = |\boldsymbol{\omega}_m|$.

Towards the end of each aggregation round, the selected clients upload their local models using an OFDMA protocol. Let $\mathbf{x}_k^\tau \triangleq \left[x_{1,k}^\tau, \cdots, x_{M,k}^\tau\right]^T \in \mathbb{C}^{M \times 1}$ represent the signals that each of the M clients would send in subchannel k, if it is allocated with the subchannel at the τ-th aggregation round. Consider Rayleigh fading, the channel gain of client m in subchannel k is

$$h_{m,k}^{\tau} = \sqrt{\epsilon_o (d_m)^{-\alpha}}\tilde{h}^{\tau}, \ \forall m, k, \tag{7.4}$$

where $\tilde{h}^{\tau}$ is a random scattering element captured by zero-mean and unit-variance circularly symmetric complex Gaussian (CSCG) variables. ϵ_o is the path loss at the reference distance $d_0 = 1$ m. α is the path loss exponent. d_m is the distance between client m and the BS.

Assume that the channels experience block fading, i.e., the channels stay the same within a global aggregation round and vary independently between aggregation rounds [36]. At the τ-th aggregation round, the received signal to noise ratio (SNR) at the BS from client m in subchannel k is given by

$$\gamma_{m,k}^{\tau} = \frac{p_{m,k}^{\tau}|h_{m,k}^{\tau}|^2}{\sigma^2}, \tag{7.5}$$

where $p_{m,k}^{\tau}$ is the transmit power of client m in subchannel k, and σ^2 is the variance of the zero-mean CSCG noise $n_{m,k}^{\tau}$, i.e., $n_{m,k}^{\tau} \in \mathcal{CN}\left(0, \sigma^2\right)$.

Each client can take a modulation mode from a discrete set of L modulation modes. The set is denoted by $\mathcal{L} = \{0, 1, \cdots, L\}$, where $L = |\mathcal{L}|$ and $|\cdot|$ stands for cardinality. The data rate of modulation l is represented by r_l. $l = 0$ indicates no transmission, i.e., $r_0 = 0$. At the BS, the bit error rate (BER) of the local model from client m using subchannel k and modulation l is [12]

$$\chi_{m,k,l}^{\tau} = \beta_1 \exp\left(-\frac{\beta_2 \gamma_{m,k}^{\tau}}{2^{r_l} - 1}\right), \tag{7.6}$$

where β_1 and β_2 are constants relying on the modulation.

To ensure the local model parameters are transmitted from client m to the BS using subchannel k and modulation l achieve the required BER χ_0, the minimum transmit power required is [27]

$$p_{m,k,l}^{\tau} = \frac{(2^{r_l} - 1)\ln\left(\frac{\beta_1}{\chi_0}\right)\sigma^2}{\beta_2 \left|h_{m,k}^{\tau}\right|^2}, \tag{7.7}$$

which is acquired by outplacing $\chi_{m,k,l}^{\tau}$ with χ_0 in (7.6) and then substituting (7.5) into (7.6), followed by rearranging (7.6). Here, we choose a sufficiently small value of χ_0 (e.g., $\leq 10^{-6}$). In other words, the BS receives reasonably accurate and reliable local models.

Let $\lambda_{m,k,l}^{\tau} = 1$ denote the assignment of subchannel k and modulation l to client m to upload its local model, given $|h_{m,k}^{\tau}|^2$; and $\lambda_{m,k,l}^{\tau} = 0$ denotes otherwise. Let $\boldsymbol{\lambda}^{\tau} := \left\{\lambda_{m,k,l}^{\tau}, \ \forall m \in \mathcal{M}, \ k \in \mathcal{K}, \ l \in \mathcal{L}\right\}$ represent the collection of all indicators to be optimized in the τ-th aggregation round. The data rate of client m is obtained as

$$R_m^{\mathrm{UL},\tau} = \sum_{k=1}^{K} \sum_{l=0}^{L} \lambda_{m,k,l}^{\tau} \times r_l. \tag{7.8}$$

According to the minimum transmit power specified in (7.7), the transmit power of client m is given by

$$P_m^{\tau} = \sum_{k=1}^{K} \sum_{l=0}^{L} \lambda_{m,k,l}^{\tau} \times p_{m,k,l}^{\tau}. \tag{7.9}$$

The delay for client m to upload its local model is given by

$$T_m^{\mathrm{UL},\tau} = \frac{|\boldsymbol{\omega}_m|}{R_m^{\mathrm{UL},\tau}} = \frac{|\boldsymbol{\omega}_m|}{\sum_{k=1}^{K} \sum_{l=0}^{L} \lambda_{m,k,l}^{\tau} \times r_l}, \tag{7.10}$$

where $|\boldsymbol{\omega}_m|$ is the size of client m's local model (in bits).

7.2.3 Computing Model

We adopt the number of floating-point operations (FLOPs) to reflect the computational capability of the clients. μ denotes the number of FLOPs required to train a local model relying on a mini-batch of a local dataset. In the τ-th aggregation round, the number of FLOPs required for client m is $\mu_m^{\tau} = \mu I_m^{\tau}$. The local training delay of client m is given by

$$T_m^{\mathrm{C},\tau} = \frac{\mu_m^{\tau}}{\beta_m} = \frac{\mu I_m^{\tau}}{\beta_m}. \tag{7.11}$$

Here, β_m is the computational (or processing) speed of client m.

As per any global aggregation τ, the delay undergone by client m is given by

$$T_m^{\tau} = T_m^{\mathrm{C},\tau} + T_m^{\mathrm{UL},\tau} + T^{\mathrm{DL}}, \tag{7.12}$$

which considers both the transmission and computation delays. Given the equal duration of each FL round, if the uplink and downlink delays are longer, there is a shorter time for local model training. If the computing delay is longer per iteration, the client can only complete fewer iterations. This impacts the convergence and accuracy of the training.

7.3 Problem Statement

We propose to jointly design the selection of clients, subchannels and modulations, i.e., $\boldsymbol{\lambda}^\tau$, to minimize the global loss function of the OFDMA-F^{2}L system per aggregation round, i.e., the τ-th round. The problem is formulated as

$$\mathbf{P7.1}: \min_{\{\boldsymbol{\lambda}^\tau, \boldsymbol{\omega}\}} \sum_{m \in \mathcal{M}} \frac{D_m \zeta_m^\tau F_m(\boldsymbol{\omega})}{D} \tag{7.13a}$$

$$\text{s.t. } \zeta_m^\tau = \operatorname{sgn}\left(\sum_{k=1}^{K}\sum_{l=0}^{L} \lambda_{m,k,l}^\tau\right), \tag{7.13b}$$

$$P_m^\tau \leq P_m^{\max}, \forall m, \tag{7.13c}$$

$$\sum_{k=1}^{K}\sum_{m=1}^{M}\sum_{l=0}^{L} \lambda_{m,k,l}^\tau \leq K, \tag{7.13d}$$

$$\sum_{m=1}^{M}\sum_{l=0}^{L} \lambda_{m,k,l}^\tau \leq 1, \ \forall k, \tag{7.13e}$$

$$\lambda_{m,k,l}^\tau \in \{0, 1\}, \tag{7.13f}$$

$$T_m^\tau \leq T_{\text{th}}, \tag{7.13g}$$

where $\operatorname{sgn}(\cdot)$ is the sign function. The objective (7.13a) stems from (7.1). Specifically, (7.1) minimizes the global loss function over all aggregation rounds. (7.13a) decouples the minimization over time and optimizes the variables and parameters for each round, which is in line with practical FL implementations resulting from causality. Hence, constraint (7.13b) yields

$$\zeta_m^\tau = \begin{cases} 1, \text{ if } \sum_{k=1}^{K}\sum_{l=0}^{L} \lambda_{m,k,l}^\tau \geq 1; & (7.14a) \\ 0, \text{ if } \sum_{k=1}^{K}\sum_{l=0}^{L} \lambda_{m,k,l}^\tau = 0. & (7.14b) \end{cases}$$

In other words, client m is selected if at least one subchannel and modulation are allocated to the client. Constraint (7.13c) indicates that the sum transmit power of each client is upper bounded by $P_m^{\max}$. Constraint (7.13d) specifies that the number of subchannels allocated to all clients is no greater than K. Constraint (7.13e) ensures that a subchannel is allocated to at most one client to avoid inter-client interference. Constraint (7.13g) indicates that the delay (including communication and computation delay) required to conduct the F^{2}L is upper bounded by a pre-specified threshold, i.e., T_{th}, for each global aggregation.

The determination of $\lambda_{m,k,l}^\tau$ relies not only on the transmission rate (or time), but also on the computing capability of each client. This indicates that the computing time

$T_m^{\mathrm{C},\tau}$ or the number of iterations per selected client I_m^τ is also optimized implicitly by solving problem **P7.1** per round τ. Once $\lambda_{m,k,l}^\tau$ is decided, the transmit power of client m using subchannel k and modulation l, that is, (7.7), is set to satisfy the BER requirement.

Problem **P7.1** is new, yet mathematically intractable. It is non-convex in $\boldsymbol{\lambda}^\tau$ and $\boldsymbol{\omega}$, and has a mixed integer programming nature resulting from the client selection and subchannel assignment. Moreover, the global loss function $F(\boldsymbol{\lambda}^\tau, \boldsymbol{\omega})$ is non-convex and has a large number of parameters to be optimized. As a consequence, problem **P7.1** cannot be directly solved using conventional optimization methods, such as alternating optimization and successive convex approximation.

We decouple problem **P7.1** between the global model learning, and the assignments of the client and modulation mode per subchannel. We first derive the convergence upper bound of the global loss function of OFDMA-F^{2}L, i.e., $F(\boldsymbol{\omega})$, and then derive the almost surely optimal assignments of clients, channels, and modulations to minimize the upper bound.

7.4 Convergence Analysis of OFDMA-F^{2}L

This section starts by deriving the optimality gap of OFDMA-F^{2}L. The optimality gap is minimized per communication by optimally selecting clients, channels, and modulations, to encourage the convergence of the OFDMA-F^{2}L, as delineated in Sect. 7.4.2.

7.4.1 Definitions and Assumptions

This section establishes the upper bound for the optimality gap of OFDMA-F^{2}L, i.e., **Algorithm** 4. We start by presenting the definitions and assumptions used.

Definition 7.1 (Gradient Divergence) Let γ_m denote an upper bound of the gradient difference between the local and global loss functions, i.e., $\|\nabla F_m(\boldsymbol{\omega}) - \nabla F(\boldsymbol{\omega})\| \leq \gamma_m, \forall m, \boldsymbol{\omega}$. The global gradient divergence is $\gamma \triangleq \frac{\sum_m D_m \gamma_m}{D}$.

Assumption 7.1 $\forall m \in \mathcal{M}$, we make the following assumptions:

1. The loss function of FL is smooth. Specifically, the gradient of $F_m(\boldsymbol{\omega})$ is L_c-Lipschitz continuous [31], that is, $\left\|\nabla F_m(\boldsymbol{\omega}^{\tau+1}) - \nabla F_m(\boldsymbol{\omega}^\tau)\right\| \leq L_c \left\|\boldsymbol{\omega}^{\tau+1} - \boldsymbol{\omega}^\tau\right\|$, $\forall \boldsymbol{\omega}^\tau, \boldsymbol{\omega}^{\tau+1}$, with L_c being a constant depending on the loss function, so as the gradient of $F(\boldsymbol{\omega})$ is also L_c-Lipschitz continuous;
2. The learning rate is $\eta \leq \frac{1}{L_c}$;
3. The mean squared norm of the stochastic gradients at each edge client m is uniformly bounded by $\mathbb{E}_{\mathcal{B}_m^\tau} \left\|\nabla F_m(\boldsymbol{\omega}_m^\tau(i); \mathcal{B}_m^\tau(i))\right\|^2 \leq \kappa_1 + \kappa_2 \|\nabla F(\boldsymbol{\omega}^\tau)\|^2$, $\forall m$ and τ with $\kappa_1, \kappa_2 \geq 0$;

4. $F_m(\boldsymbol{\omega})$ meets the Polyak-Lojasiewicz condition [20] with a constant ρ, implying that $F(\boldsymbol{\omega}) - F(\boldsymbol{\omega}^*) \leq \frac{1}{2\rho} \|\nabla F(\boldsymbol{\omega})\|^2$;
5. $F(\boldsymbol{\omega}^0) - F(\boldsymbol{\omega}^*) = \Theta$, in which Θ is a constant; in other words, the initial optimality gap is bounded.

These assumptions have been extensively adopted in the literature, e.g., [26, 32, 46]. Many modern ML models have multiple layers and non-convex loss functions (e.g., CNN). Nonetheless, a considerable number of ML tasks, e.g., logistic regression, have (strongly) convex loss functions, e.g., cross-entropy, ordinary least, etc. These convex loss functions satisfy **Assumption** 7.1. Many existing algorithms developed under the (strong) convexity assumptions were experimentally validated for ML models with non-convex loss functions, e.g., [26, 46], and references therein.

7.4.2 Convergence Analysis

Based on Definition 7.1 and Assumption 7.1, each local loss function in (7.1) is strongly convex and hence, OFDMA-F^{2}L converges. We establish the ensuing theorem to evaluate the upper bound for the optimality gap of OFDMA-F^{2}L, i.e., the gap between $\boldsymbol{\omega}^\tau$ and $\boldsymbol{\omega}^*$.

Theorem 7.1 *After τ global aggregations, the convergence upper bound of OFDMA-F^{2}L is given by*

$$F(\boldsymbol{\omega}^\tau) - F(\boldsymbol{\omega}^*) \leq (1 - 2\rho\phi_2)^\tau \Theta + \phi_1 \frac{1 - (1 - 2\rho\phi_2)^\tau}{2\rho\phi_2}, \tag{7.15}$$

where $\phi_1 \triangleq \frac{\eta^2 L_c}{2} \sum_{m \in \mathcal{M}} \frac{A^2 \kappa_1 D_m^2 \zeta_m^\tau}{D^2 I_m^\tau}$ *and* $\phi_2 \triangleq \eta - \frac{\eta^2 L_c}{2} \sum_{m \in \mathcal{M}} \frac{A^2 \kappa_2 D_m^2 \zeta_m^\tau}{D^2 I_m^\tau}$.

Proof See Appendix 7.8. □

To ensure the convergence of OFDMA-F^{2}L, the following proposition is developed based on **Theorem** 7.1.

Proposition 7.1 *To guarantee the convergence of OFDMA-F^{2}L, it must hold that* $1 - 2\rho\phi_2 < 1$. *Hence,* κ_2 *must satisfy*

$$\kappa_2 < \frac{2}{\eta L_c A^2 \sum_{m \in \mathcal{M}} \frac{D_m^2 \zeta_m^\tau}{D^2 I_m^\tau}} \overset{\eta = \frac{1}{L_c}}{=} \frac{2}{A^2 \sum_{m \in \mathcal{M}} \frac{D_m^2 \zeta_m^\tau}{D^2 I_m^\tau}}, \tag{7.16}$$

which provides a sufficient condition for the convergence of OFDMA-F^{2}L.

Corollary 7.1 *In the case of* $1 - 2\rho\phi_2 < 1$, *as* $\tau \to \infty$, *the convergence upper bound of OFDMA-F*^{2}*L is obtained as*

$$F(\boldsymbol{\omega}^{\tau}) - F(\boldsymbol{\omega}^{*}) \leq \frac{\phi_1}{2\rho\phi_2} \tag{7.17a}$$

$$= \frac{\eta^2 L_c}{4\rho} \frac{\sum_{m\in\mathcal{M}} \frac{A^2\kappa_1 D_m^2 \zeta_m^{\tau}}{D^2 I_m^{\tau}}}{\eta - \frac{\eta^2 L_c}{2}\sum_{m\in\mathcal{M}} \frac{A^2\kappa_2 D_m^2 \zeta_m^{\tau}}{D^2 I_m^{\tau}}} \tag{7.17b}$$

$$\leq \frac{1}{2\rho} \sum_{m\in\mathcal{M}} \frac{D_m^2 \zeta_m^{\tau}}{D^2 I_m^{\tau}} \propto \sum_{m\in\mathcal{M}} \frac{D_m^2 \zeta_m^{\tau}}{D^2 I_m^{\tau}}. \tag{7.17c}$$

The upper bound is proportional to $\sum_{m\in\mathcal{M}} \frac{D_m^2 \zeta_m^{\tau}}{D^2 I_m^{\tau}}$.

According to Corollary 7.1, minimizing the optimality gap of OFDMA-F^{2}L is equivalent to minimizing $\sum_{m\in\mathcal{M}} \frac{D_m^2 \zeta_m^{\tau}}{D^2 I_m^{\tau}}$ by optimally selecting clients, channels, and modulations. Consequently, we can reformulate the problem of interest to facilitate the problem-solving in Sect. 7.5.

7.5 Optimal Client, Channel, and Modulation Selection for OFDMA-F^{2}L

Since the selections of the clients, subchannels, and modulations are performed independently in every aggregation round, i.e., the τ-th aggregation round, we suppress the superscript "τ" in the rest of this chapter for brevity of notation.

According to Theorem 7.1, minimizing the global loss function can be transformed into minimizing the optimality gap of OFDMA-F^{2}L, i.e., between $F(\boldsymbol{\omega})$ and $F(\boldsymbol{\omega}^*)$. Since the optimality gap is proportional to $\sum_{m\in\mathcal{M}} \frac{D_m^2 \zeta_m}{D^2 I_m}$ per aggregation round τ, as stated in Corollary 7.1, problem **P7.1** can be rewritten in a much simpler form, as given by

$$\textbf{P7.2}: \min_{\{\boldsymbol{\lambda}\}} \sum_{m\in\mathcal{M}} \frac{D_m^2 \zeta_m}{D^2 I_m} \tag{7.18a}$$

$$\text{s.t. } 1 \leq I_m \leq A, \tag{7.18b}$$

$$(7.13\text{b}) - (7.13\text{g}).$$

Problem **P7.2** is still difficult to tackle, as it is a mixed integer program. Moreover, the feasible solution region of the problem is non-convex since the optimization variable $\boldsymbol{\lambda}$ is discrete. In the ensuing sections, we first convert the objective function to an explicit function of $\boldsymbol{\lambda}$, and then obtain the almost surely optimal solution to the problem.

It is noted that in Problem **P7.2**, the number of training iterations per client per round, I_m, is also optimized while maximizing the objective since it can be uniquely determined given T_m^{C} and T_m^{UL} (or $\boldsymbol{\lambda}$). This indicates that the selections of clients, channels, and modulations are based on their available computing powers. In other words, OFDMA-F^{2}L is flexible in the number of training iterations accomplished by each selected client by taking the potentially different and even time-varying computational capabilities of the clients into account. This flexibility enables the algorithm to achieve better efficiency in both communication and computation.

7.5.1 Problem Transformation

We proceed to show that optimizing the local iteration numbers I_m can be achieved by optimizing the uplink transmission delay T_m^{UL}. From (7.11) and (7.12), we have

$$I_m = T_m^C \beta_m / \mu, \tag{7.19a}$$

$$T_m^C = T_m - T_m^{\mathrm{UL}} - T^{\mathrm{DL}}. \tag{7.19b}$$

Substituting (7.19b) into (7.19a), we have

$$I_m = (T_m - T_m^{\mathrm{UL}} - T^{\mathrm{DL}})\beta_m / \mu. \tag{7.20}$$

Here, I_m increases with T_m; T^{DL} and μ are given. For the set of selected clients, i.e., $\mathcal{M}' \in \mathcal{M}$ ($|\mathcal{M}'| = M'$), we have $\zeta_m = 1$, $\forall m \in \mathcal{M}'$. The objective function in (7.18a) can be recast as $\sum_{m\in\mathcal{M}'} \frac{D_m^2}{D^2 I_m}$. By exploiting the harmonic-arithmetic mean inequality (i.e., $\frac{n}{\sum_{i=1}^n \frac{1}{x_i}} \le \frac{\sum_{i=1}^n x_i}{n}$, $\forall x_i > 0$), it follows that

$$\sum_{m\in\mathcal{M}'} \frac{D_m^2}{D^2 I_m} \ge \frac{M'^2}{\sum_{m\in\mathcal{M}'} \frac{D^2 I_m}{D_m^2}} \tag{7.21a}$$

$$= \frac{M'^2}{\sum_{m\in\mathcal{M}'} (T_m - T_m^{\mathrm{UL}} - T^{\mathrm{DL}})\beta_m D^2 / \mu D_m^2} \tag{7.21b}$$

$$\ge \frac{M'^2}{\sum_{m\in\mathcal{M}'} (T_{\mathrm{th}} - T_m^{\mathrm{UL}} - T^{\mathrm{DL}})\beta_m D^2 / \mu D_m^2}, \tag{7.21c}$$

where (7.21b) is obtained by plugging (7.20) into (7.21a), and (7.21c) is obtained by substituting (7.13g) into (7.21b).

We can minimize $\sum_{m\in\mathcal{M}'} \frac{D_m^2}{D^2 I_m}$ by minimizing its lower bound, i.e., (7.21c), which, in turn, is equivalent to maximizing the denominator of (7.21c), i.e., $\sum_{m\in\mathcal{M}'} (T_{\mathrm{th}} - T_m^{\mathrm{UL}} - T^{\mathrm{DL}})\beta_m D^2 / \mu D_m^2$. Since D_m, D, T_{th}, T^{DL}, and μ are given constants, we only need to minimize $\sum_{m\in\mathcal{M}'} T_m^{\mathrm{UL}} \beta_m D^2 / D_m^2$, or $\sum_{m\in\mathcal{M}'} \beta_m D^2 / R_m^{\mathrm{UL}} D_m^2$ based on (7.10). By applying the harmonic-arithmetic mean inequality again,

$$\sum_{m\in\mathcal{M}'}\frac{\beta_m D^2}{R_m^{\mathrm{UL}}D_m^2}\geq\frac{M'^2}{\sum_{m\in\mathcal{M}'}\frac{R_m^{\mathrm{UL}}D_m^2}{\beta_m D^2}},\tag{7.22}$$

and minimizing $\sum_{m\in\mathcal{M}'}\beta_m D^2/R_m^{\mathrm{UL}}D_m^2$ can be done by minimizing its minimum, i.e., $\frac{M'^2}{\sum_{m\in\mathcal{M}'}\frac{R_m^{\mathrm{UL}}D_m^2}{\beta_m D^2}}$, or maximizing $\sum_{m\in\mathcal{M}'}\frac{R_m^{\mathrm{UL}}D_m^2}{\beta_m D^2}$. Problem **P7.2** can be rewritten as

$$\begin{aligned}\mathbf{P7.3}:\ \max_{\{\boldsymbol{\lambda}\}}\ &\sum_{m\in\mathcal{M}}\frac{R_m^{\mathrm{UL}}D_m^2\zeta_m}{\beta_m D^2} &\text{(7.23a)}\\ \text{s.t. }&T_{\mathrm{th}}-T^{\mathrm{DL}}-\frac{A\mu}{\beta_m}\leq T_m^{\mathrm{UL}}\leq T_{\mathrm{th}}-T^{\mathrm{DL}}-\frac{\mu}{\beta_m},\forall m, &\text{(7.23b)}\\ &(7.13\mathrm{b})-(7.13\mathrm{f}),\end{aligned}$$

where (7.23b) is obtained by substituting (7.20) into (7.18b). Here, the objective in (7.23a) is equivalent to $\sum_{m\in\mathcal{M}'}\frac{R_m^{\mathrm{UL}}D_m^2}{\beta_m D^2}$ since ζ_m indicates the selection of clients; i.e., $\zeta_m=1$ if client m is selected, or $\zeta_m=0$, otherwise.

Problem **P7.3** can be further rewritten as

$$\begin{aligned}\mathbf{P7.4}:\ \max_{\{\boldsymbol{\lambda}\}}\ &\sum_{m\in\mathcal{M}}\frac{R_m^{\mathrm{UL}}D_m^2\zeta_m}{\beta_m D^2} &\text{(7.24a)}\\ \text{s.t. }&\frac{N}{T_{\mathrm{th}}-T^{\mathrm{DL}}-\frac{\mu}{\beta_m}}\leq R_m^{\mathrm{UL}}\leq\frac{N}{T_{\mathrm{th}}-T^{\mathrm{DL}}-\frac{A\mu}{\beta_m}},\forall m, &\text{(7.24b)}\\ &(7.13\mathrm{b})-(7.13\mathrm{f}),\end{aligned}$$

where (7.24b) is obtained by substituting (7.10) into (7.23b).

From (7.8) and (7.14), it follows that $R_m^{\mathrm{UL}}=0$ when $\zeta_m=0$, and $R_m^{\mathrm{UL}}=\sum_{k=1}^{K}\sum_{l=0}^{L}\lambda_{m,k,l}\times r_l$ when $\zeta_m=1$. As a result, $R_m^{\mathrm{UL}}\zeta_m$ can be replaced by R_m^{UL}, and (7.13b) can be suppressed. As a result, problem **P7.4** is equivalent to

$$\begin{aligned}\mathbf{P7.5}:\ \max_{\{\boldsymbol{\lambda}\}}\ &\sum_{m\in\mathcal{M}}R_m^{\mathrm{UL}}D_m^2/\beta_m D^2\\ \text{s.t. }&(7.13\mathrm{c})-(7.13\mathrm{f}),\ (7.24\mathrm{b}).\end{aligned}\tag{7.25}$$

Problem **P7.5** is in the form of weighted sum rate maximization, where the data rate of each client m is weighted by the weighting coefficient $D_m^2/\beta_m D^2$. However, the problem is still intractable for classic convex optimization tools, because it is a mixed-integer program as the result of constraint (7.13f). To solve problem **P7.5**, we propose a Lagrange-dual-based method, as delineated in the next subsection.

Note that Problem **P7.5** has a form of weighted sum rate maximization, where the weighting coefficient is $D_m^2/\beta_m D^2$ for each client m. This is reasonable because, given the delay threshold T_{th}, the maximum number of local iterations that a client can run is bounded by its data rate in an aggregation round, as indicated in (7.24b); in other words, a higher data rate results in a shorter uplink delay and, in turn, a longer time for local training. To this end, maximizing the number of local iterations per aggregation round would require the data rate to be maximized. Moreover, the optimality gap of the FL is minimized by maximizing the number of local iterations per aggregation round; see (7.17c). As a result, the minimization of the optimality gap can be transformed to the maximization of the weighted sum rate, as stated in Problem **P7.5**.

7.5.2 Optimal Client, Subchannel, and Modulation Selection

By defining $\boldsymbol{\xi} = \{\xi_m, \forall m\}$, $\boldsymbol{\nu} = \{\nu_m, \forall m\}$, and $\boldsymbol{\iota} = \{\iota_m, \forall m\}$ as the dual variables concerning (7.13c) and (7.24b), respectively, the Lagrange function of Problem **P7.5** is obtained as

$$\begin{aligned} L(\boldsymbol{\lambda}, \boldsymbol{\xi}, \boldsymbol{\nu}, \boldsymbol{\iota}) = & \sum_{m=1}^{M} \left[R_m^{\mathrm{UL}} D_m^2 / \beta_m D^2 - \xi_m \left(P_m - P_m^{\max} \right) \right] \\ & - \sum_{m=1}^{M} \nu_m \left(-R_m^{\mathrm{UL}} + \frac{N}{T_{\mathrm{th}} - T^{\mathrm{DL}} - \frac{\mu}{\beta_m}} \right) \\ & - \sum_{m=1}^{M} \iota_m \left(R_m^{\mathrm{UL}} - \frac{N}{T_{\mathrm{th}} - T^{\mathrm{DL}} - \frac{A\mu}{\beta_m}} \right). \end{aligned} \tag{7.26}$$

Therefore, the dual problem is written as

$$\min_{\xi, \nu, \iota} D(\xi, \boldsymbol{\nu}, \boldsymbol{\iota}). \tag{7.27}$$

The Lagrange dual function is given by

$$D(\boldsymbol{\xi}, \boldsymbol{\nu}, \boldsymbol{\iota}) = \max_{\{\boldsymbol{\lambda}\}} L(\boldsymbol{\lambda}, \boldsymbol{\xi}, \boldsymbol{\nu}, \boldsymbol{\iota}). \tag{7.28}$$

By defining $\varpi_{m,k,l}(\xi_m, \nu_m, \iota_m) = -\xi_m p_{m,k,l} \left(|h_{m,k}|^2 \right) + \left(D_m^2 / \beta_m D^2 + \nu_m - \iota_m \right) r_l$ for brevity, (7.26) is rearranged as

$$\begin{aligned} L(\boldsymbol{\lambda}, \boldsymbol{\xi}, \boldsymbol{\nu}, \boldsymbol{\iota}) = & \sum_{m=1}^{M} \left(\xi_m P_m^{\max} + \sum_{k=1}^{K} \sum_{l=0}^{L} \lambda_{m,k,l} \varpi_{m,k,l}(\xi_m, \nu_m, \iota_m) \right) \\ & + \sum_{m=1}^{M} \left(\frac{\iota_m N}{T_{\mathrm{th}} - T^{\mathrm{DL}} - \frac{A\mu}{\beta_m}} - \frac{\nu_m N}{T_{\mathrm{th}} - T^{\mathrm{DL}} - \frac{\mu}{\beta_m}} \right). \end{aligned} \tag{7.29}$$

Given $\boldsymbol{\xi}$, $\boldsymbol{\nu}$, and $\boldsymbol{\iota}$, the primary variable $\boldsymbol{\lambda}$ can be acquired via resolving

$$\begin{aligned}&\max_{\boldsymbol{\lambda}} \sum_{k=1}^{K} \left\{ \sum_{m=1}^{M} \sum_{l=0}^{L} \lambda_{m,k,l} \varpi_{m,k,l} \left(\xi_m, \nu_m, \iota_m \right) \right\}, \\ &\text{s.t. (7.13e), (7.13f).}\end{aligned} \tag{7.30}$$

The optimal channel and modulation selections follow a "winner-takes-all" policy [14, 45]. For subchannel k, the m_k^*-th client is selected to upload its local model utilizing the l_k^*-th modulation, that is,

$$\left\{ m_k^*, l_k^* \right\} = \arg\max_{m,l} \varpi_{m,k,l} \left(\xi_m, \nu_m, \iota_m \right), \ \forall k \in \mathcal{K}. \tag{7.31}$$

A greedy policy is employed to optimize $\boldsymbol{\lambda}$:

$$\begin{cases} \lambda_{m,k,l}^* = 1, \text{ if } \{m, l\} = \left\{ m_k^*, l_k^* \right\}; \\ \lambda_{m,k,l}^* = 0, \text{ otherwise.} \end{cases} \tag{7.32}$$

With $\boldsymbol{\lambda}^* (\boldsymbol{\xi}, \boldsymbol{\nu}, \boldsymbol{\iota})$ acquired in (7.32), the sub-gradient descent algorithm could be employed to refresh $\boldsymbol{\xi}$, $\boldsymbol{\nu}$ and $\boldsymbol{\iota}$ via solving (7.27). Then, $\boldsymbol{\xi}$, $\boldsymbol{\nu}$ and $\boldsymbol{\iota}$ are refreshed by [2]

$$\xi_m (\ell + 1) = \left[\xi_m(\ell) + \varepsilon \left(P_m \left(\boldsymbol{\lambda}^* \left(\boldsymbol{\xi}(\ell), \boldsymbol{\nu}(\ell), \boldsymbol{\iota}(\ell) \right) \right) - P_m^{\max} \right) \right]^+; \tag{7.33a}$$

$$\nu_m (\ell + 1) = \left[\nu_m(\ell) + \varepsilon \left(R_m^{\text{UL}} \left(\boldsymbol{\lambda}^* \left(\boldsymbol{\xi}(\ell), \boldsymbol{\nu}(\ell), \boldsymbol{\iota}(\ell) \right) \right) - \frac{N}{T_{\text{th}} - T^{\text{DL}} - \frac{\mu}{\beta_m}} \right) \right]^+, \ \forall m; \tag{7.33b}$$

$$\iota_m (\ell + 1) = \left[\iota_m(\ell) + \varepsilon \left(- R_m^{\text{UL}} \left(\boldsymbol{\lambda}^* \left(\boldsymbol{\xi}(\ell), \boldsymbol{\nu}(\ell), \boldsymbol{\iota}(\ell) \right) \right) + \frac{N}{T_{\text{th}} - T^{\text{DL}} - \frac{A\mu}{\beta_m}} \right) \right]^+, \ \forall m, \tag{7.33c}$$

with ε being the step size, ℓ being the index of an iteration, and $[x]^+ = \max(0, x)$. Initially, $\boldsymbol{\xi}$, $\boldsymbol{\nu}$ and $\boldsymbol{\iota}$ are no smaller than zero, that is, $\xi_m(0) \geq 0$, $\nu_m(0) \geq 0$, and $\iota_m(0) \geq 0, \forall m$, to ensure the convergence of (7.33).

While solving (7.25) involves a non-convex mixed-integer programming problem, the zero-duality gap can still be ensured when the random channel gain $|h_{m,k}|^2$, $m \in \mathcal{M}$, $k \in \mathcal{K}$, has a differentiable cumulative distribution function (CDF), as stated in the following lemma.

Lemma 7.1 *Given any ergodic fading with a continuous CDF, Problem* **P7.5** *holds with strong duality. The solution that is almost surely optimal (i.e., with probability of one) for Problem* **P7.5** *is* $\lambda_{m,k,l}^* \left(\xi_m^*, \nu_m^*, \iota_m^* \right), \forall m, k$, *when the channel gains* $|h_{m,k}|^2$ *have a continuous CDF, where* ξ_m^*, ν_m^*, *and* ι_m^* *are obtained in (7.33) with any initial* $\xi_m(0) > 0$, $\nu_m(0) \geq 0$, *and* $\iota_m(0) \geq 0, \forall m$ *[11].*

Algorithm 5: Proposed Optimal Client, Subchannel, and Modulation Selection.

Initialization: Input $\boldsymbol{\xi}(0)$, $\boldsymbol{\nu}(0)$, $\boldsymbol{\iota}(0)$, $\ell = 0$, and the maximum iteration number of the Lagrange method $L_{\max}$.
Obtain the dual problem of Problem **P7.5** according to (7.27): $\min_{\boldsymbol{\xi},\boldsymbol{\nu},\boldsymbol{\iota}} \max_{\boldsymbol{\lambda}} L(\boldsymbol{\lambda}, \boldsymbol{\xi}, \boldsymbol{\nu}, \boldsymbol{\iota})$.
while $L(\boldsymbol{\lambda}, \boldsymbol{\xi}, \boldsymbol{\nu}, \boldsymbol{\iota})$ *does not converge, and* $\ell \leq L_{\max}$ **do**
  Obtain $\boldsymbol{\lambda}^*(\ell + 1)$ by maximizing $L(\boldsymbol{\lambda}, \boldsymbol{\xi}(\ell), \boldsymbol{\nu}(\ell), \boldsymbol{\iota}(\ell))$ using a greedy strategy as in (7.30)–(7.32).
  Update $\boldsymbol{\xi}(\ell + 1)$, $\boldsymbol{\nu}(\ell + 1)$ and $\boldsymbol{\iota}(\ell + 1)$ according to (7.33) with $\boldsymbol{\lambda}^*(\ell + 1)$.
  $\ell \leftarrow \ell + 1$.
end
Output the optimal selection of subchannels and modulations $\boldsymbol{\lambda} = \boldsymbol{\lambda}^*$.

Proof The almost sure optimality of the solution to Problem **P7.5** can be established by first ratifying the almost sure uniqueness of the solution in the following three situations.

First, if $\max_{m,l} \varpi_{m,k,l}(\xi_m, \nu_m, \iota_m) = 0$, then all clients' channels experience a deep fade in subchannel k utilizing modulation l. Even though client m is chosen for the subchannel, $l_k^*(\xi_m, \nu_m, \iota_m) = 0$, the BS should refrain from transmitting on subchannel k; see (7.32). The "winner" of the subchannel is assigned with the 0-th modulation mode (i.e., no transmission). Second, if $\max_{m,l} \varpi_{m,k,l}(\xi_m, \nu_m, \iota_m) > 0$ and a sole "winner" obtains the subchannel, then there is a unique optimal policy from (7.32). Third, if $\max_{m,l} \varpi_{m,k,l}(\xi_m, \nu_m, \iota_m) > 0$ and several $\{m, l\}$ pairs gain the k-th subchannel, the policy becomes non-unique. Nevertheless, having multiple "winners" accounts for an event of Lebesgue measure zero [22], given a continuous CDF of the stochastic channel gain. The selection of a "winner" is subject to a "measure zero", as we maximize the average net reward in (7.31). By considering all three cases, the optimal allocation strategy $\lambda_{m,k,l}^*(\xi_m, \nu_m, \iota_m)$ in (7.32) is unique with probability of 1.

Given the almost sure uniqueness, we confirm the (almost sure) optimality of the solution to Problem **P7.5** obtained by running (7.32) and (7.33). By relaxing $\lambda_{m,k,l}$ to a continuous value within [0, 1], Problem **P7.2** becomes a convex linear program (LP). The dual problem in (7.27) still holds for the relaxed LP. Since the LP enjoys strong duality, the "winner-takes-all" policy $\boldsymbol{\lambda}^*(\boldsymbol{\xi}, \boldsymbol{\nu}, \boldsymbol{\iota})$ maximizing the dual task of the LP is almost surely unique and hence optimal for the LP. As the LP offers no smaller maximum objective value than problem **P7.5**, $\boldsymbol{\lambda}^*(\boldsymbol{\xi}, \boldsymbol{\nu}, \boldsymbol{\iota})$ is almost surely optimal for (7.25). □

While future OFDMA systems have the potential to flexibly assign subchannels to a device, earlier systems, such as 3GPP Long-Term Evolution (LTE) and IEEE 802.11ax, may have more rigid constraints for subchannel allocation. For example, a device can only be assigned up to two separate clusters of consecutive subchannels in LTE/LTE-Advanced [6]. The proposed algorithm can be potentially extended to incorporate this constraint. For example, when a client is assigned with three clusters of subchannels after $\boldsymbol{\lambda}$ is obtained in (7.32), we can keep two clusters for the client and redistribute the third cluster to the clients allocated with subchannels adjacent to the cluster or those allocated with fewer than two clusters of subchannels.

7.5.3 Complexity Analysis

The algorithm for solving Problem **P7.5** is summarized in Algorithm 5. Algorithm 5 has a computational complexity of $\mathcal{O}(KML\log(1/\epsilon))$, with $\mathcal{O}(\log(1/\epsilon))$ representing iterations required to attain the accuracy requirement ϵ. During an iteration, (7.32) is employed to evaluate all M clients, and L modulations for subchannel k to select the 2-pair $\{m_k^*, l_k^*\}$ that maximizes the net reward $\varpi_{m,k,l}(\lambda, \nu_m)$, resulting in the complexity of $\mathcal{O}(ML)$. The complexity of the other steps, such as the dual variables updating in (7.33), is $\mathcal{O}(M+1)$, which is relatively negligible. Given K subchannels, the total complexity of Algorithm 2 is $\mathcal{O}(KML\log(1/\epsilon))$. The complexity scales linearly with the number of clients. This is comparable with the best-case scenario of the existing technique, where the weighted sum rate maximization problem, i.e., Problem **P7.5**, is transformed into a bipartite matching problem and solved with the Hungarian algorithm. The Hungarian algorithm can eliminate the need for gradient calculations and has a best-case complexity of $\mathcal{O}(KLM)$, but it may suffer from a worst-case complexity of $\mathcal{O}(KLM^2)$ [5].

7.6 Simulation Results

A square region that has a side length of 250 m is considered, with the BS placed at its center, i.e., (0, 0, 20) m. The sides of the region are aligned with the x- and y-axes. The edge clients are uniformly distributed within the square region, as depicted in Fig. 7.1. The position of the edge client m is $(x_m, y_m, 1.5)$ m, $\forall m \in \mathcal{M}$. We set $M = 10$ clients. The maximum number of iterations between two sequential aggregations is $A = 10$. We consider Rayleigh fading for both uplink and downlink channels. The BER requirement of each client is set to $\chi_0 = 10^{-6}$. The maximum number of aggregations is $G = 50$ for the MLP model and $G = 300$ for the CNN model; unless otherwise specified. The model sizes of the MLP and CNN are 1.018×10^5 and 4.214×10^5, respectively. The other parameters are provided in Table 7.3.

We perform the experiments on three datasets:

- *Standard MNIST dataset*, comprising of 60,000 training and 10,000 testing grayscale images. The images depict handwritten digits ranging from 0 to 9;
- *CIFAR10 dataset*, comprised of 60,000 color images of size 32 × 32, split into ten classes (6,000 per class). The dataset includes 50,000 and 10,000 images for training and testing, respectively;
- *FMNIST dataset*, comprised of Zalando's article images, which are 28 × 28 grayscale images classified into ten classes. It includes 60,000 and 10,000 images

Table 7.3 Simulation Parameters for OFDMA-F^{2}L

Category	Parameter	Value
Communication Parameters		
BS maximum transmit power	$P_{\max}$	30 dBm
Client power budget	$P_m^{\max}$	20 dBm
Number of subchannels	K	8 or 16
System bandwidth	B_w	100 MHz
Noise power density	σ^2	-169 dBm/Hz
Path loss at $d_0 = 1$ m	ϵ_0	-30 dB
Path loss exponent	α_d	2.8
Modulation and Coding		
Number of modulations	L	4
Modulation rate set	–	$\{0, 2, 4, 6\}$ bits/symbol
BER requirement	χ_0	10^{-6}
Modulation/coding coefficients	β_1, β_2	0.2, -1.6 [12]
Learning and System Parameters		
Number of clients	M	10
MLP model size	–	1.018×10^5
CNN model size	–	4.214×10^5
Aggregation time limit	T_{th}	10 s
Training FLOPs per model	μ	0.2
Client computation speed	β	$U(9, 12)$
Local data size	D_m	$U(300, 500)$
Global data size	D	$\sum_m D_m$

for training and testing, respectively. Each example corresponds to a grayscale image of size 28×28 and is labeled with one of the ten classes.

To the best of our knowledge, there is no existing study on the OFDMA-F^{2}L systems. Let alone solutions to the challenging discrete selections of the clients, channels, and modulations in the system. With due diligence, we come up with the following three baselines, which are the possible alternatives to the OFDMA-F^{2}L with the proposed optimal client, subchannel, and modulation selections.

- **Baseline 1:** This is OFDMA-F^{2}L with only two modulations, where the clients, subchannels, and modulations are optimally selected in the same way as done in Algorithm 2. $l = 0$ indicates no transmission from a selected client to the BS. $l = 1$ indicates the client transmits to the BS with a modulation rate of 4 bits/symbol.
- **Baseline 2:** This is OFDMA-F^{2}L with random client selection and optimal subchannel and modulation selections. The optimal subchannel and modulation selections are made by following the "winner-takes-all" strategy.

- **Baseline 3:** This scheme runs Sync-FL in the OFDMA system, referred to as OFDMA-Sync-FL, where the selections of the clients, channels, and modulations are jointly optimized under a persistent number of local iterations across all selected clients in each aggregation round. Specifically, by setting $I_m = A$ in problem **P7.2** when $\zeta_m = 1$ and following the analysis in Sect. 7.5.1, problem **P7.4** is updated for OFDMA-Sync-FL as

$$\begin{aligned}\textbf{P7.4'}: \max_{\{\lambda\}} & \sum_{m\in\mathcal{M}} D_m^2 \zeta_m / D^2 \\ \text{s.t. } & T_m^{\mathrm{UL}} \leq T_{\mathrm{th}} - T^{\mathrm{DL}} - \frac{A\mu}{\beta_m}, \\ & (7.13\mathrm{c}) - (7.13\mathrm{f}),\end{aligned}$$

which can be solved using the "winner-takes-all" strategy, as in Algorithm 5.

All experiments are performed on a server running on Ubuntu 20.04.5 LTS, equipped with Intel@ Xeon(R) W-2235 CPU@3.80GHz featuring 12 cores and an NVIDIA GeForce RTX 3090.

7.6.1 OFDMA-F²L on MLP Model

We first evaluate the impact of the optimal client, subchannel, and modulation selections on the learning performance (or convergence) of OFDMA-F^{2}L on the MLP model. The MLP is a feedforward neural network with fully-connected layers. The MNIST dataset is used to train the MLP with a 32-unit hidden layer. We employ the softmax with ten classes regarding the ten digits. A cross-entropy loss function is used to capture the error of the model trained on the local dataset. The accuracy of the model is defined as the ratio of the number of correct predictions made by the model to the total number of predictions.

Figures 7.2 and 7.3 show the (testing) loss (function value) and accuracy, respectively, of the MLP on the MNIST when there are $K = 8$ and 16 subchannels. The x-axis indicates the global aggregations. It is seen that the FL algorithms show fast and smooth convergence within 50 training epochs. The loss function value reaches about 0.35, and the accuracy is about 0.92 under OFDMA-F^{2}L with the optimal selections. Given the number of subchannels, i.e., $K = 8$ or $K = 16$, OFDMA-F^{2}L with the optimal selections outperforms the three baseline schemes with faster convergence speeds, smaller losses, and higher learning accuracies. In addition, Baseline 1 is better than Baseline 2 since Baseline 1 optimizes the client selection policy while Baseline 2 selects clients randomly. The performances of Baselines 2 and 3 are close since they each execute part of OFDMA-F^{2}L with the optimal selections.

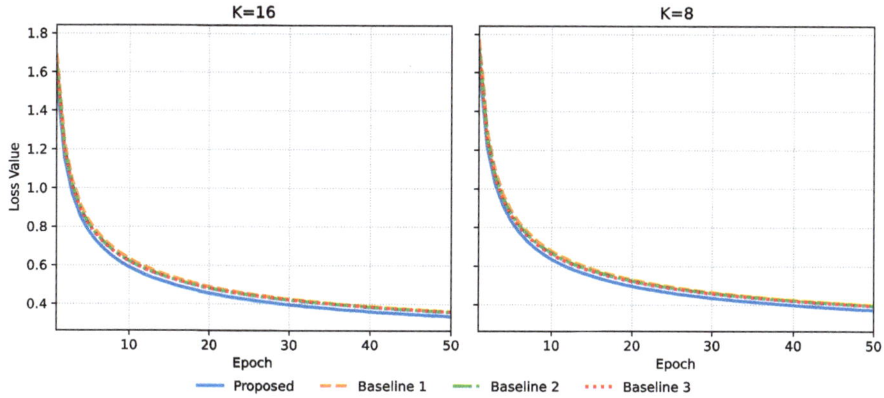

Fig. 7.2 Loss of the MLP on the MNIST under $K = 8$ and $K = 16$

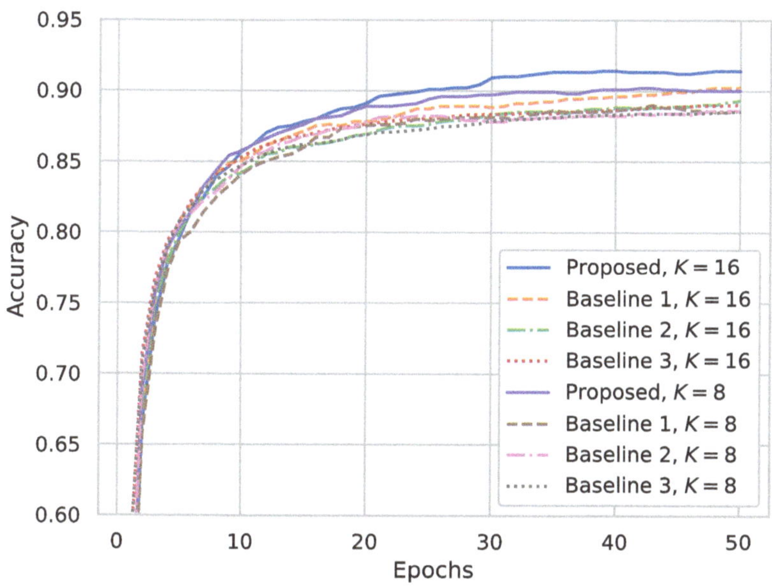

Fig. 7.3 Accuracy of the MLP on the MNIST under $K = 8$ and $K = 16$

7.6.2 OFDMA-F^{2}L on CNN Model

We also evaluate the impact of the optimal client, subchannel, and modulation selections on the learning performance (or convergence) of OFDMA-F^{2}L on the CNN model. The CNN comprises two convolutional layers that have a kernel size of five, followed by three fully-connected layers. We train the CNN using the SGD optimizer. The ReLU units are used along with a softmax function that corresponds to ten classes for the CIFAR10 and ten digits of the FMNIST.

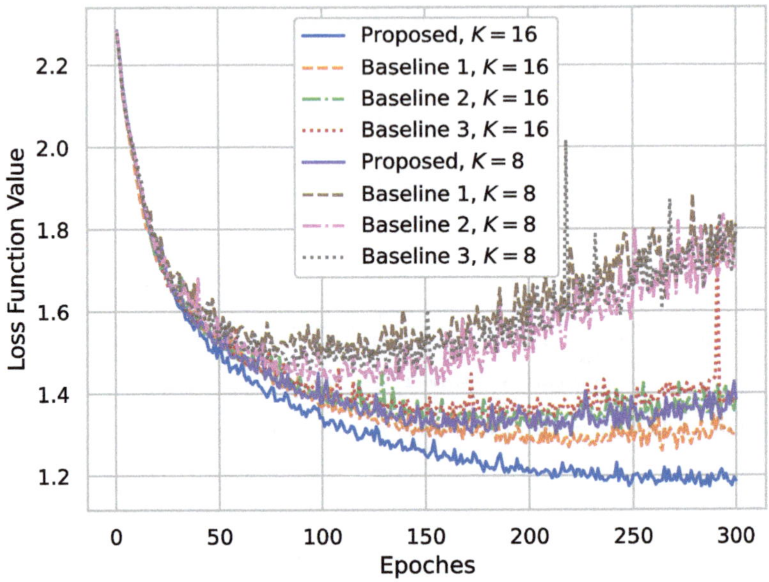

Fig. 7.4 Loss of the CNN on the CIFAR10 under $K = 8$ and $K = 16$

Figures 7.4 to 7.7 show the (testing) losses and accuracies of the CNN on the CIFAR10 and FMNIST with 300 training epochs. As the training progresses, the testing loss decreases, and the testing accuracy increases. At the beginning, the testing loss decreases rapidly, indicating that the model learns fast from the available data. Once a certain number of aggregations is reached, the rate of improvement slows down, and the testing loss converges to the minimum value, as shown in Figs. 7.4 and 7.6. Similarly, the testing accuracy increases rapidly at first, but levels off. The model reaches a plateau in its performance in Figs. 7.5 and 7.7.

It is observed in Figs. 7.4 and 7.5 that on the CIFAR10 dataset, the loss function value stabilizes around 1.2 and the accuracy is about 0.6 under OFDMA-F^{2}L with the optimal selections. In contrast, the three baseline schemes yield higher loss function values and worse accuracies. In particular, when $K = 8$, the (testing) losses of the three baselines first decrease and then increase as the training goes on, indicating a divergence of the model learned from the training data (or, in other words, the model incurs overfitting to the training data and does not generalize well to the testing data). This is because when K is small (e.g., $K = 8$), the total transmission rate is low, and the number of clients involved in the model aggregation is small, in which case, the baselines could not provide enough data for learning a good model.

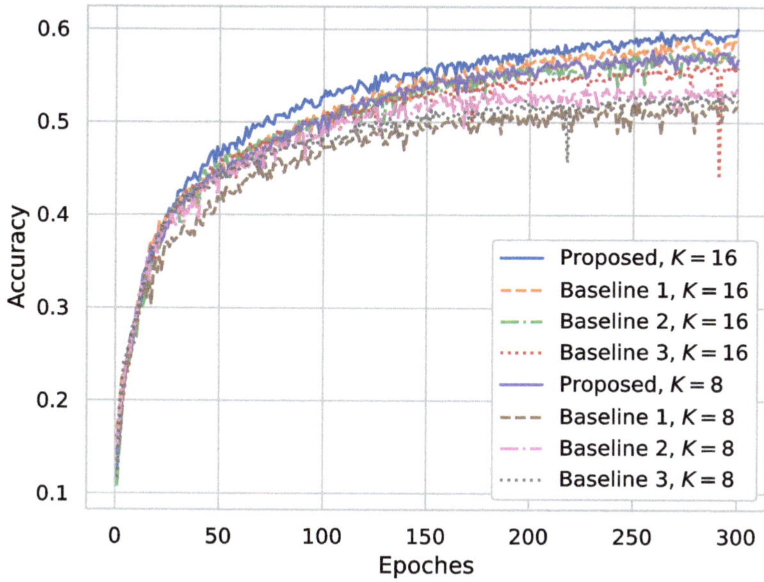

Fig. 7.5 Accuracy of the CNN on the CIFAR10 under $K = 8$ and $K = 16$

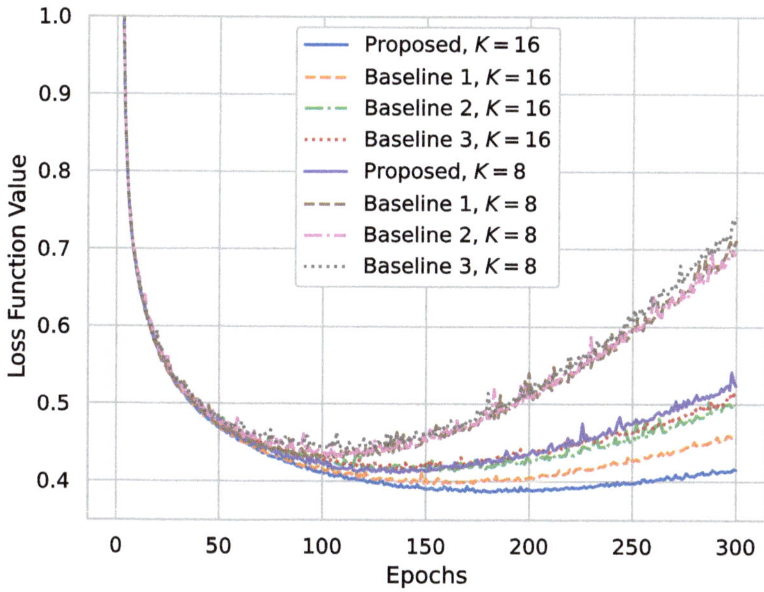

Fig. 7.6 Loss of the CNN on the FMNIST under $K = 8$ and $K = 16$

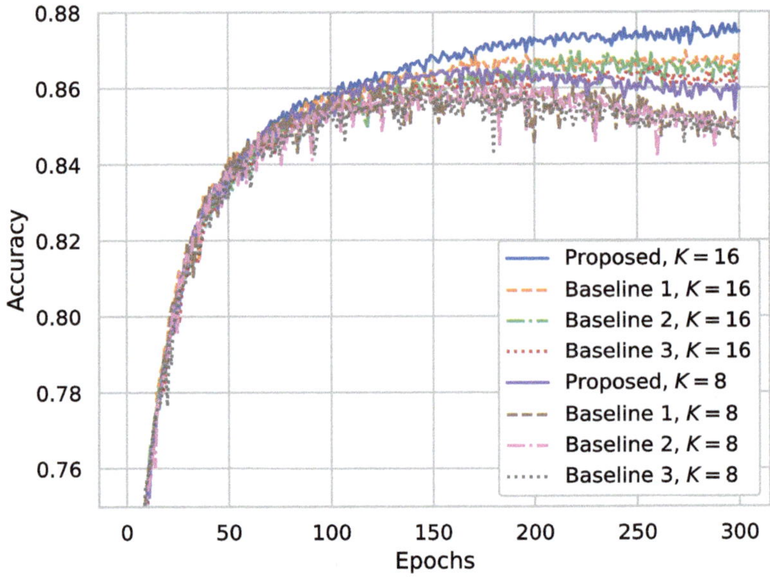

Fig. 7.7 Accuracy of the CNN on the FMNIST under $K = 8$ and $K = 16$

It is also observed in Figs. 7.4 and 7.5 that Baseline 3 is the least stable, since it necessitates all selected clients to upload their local models simultaneously after completing the same number of iterations using Sync-FL. In this case, the global training can be significantly affected by clients with poor channels or low computing powers, especially when a large dataset is concerned.

In Figs. 7.6 and 7.7, we see that the FL algorithms are much smoother than on the CIFAR10 dataset and have faster convergence. This is because the FMNIST is a dataset of images of clothing items, which is considerably smaller compared to the CIFAR10 dataset. The images in the FMNIST dataset are also simpler and have fewer variations in color and texture, compared to the CIFAR10 images. As a result, OFDMA-F^{2}L with the optimal selections may take a shorter time to converge to a satisfactory solution. When $K = 16$, the accuracy of OFDMA-F^{2}L with the optimal selections stabilizes at about 0.88, while the accuracies of the three baselines are about 0.865. In other words, OFDMA-F^{2}L with the optimal selections has better learning performance than the baselines. When $K = 8$, the (testing) loss function values of all four algorithms initially decrease and then increase during training, indicating that the models undergo overfitting. This is likely due to the low transmission rate and the limited number of clients involved in the training process, which are insufficient to capture the full complexity of the data distribution.

7.6.3 Performance Analysis of OFDMA-F^{2}L

Figure 7.8 plots the client selection results and the corresponding numbers of local iterations for each selected client at a specific training epoch of OFDMA-F^{2}L and the other baselines when $K = 8$ and $K = 16$. Compared to the three baselines, more clients are selected under OFDMA-F^{2}L with the optimal selections. Besides, OFDMA-F^{2}L with the optimal selections accommodates more local iterations per aggregation round than the three baselines, shows a better learning performance than the baselines in Sects. 7.6.1 and 7.6.2. When $K = 16$, Baseline 1 supports more local iterations per aggregation round than Baseline 2. When $K = 8$, Baseline 1 accommodates fewer local iterations per aggregation round than Baseline 2. This is consistent with the learning performance of the three baselines presented in Sects. 7.6.1 and 7.6.2.

Figures 7.9 and 7.10 show the objective value of problem **P7.5** and the (uplink) sum rate of the proposed OFDMA-F^{2}L with the optimal selections when the number of subchannels or modulation modes changes. In general, the performance metrics increase as the subchannels or modulation modes rise since the system can have a higher data rate, a shorter communication delay, and hence the lower objective function value and higher accuracy. OFDMA-F^{2}L with the optimal selections significantly outperforms the three baselines, as the subchannels or modulations increase.

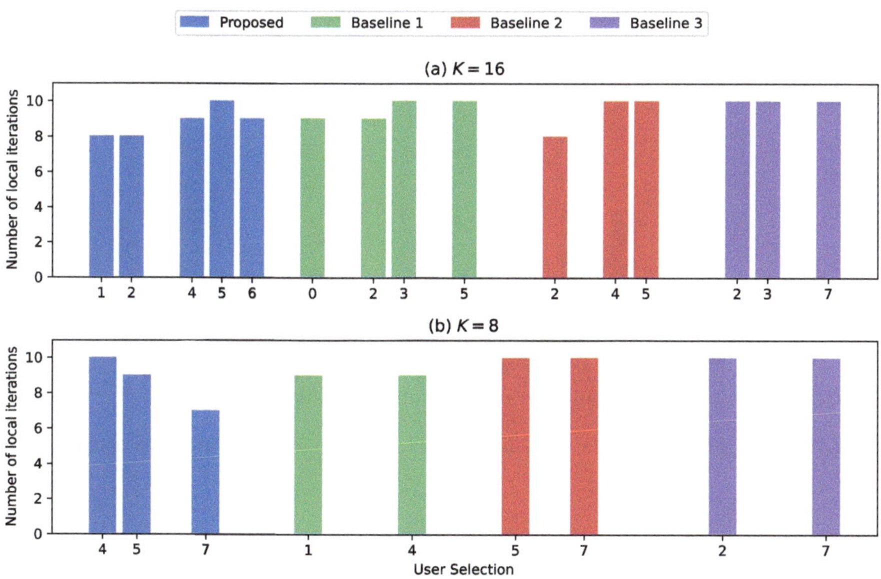

Fig. 7.8 The number of local iterations in a selected aggregation round vs. the selected clients in the round

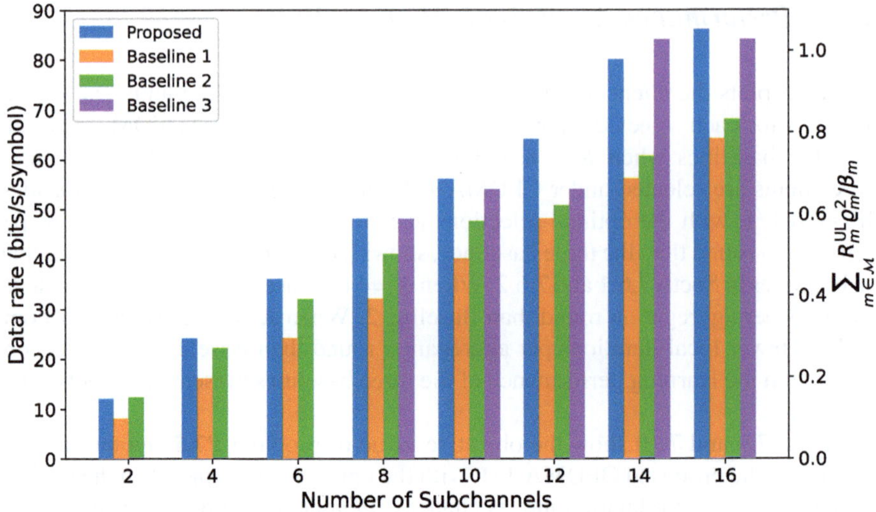

Fig. 7.9 $\sum_{m\in\mathcal{M}} R_m^{\mathrm{UL}}\varrho_m^2/\beta_m$ and sum data rate vs. the number of subchannels

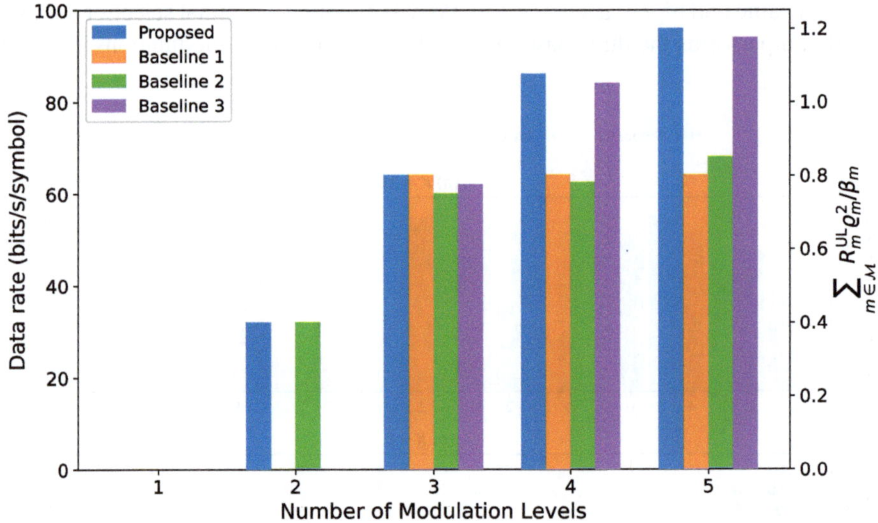

Fig. 7.10 $\sum_{m\in\mathcal{M}} R_m^{\mathrm{UL}} D_m^2/\beta_m D^2$ and sum data rate vs. the number of modulation modes

It is also observed in Fig. 7.9 that the objective value and data rate change linearly and synchronously with the number of subchannels under Baseline 1, as it has only two possible modulation modes. When the number of subchannels $K \leq 6$, the data rate is always zero under Baseline 3. This is because the total delay constraint cannot be satisfied with insufficient communication resources. It is further shown in Fig. 7.10 that when there are more than eight modulation modes, the performance of OFDMA-F^{2}L with the optimal selections does not change, as the total power required reaches the power budget $P_m^{\max}$. When there are over three modulation modes, the performance of Baseline 1 stays the same since it only considers two modulation modes. In contrast, the unstable performance of Baseline 2 results from the random client selection policy. These results corroborate the merits of OFDMA-F^{2}L with the optimal selections.

7.7 Chapter Summary

We have presented a new OFDMA-F^{2}L framework, which allows clients to train varying numbers of iterations per aggregation round and concurrently upload their local models, thereby increasing participating clients and extending their training times. A convergence upper bound on OFDMA-F^{2}L has been derived and minimized by jointly optimizing the selections of clients, subchannels, and modulations to maximize the weighted sum rate of the clients in each aggregation round of OFDMA-F^{2}L. A Lagrange-dual based method has been developed to maximize the weighted sum rate, resulting in a "winner-takes-all" strategy delivering almost surely optimal selections of clients, subchannels, and modulations. As tested experimentally on MLP and CNN, OFDMA-F^{2}L with the optimal selections can efficiently reduce communication delay, and improve the training convergence and accuracy, e.g., by about 18% and 5%, compared to its potential alternatives.

7.8 Proof of Theorem 7.1

Assuming that the gradient of the global loss function is L_c-Lipschitz continuous, and the first-order Taylor expansion can be conducted to obtain the upper bound of $F(\omega^{(\tau+1)}) - F(\omega^{\tau})$, as provided by

$$F(\omega^{(\tau+1)}) - F(\omega^{\tau}) \leq \nabla F\left(\omega^{\tau}\right)\left(\omega^{(\tau+1)} - \omega^{\tau}\right) + \frac{L_c}{2}\left\|\omega^{(\tau+1)} - \omega^{\tau}\right\|^2. \tag{7.34}$$

By substituting (7.2) and (7.3) into (7.34), we have

$$F\left(\boldsymbol{\omega}^{(\tau+1)}\right) - F\left(\boldsymbol{\omega}^{\tau}\right) \leq -\nabla F\left(\boldsymbol{\omega}^{\tau}\right) \sum_{m\in\mathcal{M}} \varrho_m^{\tau} \zeta_m^{\tau} \left(\sum_{i=0}^{I_m^{\tau}-1} \eta \nabla F_m(\boldsymbol{\omega}_m^{\tau}(i)) \right)$$

$$+ \frac{L_c}{2} \left\| - \sum_{m\in\mathcal{M}} \varrho_m^{\tau} \zeta_m^{\tau} \left(\sum_{i=0}^{I_m^{\tau}-1} \eta \nabla F_m(\boldsymbol{\omega}_m^{\tau}(i)) \right) \right\|^2 \tag{7.35a}$$

$$= \frac{\eta^2 L_c}{2} \left\| \sum_{m\in\mathcal{M}} \varrho_m^{\tau} \zeta_m^{\tau} \left(\sum_{i=0}^{I_m^{\tau}-1} \nabla F_m(\boldsymbol{\omega}_m^{\tau}(i)) \right) \right\|^2$$

$$- \eta \nabla F\left(\boldsymbol{\omega}^{\tau}\right) \sum_{m\in\mathcal{M}} \varrho_m^{\tau} \zeta_m^{\tau} \left(\sum_{i=0}^{I_m^{\tau}-1} \nabla F_m(\boldsymbol{\omega}_m^{\tau}(i)) \right). \tag{7.35b}$$

We proceed by computing the expected value of both sides of (7.35) regarding $\mathcal{B}^{\tau}$, representing the sampled mini-batches from $\mathcal{D}$ at the τ-th global aggregation. It follows that

$$\mathbb{E}_{\mathcal{B}^{\tau}}\left\{F\left(\boldsymbol{\omega}^{(\tau+1)}\right)\right\} \leq F\left(\boldsymbol{\omega}^{\tau}\right) - \eta \left\|\nabla F\left(\boldsymbol{\omega}^{\tau}\right)\right\|^2$$

$$+ \frac{\eta^2 L_c}{2} \mathbb{E}_{\mathcal{B}^{\tau}} \left\{ \left\| \sum_{m\in\mathcal{M}} \varrho_m^{\tau} \zeta_m^{\tau} \left(\sum_{i=0}^{I_m^{\tau}-1} \nabla F_m(\boldsymbol{\omega}_m^{\tau}(i); \mathcal{B}_m^{\tau}(i)) \right) \right\|^2 \right\}. \tag{7.36}$$

Since $\varrho_m^{\tau} = \frac{A}{I_m^{\tau}} \cdot \frac{D_m}{D}$, $\mathbb{E}_{\mathcal{B}^{\tau}} \left\| \sum_{m\in\mathcal{M}} \varrho_m^{\tau} \left(\sum_{i=0}^{I_m^{\tau}-1} \nabla F_m(\boldsymbol{\omega}_m^{\tau}(i); \mathcal{B}_m^{\tau}(i)) \right) \right\|^2$ in (7.36) is rewritten as

$$\mathbb{E}_{\mathcal{B}^{\tau}} \left\{ \left\| \sum_{m\in\mathcal{M}} \frac{A D_m \zeta_m^{\tau}}{D I_m^{\tau}} \left(\sum_{i=0}^{I_m^{\tau}-1} \nabla F_m(\boldsymbol{\omega}_m^{\tau}(i); \mathcal{B}_m^{\tau}(i)) \right) \right\|^2 \right\} \tag{7.37a}$$

$$= \sum_{m\in\mathcal{M}} \frac{A^2 D_m^2 \zeta_m^{\tau}}{D^2} \mathbb{E}_{\mathcal{B}_m^{\tau}} \left\{ \left\| \frac{1}{I_m^{\tau}} \sum_{i=0}^{I_m^{\tau}-1} \nabla F_m(\boldsymbol{\omega}_m^{\tau}(i); \mathcal{B}_m^{\tau}(i)) \right\|^2 \right\} \tag{7.37b}$$

$$= \sum_{m\in\mathcal{M}} \frac{A^2 D_m^2 \zeta_m^{\tau}}{D^2 \left(I_m^{\tau}\right)^2} \sum_{i=0}^{I_m^{\tau}-1} \mathbb{E}_{\mathcal{B}_m^{\tau}(i)} \left\| \nabla F_m(\boldsymbol{\omega}_m^{\tau}(i); \mathcal{B}_m^{\tau}(i)) \right\|^2 \tag{7.37c}$$

$$\leq \sum_{m\in\mathcal{M}} \frac{A^2 D_m^2 \zeta_m^{\tau}}{D^2 I_m^{\tau}} \left(\kappa_1 + \kappa_2 \|\nabla F(\boldsymbol{\omega}^{\tau})\|^2 \right), \tag{7.37d}$$

where (7.37b) is due to the i.i.d. datasets of all clients; (7.37c) is because the sampled mini-batches are considered to be i.i.d. between the two consecutive aggregations; and (7.37d) is due to the assumption of $\mathbb{E}_{\mathcal{B}_m^\tau} \left\| \nabla F_m(\boldsymbol{\omega}_m^\tau(i); \mathcal{B}_m^\tau(i)) \right\|^2 \leq \kappa_1 + \kappa_2 \|\nabla F(\boldsymbol{\omega}^\tau)\|^2, \forall m$ and τ.

Since $\mathbb{E}_{\mathcal{B}^\tau}\{F(\boldsymbol{\omega}^{\tau+1})\} = F(\boldsymbol{\omega}^{\tau+1})$, we substitute (7.37) into (7.36), yielding

$$F(\boldsymbol{\omega}^{(\tau+1)}) \leq F(\boldsymbol{\omega}^\tau) - \phi_2 \|\nabla F(\boldsymbol{\omega}^\tau)\|^2 + \phi_1, \tag{7.38}$$

where $\phi_1 \triangleq \frac{\eta^2 L_c}{2} \sum_{m \in \mathcal{M}} \frac{A^2 \kappa_1 D_m^2 \zeta_m^\tau}{D^2 I_m^\tau}$ and $\phi_2 \triangleq \eta - \frac{\eta^2 L_c}{2} \sum_{m \in \mathcal{M}} \frac{A^2 \kappa_2 D_m^2 \zeta_m^\tau}{D^2 I_m^\tau}$.

Detracting $F(\boldsymbol{\omega}^*)$ from both sides of (7.38), it follows

$$F(\boldsymbol{\omega}^{(\tau+1)}) - F(\boldsymbol{\omega}^*) \leq F(\boldsymbol{\omega}^\tau) - F(\boldsymbol{\omega}^*) - \phi_2 \|\nabla F(\boldsymbol{\omega}^\tau)\|^2 + \phi_1. \tag{7.39}$$

Considering the Polyak-Lojasiewicz condition, we have

$$\|\nabla F(\boldsymbol{\omega}^\tau)\|^2 \geq 2\rho \left(F(\boldsymbol{\omega}^\tau) - F(\boldsymbol{\omega}^*) \right). \tag{7.40}$$

By substituting (7.40) into (7.39), we have

$$F(\boldsymbol{\omega}^{(\tau+1)}) - F(\boldsymbol{\omega}^*) \leq (1 - 2\rho\phi_2) \left[F(\boldsymbol{\omega}^\tau) - F(\boldsymbol{\omega}^*) \right] + \phi_1. \tag{7.41}$$

Based on (7.41), we can use mathematical induction and finally obtain

$$\begin{aligned} F(\boldsymbol{\omega}^\tau) - F(\boldsymbol{\omega}^*) \leq & \left[F(\boldsymbol{\omega}^0) - F(\boldsymbol{\omega}^*) \right] (1 - 2\rho\phi_2)^\tau \\ & + \phi_1 \frac{1 - (1 - 2\rho\phi_2)^\tau}{2\rho\phi_2}, \end{aligned} \tag{7.42}$$

which concludes this proof.

References

1. Abutuleb, A., Sorour, S., Hassanein, H.S.: Joint task and resource allocation for mobile edge learning. In: Proc. IEEE Globecom Conf. (GLOBECOM), pp. 1–6 (Dec. 2020). https://doi.org/10.1109/GLOBECOM42002.2020.9322399
2. Boyd, S., Vandenberghe, L.: Convex Optimization. Cambridge university press (2004)
3. Chai, Z., Chen, Y., Anwar, A., Zhao, L., Cheng, Y., Rangwala, H.: FedAT: A high-performance and communication-efficient federated learning system with asynchronous tiers. In: Proc. Int. Conf. High Perfor. Comput., Networking, Storage Analysis, pp. 1–17 (2021). https://doi.org/10.1145/3458817.3476211
4. Chen H, Huang S, Zhang D, Xiao M, Skoglund M, Poor HV (2022) Federated learning over wireless iot networks with optimized communication and resources. IEEE Internet Things J 9(17):16592–16605. https://doi.org/10.1109/JIOT.2022.3151193. Sep.
5. Chen M, Yang Z, Saad W, Yin C, Poor HV, Cui S (2021) A joint learning and communications framework for federated learning over wireless networks. IEEE Trans Wireless Commun

20(1):269–283. https://doi.org/10.1109/TWC.2020.3024629. Jan.
6. Chen X, Cui J, Ni W, Wang X, Zhu Y, Zhang J, Xu S (2020) DFT-s-OFDM: Enabling flexibility in frequency selectivity and multiuser diversity for 5G. IEEE Consumer Electron Mag 9(6):15–22. https://doi.org/10.1109/MCE.2020.2988045. Nov.
7. Chen, X., Zhou, Z., Ni, W., Pan, G., Wang, X., Zhang, S., Sun, Y.: Privacy-preserving resource allocation for asynchronous federated learning. In: 2024 IEEE 99th Vehicular Technology Conference (VTC2024-Spring), pp. 1–6 (2024). https://doi.org/10.1109/VTC2024-Spring62846.2024.10683194
8. Chen Y, Sun X, Jin Y (2020) Communication-efficient federated deep learning with layerwise asynchronous model update and temporally weighted aggregation. IEEE Trans Neural Netw Learn Syst 31(10):4229–4238. https://doi.org/10.1109/TNNLS.2019.2953131. Oct.
9. Cui Y, Lv T, Li W, Ni W, Hossain E (2024) Over-the-air federated learning in user-centric cell-free networks. IEEE Wireless Communications Letters 13(12):3683–3687. https://doi.org/10.1109/LWC.2024.3486201
10. Du J, Jiang B, Jiang C, Shi Y, Han Z (2023) Gradient and channel aware dynamic scheduling for over-the-air computation in federated edge learning systems. IEEE J Sel Areas Commun 41(4):1035–1050. https://doi.org/10.1109/JSAC.2023.3242727. Apr.
11. Gao N, Wang X (2011) Optimal subcarrier-chunk scheduling for wireless OFDMA systems. IEEE Trans Wireless Commun 10(7):2116–2123. https://doi.org/10.1109/TWC.2011.050511.100458
12. Goldsmith AJ, Chua S (1998) Adaptive coded modulation for fading channels. IEEE Trans Commun 46(5):595–602. https://doi.org/10.1109/26.668727
13. Gu B, Xu A, Huo Z, Deng C, Huang H (2022) Privacy-preserving asynchronous vertical federated learning algorithms for multiparty collaborative learning. IEEE Trans Neural Netw Learn Syst 33(11):6103–6115. https://doi.org/10.1109/TNNLS.2021.3072238. Nov.
14. He T, Wang X, Ni W (2014) Optimal chunk-based resource allocation for OFDMA systems with multiple BER requirements. IEEE Trans Veh Technol 63(9):4292–4301. https://doi.org/10.1109/TVT.2014.2313354
15. He Y, Yang M, He Z, Guizani M (2023) Resource allocation based on digital twin-enabled federated learning framework in heterogeneous cellular network. IEEE Trans Veh Tech 72(1):1149–1158. https://doi.org/10.1109/TVT.2022.3205778. Jan.
16. Hu, H., Salcic, Z., Sun, L., Dobbie, G., Zhang, X.: Source inference attacks in federated learning. In: 2021 IEEE International Conference on Data Mining (ICDM), pp. 1102–1107. IEEE (2021)
17. Hu, S., Chen, X., Ni, W., Hossain, E., Wang, X.: Distributed machine learning for wireless communication networks: Techniques, architectures, and applications. IEEE Commun. Surveys Tuts. **23**(3), 1458–1493 (3rd Quart., 2021)
18. Ji, Z., Qin, Z.: Federated learning for distributed energy-efficient resource allocation. In: Proc. IEEE Int. Conf. Commun. (ICC), pp. 1–6 (May 2022). https://doi.org/10.1109/ICC45855.2022.9882281
19. Jiao Y, Wang P, Niyato D, Lin B, Kim DI (2021) Toward an automated auction framework for wireless federated learning services market. IEEE Trans Mobile Comput 20(10):3034–3048. https://doi.org/10.1109/TMC.2020.2994639. Oct.
20. Karimi, H., Nutini, J., Schmidt, M.: Linear convergence of gradient and proximal-gradient methods under the Polyak-Lojasiewicz condition. In: Joint European Conference on Machine Learning and Knowledge Discovery in Databases, pp. 795–811. Springer (2016)
21. Kurunathan H, Huang H, Li K, Ni W, Hossain E (2024) Machine learning-aided operations and communications of unmanned aerial vehicles: A contemporary survey. IEEE Communications Surveys & Tutorials 26(1):496–533. https://doi.org/10.1109/COMST.2023.3312221
22. Lenz D (2002) Singular spectrum of Lebesgue measure zero for one-dimensional quasicrystals. Commun Math Phys 227(1):119–130
23. Li W, Lv T, Cao Y, Ni W, Peng M (2023) Multi-carrier NOMA-empowered wireless federated learning with optimal power and bandwidth allocation. IEEE Trans Wireless Commun 22(12):9762–9777. https://doi.org/10.1109/TWC.2023.3273312

24. Lim, W.Y.B., Luong, N.C., Hoang, D.T., Jiao, Y., Liang, Y.C., Yang, Q., Niyato, D., Miao, C.: Federated learning in mobile edge networks: A comprehensive survey. IEEE Commun. Surveys Tuts. **22**(3), 2031–2063 (3rd Quart. 2020). https://doi.org/10.1109/COMST.2020.2986024
25. Lin X, Wu J, Li J, Zheng X, Li G (2023) Friend-as-learner: Socially-driven trustworthy and efficient wireless federated edge learning. IEEE Trans Mobile Comput 22(1):269–283. https://doi.org/10.1109/TMC.2021.3074816. Jan.
26. Liu J, Xu H, Wang L, Xu Y, Qian C, Huang J, Huang H (2023) Adaptive asynchronous federated learning in resource-constrained edge computing. IEEE Trans Mobile Comput 22(2):674–690. https://doi.org/10.1109/TMC.2021.3096846. Feb.
27. Malik, H., Alam, M.M., Le Moullec, Y., Ni, Q.: Interference-aware radio resource allocation for 5G ultra-reliable low-latency communication. In: Proc. IEEE Globecom Workshops (GC Wkshps), pp. 1–6 (2018). https://doi.org/10.1109/GLOCOMW.2018.8644301
28. McMahan, B., Moore, E., Ramage, D., Hampson, S., y Arcas, B.A.: Communication-efficient learning of deep networks from decentralized data. In: Proc. Int. Conf. Artif. Intell. Statist., pp. 1273–1282. PMLR (2017)
29. Nguyen VD, Sharma SK, Vu TX, Chatzinotas S, Ottersten B (2021) Efficient federated learning algorithm for resource allocation in wireless iot networks. IEEE Internet Things J 8(5):3394–3409. https://doi.org/10.1109/JIOT.2020.3022534. Mar.
30. Ni, F., Zhou, Z., Ni, W., Chen, X., Pan, G., Sun, Y., Zhang, S., Jamalipour, A.: Scheduling and securing asynchronous federated learning through cooperative jamming. IEEE Transactions on Cognitive Communications and Networking pp. 1–1 (2025, early access). https://doi.org/10.1109/TCCN.2025.3623377
31. O'Searcoid, M.: Metric Spaces. Springer Undergraduate Mathematics Series. Springer London (2006). https://books.google.com.au/books?id=aP37I4QWFRcC
32. Qu Z, Guo S, Wang H, Ye B, Wang Y, Zomaya AY, Tang B (2022) Partial synchronization to accelerate federated learning over relay-assisted edge networks. IEEE Trans Mobile Comput 21(12):4502–4516. https://doi.org/10.1109/TMC.2021.3083154. Dec.
33. Ruan, Y., Zhang, X., Liang, S.C., Joe-Wong, C.: Towards flexible device participation in federated learning. In: Proc. Int. Conf. Artif. Intell. Statist., vol. 130, pp. 3403–3411. PMLR (Apr. 2021)
34. Salehi M, Hossain E (2021) Federated learning in unreliable and resource-constrained cellular wireless networks. IEEE Trans Commun 69(8):5136–5151 Aug.
35. Sutton GJ, Zeng J, Liu RP, Ni W, Nguyen DN, Jayawickrama BA, Huang X, Abolhasan M, Zhang Z, Dutkiewicz E, Lv T (2019) Enabling technologies for ultra-reliable and low latency communications: From phy and mac layer perspectives. IEEE Communications Surveys & Tutorials 21(3):2488–2524. https://doi.org/10.1109/COMST.2019.2897800
36. Viswanathan H (1999) Capacity of Markov channels with receiver CSI and delayed feedback. IEEE Trans Inf Theory 45(2):761–771. https://doi.org/10.1109/18.749027
37. Wang Z, Zhang Z, Tian Y, Yang Q, Shan H, Wang W, Quek TQS (2022) Asynchronous federated learning over wireless communication networks. IEEE Trans Wireless Commun 21(9):6961–6978. https://doi.org/10.1109/TWC.2022.3153495. Sep.
38. Wu Y, Song Y, Wang T, Qian L, Quek TQS (2022) Non-orthogonal multiple access assisted federated learning via wireless power transfer: A cost-efficient approach. IEEE Trans Commun 70(4):2853–2869. https://doi.org/10.1109/TCOMM.2022.3153068. Apr.
39. Xiao B, Zhang J, Ni W, Wang X (2025) FLARE: A new federated learning framework with adjustable learning rates over resource-constrained wireless networks. IEEE Trans Wireless Commun 24(8):6674–6689. https://doi.org/10.1109/TWC.2025.3555136
40. Xie, C., Koyejo, S., Gupta, I.: Asynchronous federated optimization. In: Annual Wrkshps Optim. Mach. Learn. (2020)
41. Xu J, Wang H (2021) Client selection and bandwidth allocation in wireless federated learning networks: A long-term perspective. IEEE Trans Wireless Commun 20(2):1188–1200. https://doi.org/10.1109/TWC.2020.3031503. Feb.
42. Yang, S., Liu, Y.: Training efficiency of federated learning: A wireless communication perspective. In: Proc. Int. Conf. Wireless Commun. Signal Process. (WCSP), pp. 922–926 (Oct. 2020). https://doi.org/10.1109/WCSP49889.2020.9299704

43. Yang Z, Chen M, Saad W, Hong CS, Shikh-Bahaei M (2021) Energy efficient federated learning over wireless communication networks. IEEE Trans Wireless Commun 20(3):1935–1949. https://doi.org/10.1109/TWC.2020.3037554. Mar.
44. Yu X, Xiao B, Ni W, Wang X (2023) Optimal adaptive power control for over-the-air federated edge learning under fading channels. IEEE Trans Commun 71(9):5199–5213. https://doi.org/10.1109/TCOMM.2023.3283790. Sep.
45. Yuan X, Hu S, Ni W, Liu RP, Wang X (2023) Joint user, channel, modulation-coding selection, and RIS configuration for jamming resistance in multiuser OFDMA systems. IEEE Trans Commun 71(3):1631–1645 Mar.
46. Yuan X, Ni W, Ding M, Wei K, Li J, Vincent Poor H (2023) Amplitude-varying perturbation for balancing privacy and utility in federated learning. IEEE Trans Inf Forensics Security 18:1884–1897. https://doi.org/10.1109/TIFS.2023.3258255
47. Zeng, Q., Du, Y., Huang, K., Leung, K.K.: Energy-efficient radio resource allocation for federated edge learning. In: Proc. IEEE Int. Conf. Commun. Wrkshps. (ICC Wrkshps), pp. 1–6 (Jun. 2020). https://doi.org/10.1109/ICCWorkshops49005.2020.9145118
48. Zhang, J., Simeone, O.: LAGC: Lazily aggregated gradient coding for straggler-tolerant and communication-efficient distributed learning. IEEE Trans. Neural Netw. Learn. Syst. pp. 1–13 (2020)
49. Zhang Y, Liu D, Duan M, Li L, Chen X, Ren A, Tan Y, Wang C (2023) FedMDS: An efficient model discrepancy-aware semi-asynchronous clustered federated learning framework. IEEE Trans Parall Distr Syst 34(3):1007–1019. https://doi.org/10.1109/TPDS.2023.3237752. Mar.
50. Zhong C, Yang H, Yuan X (2023) Over-the-air federated multi-task learning over mimo multiple access channels. IEEE Trans Wireless Commun 22(6):3853–3868. https://doi.org/10.1109/TWC.2022.3221797. Jun.
51. Zhu G, Du Y, Gündüz D, Huang K (2021) One-bit over-the-air aggregation for communication-efficient federated edge learning: Design and convergence analysis. IEEE Trans Wireless Commun 20(3):2120–2135. https://doi.org/10.1109/TWC.2020.3039309. Mar.
52. Zhu G, Wang Y, Huang K (2020) Broadband analog aggregation for low-latency federated edge learning. IEEE Trans Wireless Commun 19(1):491–506. https://doi.org/10.1109/TWC.2019.2946245. Jan.

Chapter 8
Exploring Visual Explanations for Attack Detection

FL has surfaced as a notably distributed ML framework, where user devices collaboratively engage in the training of a shared ML model, supervised by a server. User devices in FL consecutively train local model updates (e.g., weight parameters or gradients) utilizing their proprietary data. Rather than transmitting raw, private data, user devices forward model updates to a server for amalgamation. In response, the server integrates local model updates to generate a comprehensive global model that is then dispatched to the devices for updating their respective local models [12, 23]. Such a communication cycle repeats until the model achieves a satisfactory accuracy level. FL prevents the potential unauthorized dissemination of private data [46]. For instance, FL allows multiple medical institutions to collaboratively train a unified ML model without directly sharing sensitive patient data. By retaining patient data locally, hospitals can contribute to model improvements by processing data on-site and sharing only model updates, thereby enhancing diagnostic tools or treatment algorithms while adhering to data protection regulations [15].

The distributed architecture of FL makes it particularly susceptible to poisoning attacks. User devices compromised by adversarial actors can alter model update parameters, subsequently contaminating the global FL models [3, 8, 27]. Existing countermeasures, leveraging Euclidean distance-based metrics to discern deviations between malicious and benign models, have demonstrated efficacy against such attacks [1, 39]. However, sophisticated adversaries can craft malicious model updates such that the Euclidean distances to benign counterparts remain below a designated threshold, thereby eluding detection by defenses reliant on this metric.

K. Li et al., *Security and Resilience in Distributed Machine Learning*, Springer Series in Reliability Engineering, https://doi.org/10.1007/978-3-032-23959-4_8

8.1 Introduction

This chapter presents a new defense mechanism, dubbed GradCAM-AE, against MP attacks on FL, which employs a Gradient-weighted Class Activation Mapping (GradCAM [26])-based approach in coupling with an AE to offer a substantially more powerful detection capability compared to existing Euclidean distance-based approaches. Specifically, GradCAM is applied at the server to create GradCAM heat maps for every uploaded model update. An AE is applied to reconstruct the heat maps, while magnifying the discernible features of the heat maps. The reconstruction errors of the GradCAM heat maps are measured, and a threshold is created based on the statistics of the reconstruction errors. A reconstructed GradCAM heat map with a reconstruction error surpassing the threshold is categorized as atypical, and the corresponding model update as malicious.

The key contributions are summarized as follows:

- We propose a novel defense method against MP attacks on FL, where GradCAM and an AE are orchestrated for the successful detection of subtle attacks.
- GradCAM is adopted to produce heat maps for each uploaded local model, hence transforming each local model into a lower-dimensional, visual representation. This provides a conduit for pinpointing malicious model updates by singling out anomalous GradCAM heat maps.
- An AE is used to remap the GradCAM heat maps to accentuate the hidden features of the heat maps, and improve the distinguishability of the heat maps and the success rate of identifying anomalous heat maps and malicious local models.
- To mitigate potential misclassifications, we also design a voting algorithm, where model updates that consistently manifest as potentially malicious over several communication rounds are conclusively identified as malicious and removed from FL.

We conducted a comprehensive assessment of the proposed GradCAM-AE framework using two public datasets, CIFAR-10 and German traffic sign recognition benchmark (GTSRB), under both IID and non-IID settings. Our assessment encompasses two prominent deep learning models, i.e., ResNet-18 [10] and MobileNetV3-Large [11]. Our approach offers superior detection rates and test accuracy of the FL global model, compared to the state-of-the-art methods.

The rest of this chapter is organized as follows. Section 8.2 discusses the literature on defense models against poisoning attacks on FL. Section 8.3 presents FL for image classification with DNNs. The threat model is presented in Sect. 8.4. In Sect. 8.5, we detail the proposed GradCAM-AE against MP attacks. Performance evaluation is presented in Sect. 8.6.

8.2 Status Quo

This section reviews existing defense models against poisoning attacks on FL.

8.2.1 *Euclidean Distance-Based Defense*

Several extant solutions, predicated upon Euclidean distance, have been devised to detect the poisoning attack within FL contexts. Within the framework, Krum [1, 40] or Multi-Krum [1, 40] computes a score for each local model update, the score is the sum of its Euclidean distance from its neighbors. Multi-Krum considers those with high scores as malicious model updates, which are excluded. In stark contrast, Trimmed-mean methodology [36] is a coordinate-wise aggregation rule that aggregates each coordinate of the model update separately. For each coordinate, first sort the values of the corresponding coordinates in the users' model updates. The largest and the smallest k values are removed, then the trimmed mean calculates the average of the remaining values, as the coordinate of the aggregated model update. To defend against poisoning attacks with a large number of malicious users, FLDetector [39] is developed to predict a device's model update in each communication round based on historical model updates. The received model update from the user and the predicted model update are inconsistent in multiple communication rounds and are flagged as malicious. A defense strategy is developed against poisoning attacks on FL [4], where users are categorized into distinct groups. A global model is trained for each user group according to an existing FL aggregation rule. Based on the global models of all the groups, a majority vote mechanism is used to identify whether a test input is poisoned by the attacker.

8.2.2 *Cosine-Similarity-Based Defense*

The study in [2] uses cosine similarity to measure the alignment of local model updates with the trusted global model update that is trained on a clean dataset. The local model updates with low similarity are scaled down or ignored. The work [24] utilizes cosine similarity to measure the angular similarity between local model updates and a reference direction (e.g., the aggregated global model or the average of local model updates). The local model updates are clustered based on their angular similarity. The clusters with low cosine similarities to the majority of local model updates are flagged as suspicious and excluded from aggregation. Similarly, the study in [47] keeps the underlying assumption that the local gradients of benign devices align closely with the direction of the global gradient, whereas malicious gradients deviate significantly. The authors use cosine similarity to measure the angular alignment between the local gradients of devices and the global gradients maintained by the server. The local gradients with obtuse angles (cosine similarity is less than zero) are flagged as malicious because they indicate opposing or disruptive directions. The server discards these local gradients to prevent them from affecting the global model's update.

8.2.3 ML-Based Defense

To detect the malicious local devices while generating an accurate model, a statistical mechanism called Auror is introduced [28]. AUROR is based on the observation that the most important model features from the majority of honest devices will exhibit a similar distribution, while those from malicious devices will exhibit an anomalous distribution. AUROR then uses k-means to cluster uploaded local model updates over training rounds and discards the malicious model updates, i.e., contributions from small clusters that exceed a threshold distance are considered malicious.

The authors of [29] presented the Federated Anomaly Analytics enhanced Distributed Learning (FAA-DL) framework, which is a lightweight, unsupervised anomaly detection method based on a support vector machine (SVM). The main idea of FAA-DL is to utilize an appropriate kernel function and soft margins to estimate a nonlinear decision boundary and separate the benign and malicious local model updates. Detecting network attacks is crucial for preventing and alleviating the consequences of network attacks. To detect and filter malicious local model updates before global model aggregation, the authors of [18] developed the FL framework, which is an integrated isolation forest algorithm. The authors argue that the leaf nodes representing the malicious model are closer to the root. The malicious local model updates can be detected by employing DRL methods to adjust the detection threshold dynamically. To resist generative adversarial network (GAN) attacks, the authors of [5] developed a system that isolates local model update parameters of the network from all devices, derailing any chance that the attacker can set up a GAN to carry out attacks.

The existing defense mechanisms for malicious local model updates, particularly those predicated on Euclidean distances, notably encounter challenges pertaining to the "curse of dimensionality" within the context of DNNs, especially considering that local model update parameters encompass millions or potentially billions of parameters (encompassing both weights and biases). In the milieu of high-dimensional spaces, Euclidean distances may inadvertently overestimate the distances between model updates, thereby diminishing the efficacy of detecting malicious model updates. Predominantly, machine-learning-based detection mechanisms significantly hinge upon the meticulous tuning of hyperparameters and the astute setting of thresholds. Recent experimental findings [9, 42] contend that this approach potentially verges on ineffective anomaly detection.

8.3 Formulation of DNN-Based Federated Learning

In this section, we consider image classification as an example to depict the training process of FL. The system consists of a server and K benign users where $\mathcal{K} \triangleq \{1, \ldots, K\}$, as shown in Fig. 8.1. Each benign user device trains a DNN that is particularly well-suited for image classification. The DNN analyzes an input image (e.g., deer), extracts features (e.g., edges, corners, and textures) through several convolutional layers, feeds the feature maps into fully connected (FC) layers, and assigns one of the labels (such as "bird", "dog", "horse", or "deer"). The DNN model is learned based on training data and represented by the parameter vector $\mathbf{W}$ (weights and biases). The user device possesses D_k training data pairs $\{(\mathbf{x}_s, y_s)\}_{s=1}^{D_k}$, consisting of a feature vector $\mathbf{x}_s$ (e.g., the pixels of an image) as the input of the model and a corresponding scalar value y_s (e.g., the real label of the image) that is the desired output of the model [33]. The dataset of the user device is denoted by $\mathcal{D}_k$, and we have

$$\mathcal{D}_k = \{(\mathbf{x}_1, y_1), \ldots, (\mathbf{x}_s, y_s), \ldots, (\mathbf{x}_{D_k}, y_{D_k})\}, \tag{8.1}$$

where D_k denotes the cardinality of $\mathcal{D}_k$, i.e., $D_k \triangleq |\mathcal{D}_k|$.

Let $f_s(\mathbf{W}, \mathbf{x}_s, y_s)$ denote the loss function of each training data sample s, which captures the approximation errors between the input $\mathbf{x}_s$ and desired output y_s.

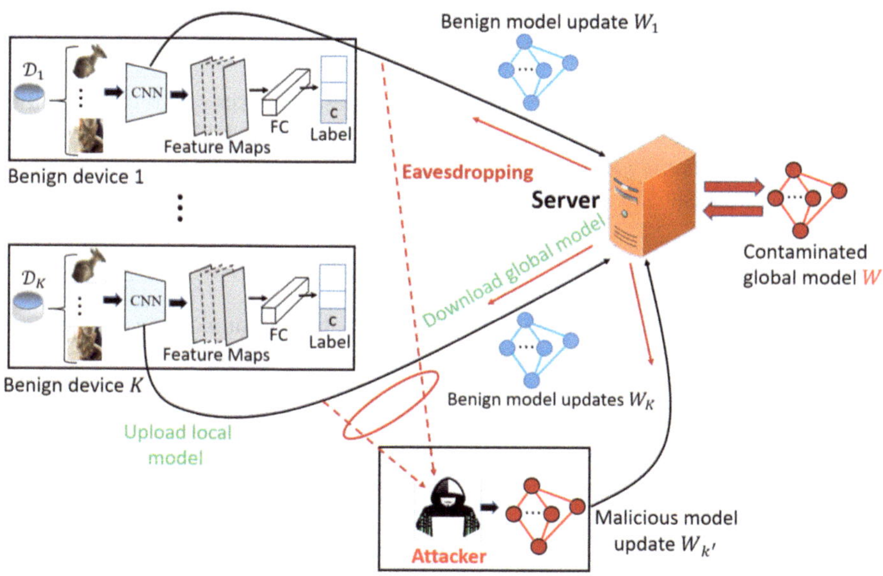

Fig. 8.1 FL with poisoning attack, where the attacker eavesdrops on the benign local models uploaded by the devices. The attacker fabricates a malicious local model update based on its eavesdropped local model updates. The Euclidean distances between the malicious and benign local models fall within a preset threshold distance

For instance, $f_s(\mathbf{W}, \mathbf{x}_s, y_s) = -y_s \times \log\left(\frac{1}{1+\exp(-\mathbf{W}^\mathbf{T}\mathbf{x}_s)}\right) - (1-y_s)\log\left(1 - \frac{1}{1+\exp(-\mathbf{W}^\mathbf{T}\mathbf{x}_s)}\right)$ accounts for cross entropy function. Here, $\mathbf{W}^\mathbf{T}$ is the transpose of $\mathbf{W}$. For each device k, given the dataset $\mathcal{D}_k$, the local loss function of $\mathbf{W}$ on the collection of data samples is defined as

$$F_k(\mathbf{W}) \triangleq \frac{1}{D_k} \sum_{s \in \mathcal{D}_k} f_s(\mathbf{W}, \mathbf{x}_s, y_s). \tag{8.2}$$

Accordingly, the global loss function on all the distributed datasets evaluated the parameter vector $\mathbf{W}$ is given by

$$F(\mathbf{W}) = \sum_{k \in \mathcal{K}} \frac{D_k}{D} F_k(\mathbf{W}), \tag{8.3}$$

where $D = \sum_{k \in \mathcal{K}} D_k$. The goal of the training process is to find the optimal parameter $\mathbf{W}^*$ such that the global loss function $F(\mathbf{W})$ is minimized, i.e.,

$$\mathbf{W}^* = \arg\min_{\mathbf{W}} F(\mathbf{W}). \tag{8.4}$$

To solve problem (8.4) while preserving the data privacy for each device, a canonical gradient-descent technique is widely used in state-of-the-art FL systems [20, 23, 37], as follows.

Let T_{FL} denote the total number of communication rounds. $\mathcal{T} \triangleq \{1, \ldots, T_{FL}\}$. In each round $t \in \mathcal{T}$, there are three steps:

- *Device selection and broadcast:* The server samples K devices satisfying eligibility requirements. The selected devices download the global model $\mathbf{W}^t$ from the server.
- *Local model training and uploading:* Each user device trains its local model based on its local dataset $\mathcal{D}_k$ for L local training epochs, starting with the global model $\mathbf{W}^t$, i.e.,

$$\mathbf{W}^t_{k,\ell+1} = \mathbf{W}^t_{k,\ell} - \eta_t \nabla F_k(\mathbf{W}^t_{k,\ell}, \mathcal{D}_k), \ \ell \in \{1, \ldots, L\}, \tag{8.5}$$

  where η_t denotes the learning rate in the t-th communication round, and $\nabla F_k(\mathbf{W}^t_{k,\ell}, \mathcal{D}_k)$ defines the local gradient estimate over $\mathcal{D}_k$ at local training epoch ℓ. After L local training epochs, each user device k uploads its updated local model $\mathbf{W}^t_k = \mathbf{W}^t_{k,L}$ to the server.
- *Global model aggregation:* The server aggregates the K devices' updated local models $\{\mathbf{W}^t_k\}_{k=1}^K$ to obtain the updated global model, given by

$$\mathbf{W}^{t+1} = \sum_{k \in \mathcal{K}} \frac{D_k}{D} \mathbf{W}^t_k, \tag{8.6}$$

  where $\mathbf{W}^{t+1}$ is sent back to all K devices.

The above process repeats until convergence is achieved or the maximum number of communication rounds T_{FL} is met. Without loss of generality, we assume all K user devices are involved in FL.

8.4 Threat Model

Malicious devices may attempt to compromise the global and local models by injecting poisoned data into their local models or intentionally tampering with the local model parameters during training. We consider that the MP attack is launched by the attackers whose goal is to impair or degrade the accuracy and convergence of FL. A general MP attack in [8, 27] is considered as the threat model. Specifically, the considered threat model presents a general MP attack in which the attacker may appear as a legitimate participant that receives the global model at each communication round, the attacker is assumed to have complete knowledge of the global model's architecture, parameters, and the aggregation rules (such as FedAvg), allowing the attacker to craft malicious local model updates to manipulate the benign FL process. By progressively injecting malicious parameters into the aggregation process, the attacker seeks to degrade the global model's accuracy over time.

During the FL process, the attackers collected by the set of $\mathcal{M} \triangleq \{1', \ldots, K'\}$ $(K' < K)$ maintain the same DNN structure as benign devices, as shown in Fig. 8.1. Meanwhile, the attacker eavesdrops on C $(C \leq K)$ benign local model updates during the model update uploading to the server. Inspired by projected gradient descent attacks [22, 48], an attacker may find the perturbation on the eavesdropped local model updates that maximize the loss of the global model while keeping the size of the perturbation smaller than a specified threshold [25].

If the server is unaware of the malicious local models, it unintentionally creates a contaminated global model. The total size of the local training data reported to the server is $S = \sum_{k\in\mathcal{K}} D_k + \sum_{k'\in\mathcal{M}} D_{k'}$, where $D_{k'}$ is the claimed data size by attacker k'. The contaminated global model is given by

$$\mathbf{W}_g^{t+1} = \sum_{k\in\mathcal{K}} \frac{D_k}{S}\mathbf{W}_k^t + \sum_{k'\in\mathcal{M}} \frac{D_k'}{S}\mathbf{W}_{k'}^t. \tag{8.7}$$

The server broadcasts $\mathbf{W}_g^{t+1}$ to all user devices. The FL training turns to minimizing the following global loss function:

$$\min_{\mathbf{W}_g^{t+1}} F(\mathbf{W}_g^{t+1}) = \sum_{k\in\mathcal{K}} \frac{D_k}{S} F_k(\mathbf{W}_g^{t+1}) + \sum_{k'\in\mathcal{M}} \frac{D_{k'}}{S} F_{k'}(\mathbf{W}_g^{t+1}), \tag{8.8}$$

where $F_{k'}(\cdot)$ is the claimed local loss function of the k'-th attacker.

The attackers aim to maximize $F(\mathbf{W}_g^{t+1})$, i.e., equivalent to minimizing the training accuracy of FL, while keeping $\mathbf{W}_{k'}^t$ imperceptible to the server that may adopt the state-of-the-art Euclidean distance-based [1, 36] or similarity-based detection methods [2] to detect and identify malicious local model updates. Suppose that an attacker can measure the Euclidean distance between $\mathbf{W}_{k'}^{t}$ and $\mathbf{W}_{g}^{t+1}$ and make it smaller than a threshold r to evade detection:

$$\max_{\mathbf{W}_{k'}^{t}} F(\mathbf{W}_{g}^{t+1}) \tag{8.9a}$$

$$\text{s.t}\ \mathfrak{D}\big(\mathbf{W}_{k'}^{t}, \mathbf{W}_{g}^{t+1}\big) \le r, \forall k' \in \mathcal{M}, \tag{8.9b}$$

where $\mathfrak{D}(\cdot, \cdot)$ is the Euclidean distance function $\mathfrak{D}\big(\mathbf{W}_{k'}^{t}, \mathbf{W}_{g}^{t+1}\big) = \|\mathbf{W}_{k'}^{t} - \mathbf{W}_{g}^{t+1}\|_2$. The optimal malicious local model updates are

$$\big(\mathbf{W}_{k'}^{t}\big)^{*} = \underset{\mathbf{W}_{k'}^{t}}{\text{argmax}}\ F(\mathbf{W}_{g}^{t+1}). \tag{8.10}$$

Note that the optimization solution to (8.9) can be found in the latest literature [17], and the goal here is not to solve this optimization problem. We aim to defend against such attacks. In particular, the threat model described in [17] represents a specific instance of the general MP attack considered, where the malicious update is generated to minimize the global model's accuracy. By exploiting the feature correlation between benign local and global models, the attacker in [17] introduces subtle perturbations to the local model updates, while the malicious model remains undetected given existing Euclidean distance-based or similarity-based defense strategies. While the threat model in [17] serves as a concrete example, our proposed defense model remains effective even against the broader class of general MP attacks in [8, 27].

8.5 GradCAM-AE Against MP Attacks

In this section, we elaborate on the GradCAM-AE, where the GradCAM and AE are leveraged on the server side for pinpointing the malicious local models.

The proposed GradCAM-AE pioneers the exploration of GradCAM-based malicious local model detection in FL, developing a methodology that eschews traditional Euclidean distance measurement [43]. The GradCAM-AE defense strategy against poisoning attacks in FL is capable of converting the local model parameters-ranging in the millions, into a low-dimensional GradCAM heat map. Additionally, an AE is designed with GradCAM to enhance the separability of the heat maps, thereby increasing the success rate in identifying anomalous heat maps generated by malicious model updates.

8.5.1 GradCAM-AE Architecture

On the device side, each of the K benign devices utilizes the DNN tailored specifically for image classification tasks, as shown in Fig. 8.2. The DNN model extracts relevant features from an input image (e.g., a bird) and subsequently maps them to the corresponding classes. The architecture of a typical DNN comprises multiple layers, each with a specific function.

1. The first layer is a convolutional layer, which applies a set of filters to the input image, thereby extracting intrinsic features, such as edges, corners, and textures. The output of the convolutional layer is a set of feature maps, each representing different aspects of the image.
2. The following layer is a pooling layer, which reduces the dimensionality of the feature maps while retaining essential information, such as the salient features of the bird's beak or feather. Various pooling methods can be employed, including max pooling or average pooling, all with the objective of downsampling the feature maps while retaining their salient features. Some DNN architectures may differ from traditional pooling operations. For example, models like SqueezeNet [14], ResNet [10], DenseNet [13], and MobileNet [11] employ alternative strategies.
3. Subsequent to several convolution and pooling layers, the extracted features are fed into one or multiple fully connected layers, where the final classification is performed.

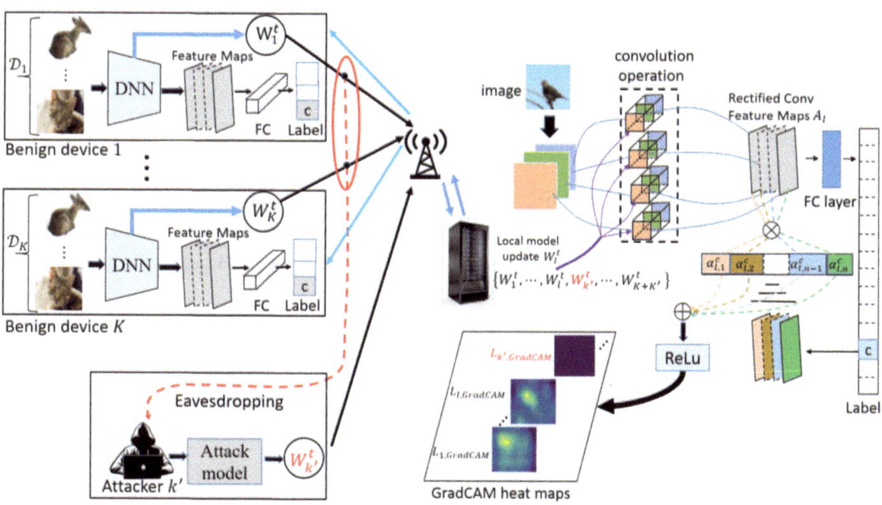

Fig. 8.2 GradCAM-assisted defense against poisoning attacks on FL. The server arbitrarily selects an image (e.g., an image with the label "bird") from the global model testing dataset to create GradCAM heat maps for every uploaded local model update. These GradCAM heat maps flow into an AE for malicious model detection

Upon receiving the local model updates from the devices, the server aggregates the local models, where the benign local models can be mingled with malicious local models. We aim to design a defense countermeasure that can pinpoint and filter malicious devices. GradCAM [26] visualization is distinguished by its high resolution and high-class discriminative ability compared to other methods, e.g., Class Activation Mapping [44]. GradCAM is leveraged in our design to detect malicious local model updates. Specifically, the server randomly picks an image from the global model testing dataset that incorporates all categories of the devices dataset as input and passes through the convolutional layers with weight and bias parameters that are replaced by each model update $\mathbf{W}_l$, $l \in \mathcal{K} \cup \mathcal{M}$ (l is the index of local model updates, including benign and malicious), obtains the feature maps[1] A_l that has n channels. The feature maps A_l are fed into a fully connected layer for final classification.

To obtain the class discriminative localization map of $L^{(c)}_{l,GradCAM} \in R^{B \times H}$ with width B and height H for any class c, GradCAM first computes the gradient of the score for class c, $Y^{(c)}$ (before softmax) with respect to each feature map A^m_l of a convolutional layer, i.e., $\frac{\partial Y^{(c)}}{\partial A^m_l}$. $m \in [1, n]$ is the index of channels. The neuron importance weights $\alpha^{(c)}_{l,m}$ can be obtained through global average pooling:

$$\alpha^{(c)}_{l,m} = \frac{1}{B \times H} \sum_{i=1}^{B} \sum_{j=1}^{H} \frac{\partial Y^{(c)}}{\partial A^m_l(i,j)}, \forall l \in \mathcal{K} \cup \mathcal{M}, \tag{8.11}$$

where $A^m_l(i,j)$ is the activation at location (i,j) of the feature map A^m_l. We apply a ReLU to the weighted linear combination of the forward activation. $L^{(c)}_{l,GradCAM}$ is given by

$$L^{(c)}_{l,GradCAM} = \text{ReLU}(\sum_{m=1}^{n} \alpha^{(c)}_{l,m} A^m_l). \tag{8.12}$$

The server uses the image and local model updates to obtain a single-channel GradCAM heat map that is input into the AE for identification. For conciseness, we suppress the indicator of classes c and rewrite $L^{(c)}_{l,GradCAM}$ as $L_{l,GradCAM}$ in what follows.

8.5.2 *Autoencoder for Malicious GradCAM Heat Map Identification*

Considering all GradCAM heat maps are unlabeled, an AE is an efficient tool in unsupervised learning to discover non-linear features across anomaly detection sys-

[1] In this chapter, we focus on extracting the feature maps after the final convolution layer of deep learning model in that can capture high-level features and hold information regarding the important regions of the input image.

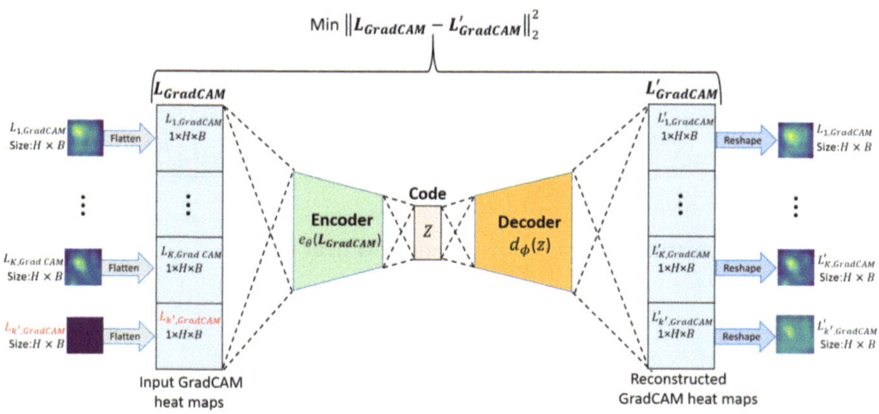

Fig. 8.3 AE-based abnormal GradCAM heat map identification. The server flattens and concatenates GradCAM heat maps as input to the encoder neural network, which compresses the GradCAM heat maps from a high dimension to a low dimension Z. The decoder neural network takes Z as its input to reconstruct the original input GradCAM heat maps

tems [45]. We use an AE to pinpoint abnormal GradCAM heat maps corresponding to malicious local model updates uploaded by the attackers.

As shown in Fig. 8.3, a canonical AE consists primarily of three components: an encoder, a code (or a latent space) and a decoder. Each GradCAM heat map $L_{l,GradCAM}$ with size of $H \times B$ is flattened into a vector with size of $1 \times H \times B$, which is further concatenated with the vectors of other GradCAM heat maps to form $\mathbf{L_{GradCAM}}$ as the input to the encoder. During the AE training, the encoder $e_\theta(\mathbf{L_{GradCAM}})$ with parameter θ compresses the GradCAM heat maps from a high-dimensional space to a low-dimensional space $z = e_\theta(\mathbf{L_{GradCAM}})$, also called the code or the latent space. The code learns the underlying features or representation of the GradCAM heat maps, which are input to the decoder $d_\phi(z)$ with parameter ϕ. The decoder further reconstructs the input GradCAM heat maps from the code, i.e., $d_\phi(z) = \mathbf{L'_{GradCAM}} = d_\phi\big(e_\theta(\mathbf{L_{GradCAM}})\big)$. After the training of (θ, ϕ), the reconstructed GradCAM heat maps are reshaped into the same size as the original GradCAM heat maps.

Loss Function. To minimize the difference between the original input GradCAM heat maps and reconstructed GradCAM heat maps, the AE loss function is defined as the mean squared error (MSE) between the encoder input GradCAM heat maps $\mathbf{L_{GradCAM}}$, and the decoder reconstructed GradCAM heat maps $\mathbf{L'_{GradCAM}}$, i.e.,

$$
\begin{aligned}
L(\theta, \phi) &= \min_{\theta,\phi} \frac{1}{|K| + |K'|} \|\mathbf{L_{GradCAM}} - \mathbf{L'_{GradCAM}}\|_2^2 \\
&= \min_{\theta,\phi} \frac{1}{K+K'} \sum_{l=1}^{K+K'} \|L_{l,GradCAM} - d_\phi(e_\theta\big(L_{l,GradCAM})\big)\|_2^2.
\end{aligned}
\tag{8.13}
$$

Once the AE completes training, the server computes the reconstruction errors between each reconstructed GradCAM heat map and its corresponding input GradCAM heat map and obtains the mean reconstruction error, i.e., $\forall l \in \mathcal{K} \cup \mathcal{M}$,

$$R_l = \frac{\sum_{i=1}^{B} \sum_{j=1}^{H} \left| L_{l,GradCAM}(i, j) - L^{'}_{l,GradCAM}(i, j) \right|}{H \times B}. \tag{8.14}$$

The average reconstruction error of all GradCAM heat maps is

$$\overline{R} = \frac{1}{R_l} \sum_{l=1}^{K+K'} R_l. \tag{8.15}$$

A threshold δ is defined as

$$\delta = \overline{R} + \alpha \times \sqrt{\frac{\sum_{l=1}^{K+K'} (R_l - \overline{R})^2}{K + K'}}, \tag{8.16}$$

where α is an empirically configured coefficient. Here, δ is used to distinguish between the benign and malicious GradCAM heat maps.

Typically, the rèconstruction errors with regard to the GradCAM heat maps follow a normal distribution, the reconstruction errors of GradCAM heat maps lie outside α standard deviation of the mean reconstruction errors are statistically rare and considered potential anomalies [31, 34]. As a result, (8.16) is used to distinguish between the benign and malicious GradCAM heat maps. If the mean reconstruction error of each GradCAM heat map is greater than the threshold δ, the corresponding input of the GradCAM heat map is considered potentially abnormal. Otherwise, it is a potential normal GradCAM heat map because the AE learns to capture variations in normal GradCAM heat maps during training.

Note that the AE is trained on both normal and abnormal (or malicious) GradCAM heat maps. The AE is designed to minimize the reconstruction error between the input heat maps and their reconstructed counterparts. The reconstruction error at the AE is a widely adopted measure of anomaly or abnormality, given that the AE can learn the distribution of normal data and thus reconstruct such data with minimal error. In contrast, out-of-distribution or abnormal inputs tend to exhibit larger reconstruction errors since the AE has not learned the patterns characteristic of such inputs, which means AE encounter difficulties in handling anomalies that do not conform to the learned patterns. This phenomenon is well-substantiated in various domains, ranging from industrial defect detection to imaging, where high reconstruction errors have repeatedly been shown to strongly correlate with anomalies [6, 41]. As a result, it is reasonable to expect that the GradCAM heat maps that can be reconstructed with low reconstruction errors are considered normal, while those with high reconstruction errors are potentially abnormal.

Also note that δ is time-varying with communication rounds as the GradCAM heat maps change over the communication rounds. The detection results are stored in a buffer:

$$O_l^t = \begin{cases} 1, & \text{if } R_l \leq \delta; \\ 0, & \text{if } R_l > \delta, \end{cases} \tag{8.17}$$

where "1" and "0" indicate a benign and a malicious local model update, respectively.

A voting mechanism is further designed to further reduce the possibility of misclassifying a benign local model update. The key idea is that the server makes final decisions every fixed ξ communication rounds to determine which devices are malicious. Specifically, the local model updates ϵ out of ξ communication rounds are detected as potentially malicious. They are determined malicious and removed from global aggregations. Finally, (8.7) is rewritten as

$$\mathbf{W}_g^{t+1} = \begin{cases} \sum_{l \in \mathcal{K} \cup \mathcal{M}} O_l^t \times \frac{D_l}{S} \times \mathbf{W}_l^t, & \text{if } t \text{ modulo } \xi \neq 0; \\ \sum_{l \in \mathcal{K} \cup \mathcal{M}} Y_l^t \times \frac{D_l}{S} \times \mathbf{W}_l^t, & \text{if } t \text{ modulo } \xi = 0, \end{cases} \tag{8.18}$$

where Y_l^t is given by

$$Y_l^t = \begin{cases} 1, & \text{if } \sum_{t=N\xi-\xi+1}^{N\xi} O_l^t < \epsilon, N = 1, \ldots, \lfloor \frac{T_{FL}}{\xi} \rfloor; \\ 0, & \text{if } \sum_{t=N\xi-\xi+1}^{N\xi} O_l^t \geq \epsilon, N = 1, \ldots, \lfloor \frac{T_{FL}}{\xi} \rfloor. \end{cases} \tag{8.19}$$

The updated global model $\mathbf{W}_g^{t+1}$ is sent back to all devices. The detailed pseudo-code is given in **Algorithm 1**.

8.6 Performance Evaluation

8.6.1 Experiment Setup

Datasets. Two datasets, i.e., CIFAR-10 [16] and GTSRB [30], are used to evaluate the performance of our proposed GradCAM-AE.

- CIFAR-10: It consists of 60000 32×32 low-resolution color images in 10 different classes, with 6000 images per class. There are 50,000 training images allocated to all benign devices for local model training, and 10,000 test images are allocated to the server for the global model testing until the end of each communication round.
- GTSRB: It contains 43 different classes of German traffic signs. The images range in size from 15×15 pixels to 250×250 pixels. There are 39209 training images, and 12630 test images are assigned to all local devices for training and the server for testing. Notice that we resize the images of different sizes to a fixed resolution of 128×128 pixels for training and testing.

Algorithm 1: The proposed GradCAM-AE for detecting malicious local model updates on FL

input : $\mathcal{K}$ - the set of benign devices, $\mathcal{M}$ - the set of attackers, T_{FL} the total number of communication round, L - the number of local epochs, $\mathcal{D}_k$ - data size of device, η_t - the learning rate of devices, $\mathbf{W}^0$ - the initialized global model, $\mathbf{M}$ - buffer with size $\xi \times (K + K')$, ξ - the interval of communication rounds, ϵ - voting threshold, I_0 - an image picked form testing dataset.
output: The final global model weight $\mathbf{W}^{T_{FL}}$

```
for t = 1, ..., T_FL do
    Server broadcasts W^t to all devices;
    M ← 0, M ∈ R^{ξ×(|K+K'|)} ;
    for k ∈ 𝒦 do
        W_k^t ← LocalUpdate(k, W^t) ;
    end
    Solve (8.9a) with constraint (8.9b);
    (W_{1'}^t)^*, ..., (W_{K'}^t)^* ← (8.10).
    if t modulo ξ ≠ 0 then
        A_l ← W_l^{t+1} and I_0 perform convolution operation ;
        L_{l,GradCAM} ← according to (8.11) and (8.12) ;
        L_GradCAM ← concatenate (L_{l,GradCAM}) ;
        L'_GradCAM ← AE(L_GradCAM) ;
        O_l^t ← (8.17) ;
        Store O_l^t into M ;
    else
        A_l ← W_l^{t+1} and I_0 perform convolution operation ;
        L_{l,GradCAM} ← according to (8.11) and (8.12) ;
        L_GradCAM ← concatenate (L_{l,GradCAM}) ;
        L'_GradCAM ← AE(L_GradCAM) ;
        O_l^t ← (8.17) ;
        Store O_l^t into M;
        Stat ← sum(M, axis = 0) ;
        Y_l^t ← (8.19)
    end
    Update global model according to (8.18) and (8.19).
end
LocalUpdate(k, W^t) ;
for ℓ = 1, ..., L do
    for b_k ∈ 𝒟_k do
        (8.5).
    end
end
```

Benchmarks. We consider the following state-of-the-art defense schemes, i.e., Multi-Krum [1], AUROR [28], Trimmed-mean [36], FAA-DL [29], FLTrust [2], FLCert [4], and FLDetector [39]. Specifically, FLTrust utilizes cosine similarity to measure the alignment of local model updates with the trusted global model update trained by a clean dataset. The local model updates with low similarity are excluded. FLCert divides devices into multiple groups and trains a global model for each group using a standard FL algorithm. FLDetector uses K-means to detect malicious devices by checking the consistency between the received and predicted model updates of the devices. In addition, we study a GradCAM-Krum anomaly detection scheme in which Krum is applied to identify the malicious GradCAM heat maps that are generated at the server.

Experimental setting. The total communication rounds $T_{FL} = 100$. For each communication round, 18 benign devices train the local models for 20 epochs using the Adam optimizer, with a batch size of 64, a learning rate of 0.0001, and a weight decay of 0. The AE trains the GradCAM heat maps for 200 epochs using the Adam optimizer, with a hidden-layer size of 128, a learning rate of 0.001, and a weight decay of 0.00001. In the voting process, the communication round interval is $\xi = 3$, and the threshold is $\epsilon = 2$.

8.6.2 Performance Under IID Datasets

Figure 8.4 shows the comparison of FL test accuracy under our proposed GradCAM-AE defense framework with existing benchmarks, where the DNN model of ResNet-18 performs image classification on CIFAR-10 and GTSRB with IID settings, respectively. GradCAM-AE can achieve the highest test accuracy of the global model as the AE can accurately exclude the abnormal GradCAM heat maps such that more benign local model updates are involved in the FL training process. While GradCAM-Krum only selects an optimal local model update as the global model to train FL, it also achieves higher test accuracy (0.79) compared with other benchmarks. The reason is that the data is IID, and the difference among the local model updates is not significant. Multi-Krum, AUROR and Trimmed-mean all directly detect millions, even many more, parameters of local model updates based on Euclidean distance. The Euclidean distance between the crafted malicious local model updates and benign ones is within the threshold designated by the server. This indicates that the malicious local model updates can elude the detection of the server and participate in the FL training process through multiple communication rounds, resulting in the global model being corrupted.

Table 8.1 shows that although the ACCs of Euclidean distance-based defense schemes are relatively high (0.859, 0.736 and 0.816), the Precision is close to zero. In other words, they can correctly classify a few benign devices but fail to identify malicious user devices. Similar performances are observed on GTSRB dataset, the detection rates of GrdaCAM-Krum still outperform the benchmarks, as shown in

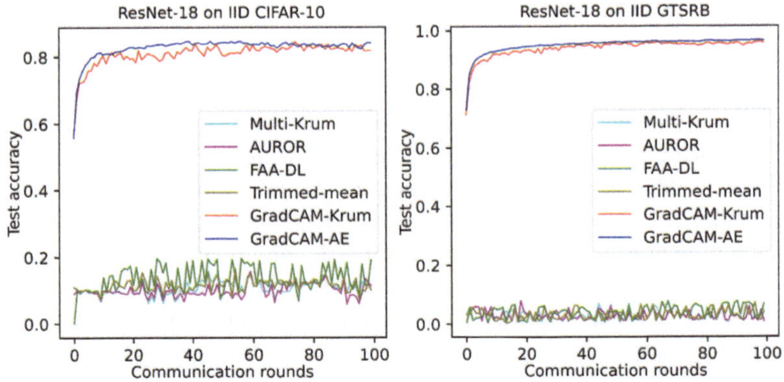

Fig. 8.4 Test accuracy versus communication rounds for **ResNet-18** on IID CIFAR-10 and GTSRB

Table 8.1 Detection rates of ResNet-18 on IID and Non-IID CIFAR-10

Method	IID CIFAR-10					Non-IID CIFAR-10				
	Rec.	Prec.	ACC	F1	AUC	Rec.	Prec.	ACC	F1	AUC
Multi-Krum	0.045	0.09	0.859	0.06	0.497	0.085	0.085	0.817	0.085	0.492
AUROR	0.015	0.006	0.736	0.009	0.416	0.035	0.017	0.750	0.023	0.432
Trimmed-mean	0.080	0.080	0.816	0.080	0.489	0.090	0.090	0.818	0.090	0.494
FAA-DL	0.610	0.098	0.412	0.168	0.500	0.550	0.109	0.514	0.182	0.530
GradCAM-Krum	1.000	1.000	1.000	1.000	1.000	0.625	0.625	0.925	0.625	0.792
GradCAM-AE	**1.000**	**1.000**	**1.000**	**1.000**	**1.000**	**1.000**	**0.943**	**0.986**	**0.961**	**0.992**

Table 8.2 Detection rates of ResNet-18 on IID and Non-IID GTSRB

Method	IID GTSRB					Non-IID GTSRB				
	Rec.	Prec.	ACC	F1	AUC	Rec.	Prec.	ACC	F1	AUC
Multi-Krum	0.045	0.090	0.859	0.060	0.497	0.045	0.080	0.857	0.060	0.497
AUROR	0.035	0.016	0.739	0.020	0.427	0.035	0.016	0.755	0.0229	0.435
Trimmed-mean	0.080	0.080	0.816	0.080	0.489	0.070	0.070	0.814	0.070	0.483
FAA-DL	0.685	0.092	0.295	0.162	0.468	0.430	0.093	0.517	0.152	0.478
GradCAM-Krum	0.625	0.625	0.925	0.625	0.792	0.550	0.550	0.910	0.550	0.750
GradCAM-AE	**1.000**	**1.000**	**1.000**	**1.000**	**1.000**	**1.000**	**0.990**	**0.998**	**0.994**	**0.999**

Table 8.2. The detailed detection rates of MobileNetV3-Large on CIFAR-10 and GTSRB with IID settings, are given in Tables 8.3 and 8.4, respectively.

FAA-DL, a one-class SVM-based method, also directly classifies the local model updates aggregated by the server as benign and malicious. However, there are two key reasons why FAA-DL fails in the experiments. First, the local model updates have the characteristics of high-dimensional feature spaces. The curse of dimensionality can lead to data sparsity, making it challenging for FAA-DL to find a suitable

Table 8.3 Detection rates of MobileNetV3-Large on IID and Non-IID CIFAR-10

Method	IID CIFAR-10					Non-IID CIFAR-10				
	Rec.	Prec.	ACC	F1	AUC	Rec.	Prec.	ACC	F1	AUC
Multi-Krum	0.050	0.100	0.860	0.060	0.500	0.035	0.070	0.857	0.047	0.492
AUROR	0.020	0.0095	0.740	0.013	0.420	0.015	0.0060	0.736	0.009	0.416
Trimmed-mean	0.100	0.100	0.820	0.100	0.500	0.500	0.080	0.816	0.080	0.489
FAA-DL	0.735	0.092	0.257	0.163	0.469	0.700	0.094	0.309	0.166	0.483
GradCAM-Krum	0.860	0.860	0.972	0.860	0.922	0.940	0.940	0.988	0.940	0.967
GradCAM-AE	**1.000**	**1.000**	**1.000**	**1.000**	**1.000**	**1.000**	**1.000**	**1.000**	**1.000**	**1.000**

Table 8.4 Detection rates of MobileNetV3-Large on IID and Non-IID GTSRB

Method	IID GTSRB					Non-IID GTSRB				
	Rec.	Prec.	ACC	F1	AUC	Rec.	Prec.	ACC	F1	AUC
Multi-Krum	0.030	0.060	0.856	0.040	0.489	0.135	0.135	0.827	0.135	0.519
AUROR	0.035	0.019	0.745	0.020	0.430	0.015	0.0083	0.7375	0.011	0.416
Trimmed-mean	0.080	0.080	0.816	0.080	0.489	0.065	0.065	0.813	0.065	0.481
FAA-DL	0.790	0.096	0.236	0.170	0.483	0.765	0.098	0.280	0.173	0.496
GradCAM-Krum	0.995	0.995	0.999	0.995	0.997	0.580	0.580	0.916	0.580	0.767
GradCAM-AE	**1.000**	**1.000**	**1.000**	**1.000**	**1.000**	**1.000**	**0.990**	**0.998**	**0.994**	**0.998**

margin that separates benign from malicious local model updates. Second, FAA-DL is sensitive to class imbalance (18 benign local model updates and 2 malicious local model updates), which means the FAA-DL biases its decision boundary towards the majority class, making it struggle to detect malicious local model updates effectively. Notwithstanding, we replace the DNN model with MobileNetV3-Large. Similar performances are observed; see Fig. 8.5. This demonstrates that GradCAM-AE is applicable to different DNN models for detecting malicious local model updates.

Table 8.5 shows the detection rates of GradCAM-AE, where the poisoning ratio of the local model (the proportion of local model parameters modified over the total number of parameters) increases from 0.2 to 0.5. Recall, Precision, ACC, F1 socre and AUC of GradCAM-AE grows from 0.955 to 1.0, 0.960 to 1.0, 0.992 to 1.0, 0.953 to 1.0, and 0.976 to 1.0, respectively, while the false positive rate (FPR) drops from 0.003 to 0. The decrease in FPR indicates that the number of benign devices mistakenly classified as malicious devices by GradCAM-AE is reduced. This is because the more the malicious local models are modified, the more it deviates from the benign local models, which makes it easier for GradCAM-AE to detect the malicious local model. When the poisoning ratio increases to 0.8, the detection rates of GradCAM-AE maintain to 1.0, as all the malicious local models have been detected by GradCAM-AE.

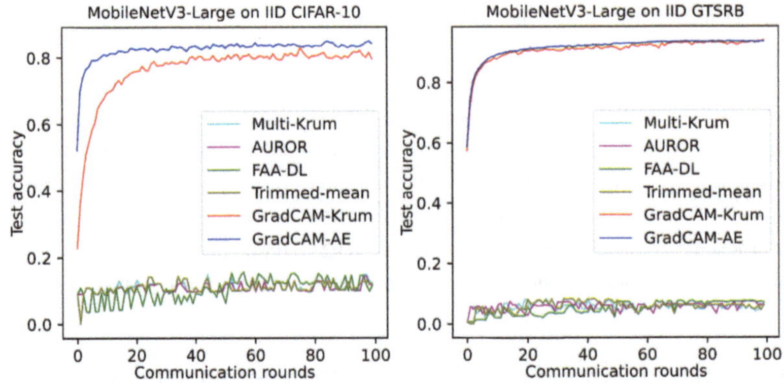

Fig. 8.5 Test accuracy versus communication rounds for **MobileNetV3-Large** on IID CIFAR-10 and GTSRB

Table 8.5 Detection rate of GradCAM-AE under different poisoning ratios

Poisoning ratio	Rec.	Prec.	FPR	ACC	F1	AUC
0.2	0.955	0.960	0.003	0.992	0.953	0.976
0.5	1.000	1.000	0.000	1.000	1.000	1.000
0.8	1.000	1.000	0.000	1.000	1.000	1.000

8.6.3 Performance Under Non-IID Datasets

Compared with IID, the Non-IID setting is much more challenging for the server to identify the malicious local model updates. This reason is that the local model updates tend to be more divergent under non-IID settings; in turn, the generated heat maps on the server are more diverse. To simulate the statistical data heterogeneity in practical FL scenarios, we adopt one of the Non-IID partition schemes, i.e., latent Dirichlet allocation (LDA), which is widely used in the latest literature [7, 19, 32, 38]. As a biased probability, the β parameter of LDA controls the distribution difference of the local training data. A larger β denotes a higher non-IID level among the local training data. Here, we set $\beta = 0.5$.

Figure 8.6 illustrates the test accuracy of ResNet-18 in non-IID CIFAR-10 and non-IID GTSRB. Under the Non-IID CIFR10 setting, the test accuracy of GradCAM-AE is much higher than that of GradCAM-Krum. GradCAM-AE can quickly converge (around the 30th communication round) as it involves more benign devices in FL training. This indicates that GradCAM-AE can accurately filter malicious model updates, as can also be confirmed by the detection indicators in Table 8.1. Although GradCAM-Krum can restrict malicious devices from participating in FL, it sacrifices accuracy and robustness, especially under non-IID CIFAR-10. The test accuracy of GradCAM-Krum fluctuates significantly and likely diverges, as it only

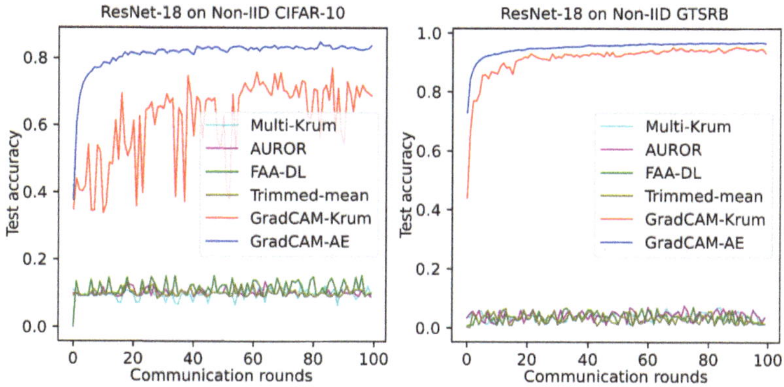

Fig. 8.6 Test accuracy versus communication rounds for **ResNet-18** on Non-IID CIFAR-10 and GTSRB

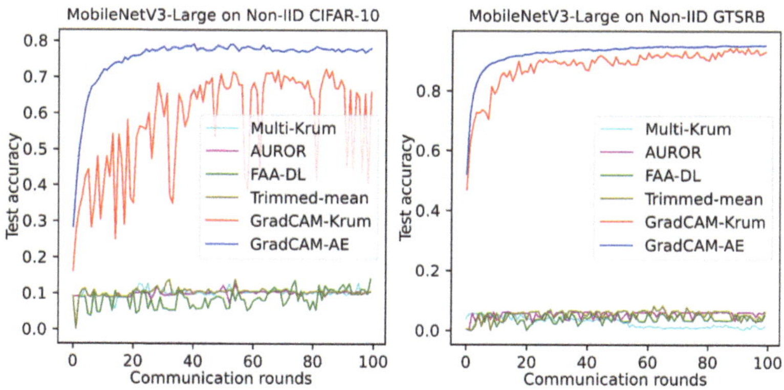

Fig. 8.7 Test accuracy versus communication rounds for **MobileNetV3-Large** on Non-IID CIFAR-10 and GTSRB

selects one local model update as the global model. The more divergent the model updates, the more diverse the GradCAM heat maps. GradCAM-Krum struggles to screen malicious model updates, which coincides with the precision of 0.625 for GradCAM-Krum on Non-IID CIFAR-10 in Table 8.1. The remaining defense methods still undergo poor anomaly detection on the IID dataset. The detection rates of MobileNetV3-Large on CIFAR-10 and GTSRB under non-IID settings are given in Tables 8.3 and 8.4, respectively.

We replace the ResNet-18 model with MobileNetV3-Large, and the changing trend of the test accuracy with the communication round is consistent with Fig. 8.6, as shown in Fig. 8.7. This means that the complexity of the DNN model itself has a limited impact on the malicious detection of GradCAM-Krum, while the complexity of local data plays a crucial role. GTSRB has fewer image features than CIFAR-10, and the local model updates learned by FL have less diversity, making GradCAM-Krum converge.

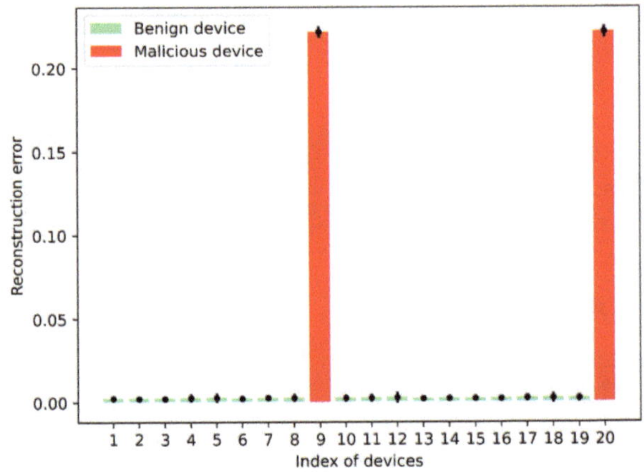

Fig. 8.8 Reconstruction errors for ResNet18 on Non-IID CIFAR-10

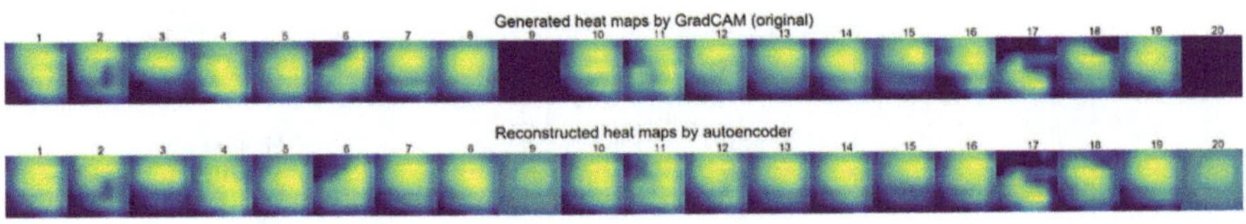

Fig. 8.9 Comparison between generated heat maps by GradCAM and reconstructed heat maps by the AE

This significant difference is quantified in Fig. 8.8, where the server averages the reconstruction errors of 20 user devices in 100 communication rounds one by one. For visualization, Fig. 8.9 only shows the original GradCAM heat maps and the reconstructed heat maps by the AE under ResNet-18 on Non-IID CIFAR-10 setting in one communication round. Strikingly, it can be noticed that the reconstruction errors of device 9 and device 20 are obviously larger than those of other user devices. In particular, we use the generated heat maps to train the AE, and after the training, the final reconstructed heat maps are used to compare with the originally generated heat maps, the heat maps with high reconstruction errors are considered abnormal. We conduct 100 communication rounds for FL training with 20 devices, and in each communication round, AE performs 200 epochs for training. Thus, we use $200 \times 20 = 4000$ heat maps to train the AE in each communication round, which does not result in overfitting.

Moreover, Fig. 8.9 shows that the reconstructed heat maps of malicious devices are significantly different from the heat maps reconstructed by benign devices. This is because the AE captures the key features of normal heat maps during the encoding process. Anomalous heat maps with different patterns, shapes, or structures result in latent representations that cannot be decoded accurately. The abnormal heat maps often fall outside the distribution of normal heat maps in the latent space, making reconstruction difficult. In particular, training an effective AE involves meeting

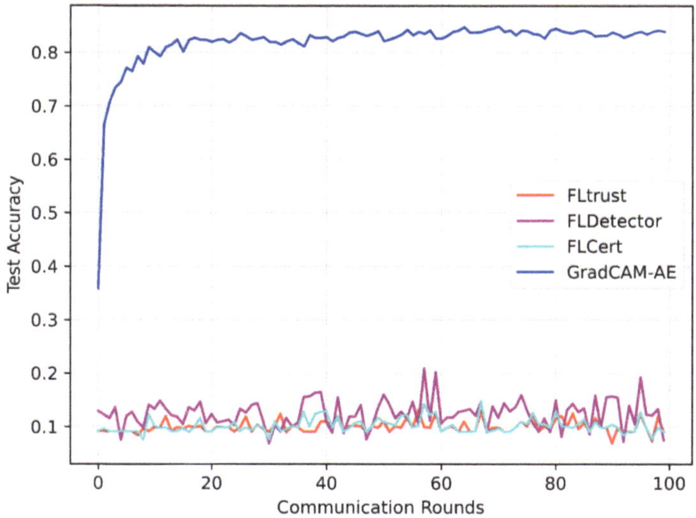

Fig. 8.10 Comparison of global model test accuracy between GradCAM-AE and benchmarks

specific criteria that ensure the model effectively learns a meaningful latent representation and reconstructs the input heat maps accurately. The primary criterion for training a high-quality AE is achieving a reconstruction error on the training heat maps that is less than the mean reconstruction error plus α times the standard deviation of all reconstruction errors.

Figure 8.10 compares GradCAM-AE, FLTrust [2], FLCert [4], and FLDetector [39]. In terms of the test accuracy of the global model, GradCAM-AE is 0.7 better than FLTrust defense. The reason is that the malicious local models generated by attackers are highly correlated with benign local models that can evade FLTrust but can be detected by GradCAM-AE. GradCAM-AE can map local models into visible heat maps, which are remapped by the AE to highlight the hidden features. The utilization of AE improves the distinguishability of the heat maps and the success rate of identifying anomalous heat maps and malicious local models. Likewise, FLCert inevitably involves malicious devices in FL training, leading to the corruption of the global model. FLDetector can prevent attackers from participating in FL training, there are still benign devices who are misclassified as malicious. FLDetector uses K-means for defense, nevertheless, it still inevitably involves malicious devices in FL training. The test accuracy of the global model under FLDetector defense is 0.68 lower than the that under our proposed GradCAM-AE defense.

Figure 8.11 shows the test accuracy of the global model, where β is set to 0.1, 0.5, or 0.9. Generally, the test accuracy of the global model increases with the growth of β. Given $\beta = 0.1$, the low detection rates of GradCAM-AE for malicious devices result in the low test accuracy of the global model, which is 0.6. When β increases from 0.5 to 0.9, the divergence degree of the local model updates is reduced, and

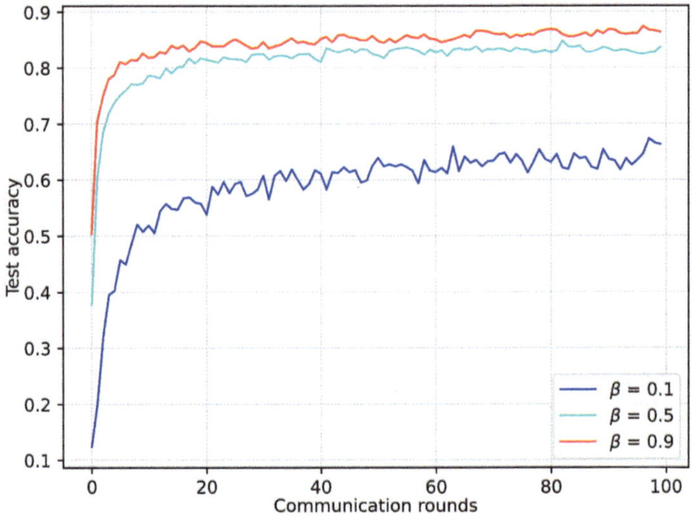

Fig. 8.11 The test accuracy of FL with GradCAM-AE, where β is set to 0.1, 0.5 or 0.9

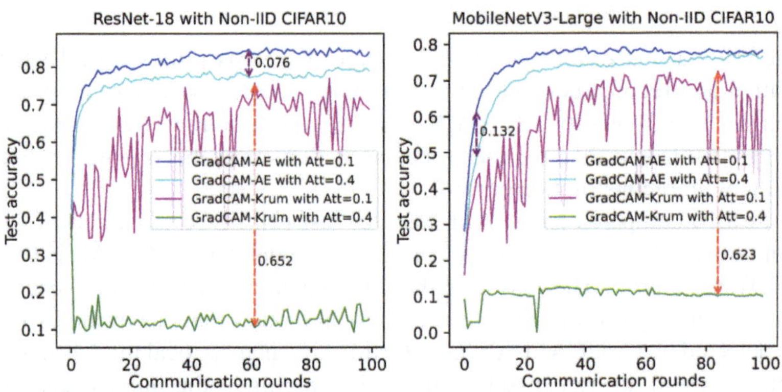

Fig. 8.12 Scalability of GradCAM-AE versus GradCAM-Krum

the detection rates of GradCAM-AE for malicious devices increase, leading to the global model's test accuracy increasing from 0.83 to 0.86.

To evaluate the scalability of GradCAM-AE, we define the attack rate as the ratio of the number of attackers to the total number of local devices, i.e., $\text{Att} = \frac{\#\text{attackers}}{\#\text{all devices}}$. GradCAM-AE and GradCAM-Krum are compared under Non-IID CIFAR-10 with 2 and 8 attackers. The total number of local devices is 20, i.e., $\text{Att} = 0.1$ and $\text{Att} = 0.4$. Figure 8.12 shows that as the attack rate increases, GradCAM-Krum fails to identify attackers. The maximum test accuracy gap of 0.652 appears in the 61st communication round. In contrast, the test accuracy of GradCAM-AE changes, and the maximum gap of 0.076 appears in the 61st communication round. Similarly, in the case of MobileNetV3-Large, the tendency of test accuracy is consistent, offering scalability

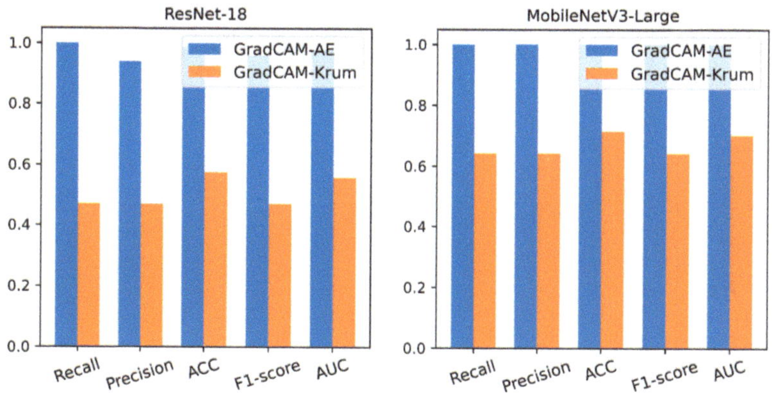

Fig. 8.13 Detection rates of ResNet-18 and MobileNetV3-Large on Non-IID CIFAR-10 with Att = 0.4

and robustness. Figure 8.13 shows detection rates of ResNet-18 and MobileNetV3-Large on Non-IID CIFAR-10 with Att = 0.4, respectively. The results indicate that the detection rates of GradCAM-AE outperform GradCAM-Krum no matter the number of attackers, offering scalability and robustness.

Figure 8.14 demonstrates that FL maintains a high test accuracy of 0.83 and 0.8 under the proposed GradCAM-AE, when the number of devices is set to 20 (18 benign devices and 2 malicious devices) and 100, respectively. The decrease in test accuracy of global model is because given the total dataset size remains same, assigning the dataset to more devices reduces the amount of data per device under non-IID (i.e., different devices have data from different distributions) scenario, smaller local datasets exacerbate this heterogeneity, which causes the detection rates of GradCAM-AE on 100 devices in detecting malicious local models to decrease.

Table 8.6 presents the detection rates of GradCAM-AE under MPAF [3] and Fang attack [8]. In particular, the MPAF [3] generates malicious local models that are aligned with an attacker-specified base model. These malicious local models are scaled up before being transmitted to the server, thereby amplifying their adversarial impact and reducing the test accuracy of the global model. Given MPFA, the Recall, Precision, ACC, F1 score, and AUC of GradCAM-AE are respectively. Moreover, our proposed GradCAM-AE is applicable to Fang attack in [8]. As shown in Table 8.6, all the malicious local models of Fang attack are detected by GradCAM-AE. This is because Fang attack alters the direction of the local model updates. As a result, the generated heat maps based on the malicious local models are different from the benign ones, which are identified as a malicious one.

Runtime. As shown in Fig. 8.15, the average runtime of our proposed GradCAM-AE (35.71 s) is higher than that of Euclidean distance-based defense methods, i.e., Multi-Krum (29.68 s) and Trimmed-mean (32.89 s). This is because the Euclidean distance-based defense does not require training, while GradCAM-AE takes a little longer time to train the AE. Nevertheless, GradCAM-AE is still faster than that of

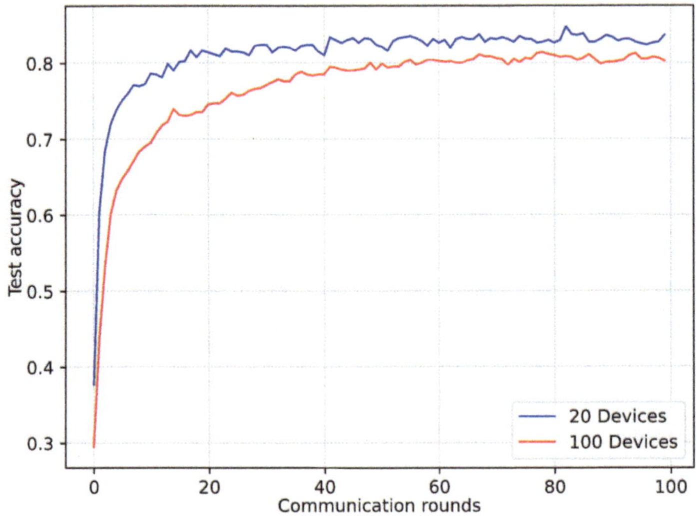

Fig. 8.14 The test accuracy of FL global model under the defense of GradCAM-AE with 20 devices vs 100 devices train ResNet-18 on Non-IID CIFAR-10 dataset

Table 8.6 Detection rates of GradCAM-AE under MPFA and Fang attack

Attack type	Rec.	Prec.	FPR	ACC	F1	AUC
MPAF [3]	0.985	0.990	0.002	0.997	0.984	0.992
Fang attack [7]	1.000	1.000	0.000	1.000	1.000	1.000

the ML-based AUROR (35.97 s) and FAA-DL (40.53 s). It is noted that although runtime of GradCAM-AE is not the fastest, detection rates of GradCAM-AE are the highest.

8.6.4 Discussion

Untargeted MP attacks are generally easier to detect compared to semantic targeted MP or backdoor attacks, as the former often leads to significant decreases in global model accuracy and can result in a denial-of-service situation. In the chapter, our primary focus is on defending against untargeted MP attacks due to their disruptive nature and potential for large-scale impact. However, our proposed GradCAM-AE has the potential to be extended to defend against more advanced attacks, such as semantic targeted MP [21], backdoor attacks, and even byzantine-robust [8] or distributed attacks [35]. At a high level, both untargeted and targeted poisoning attacks share the goal of modifying the behavior of the global model, either overtly

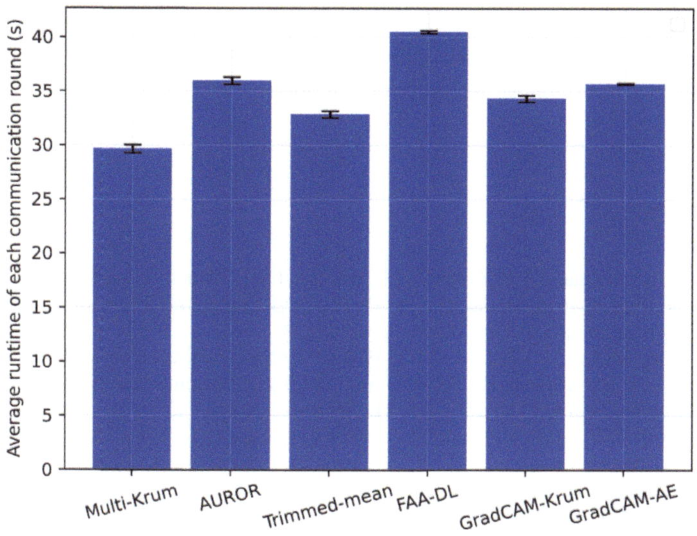

Fig. 8.15 Comparison of GradCAM-AE's average runtime with other benchmarks

or covertly. While untargeted MP attacks aim to disrupt model performance broadly, targeted and backdoor attacks focus on embedding specific malicious objectives into the model without compromising overall performance. Specifically, our method can adapt to the detection of these advanced attacks as follows.

Semantic Targeted and Backdoor Attacks: These attacks often introduce subtle, targeted changes in the local model updates that are harder to detect. GradCAM-AE, which leverages heat map generation via GradCAM and anomaly detection using an AE, is capable of identifying discrepancies in the activation patterns of malicious local updates. Even in scenarios where backdoor triggers are embedded, the heat maps produced by malicious devices will deviate significantly from those of benign devices due to the targeted intent, allowing our AE to flag such updates as anomalous. For instance, the ability to capture nuanced, target-specific activations makes GradCAM-AE a promising approach for these attack types.

Byzantine-Robust Attacks: In these attacks, adversaries aim to manipulate the direction of local model updates to mislead the aggregation process without completely disrupting the global model. GradCAM-AE can still identify such manipulations because malicious updates often result in distinct heat maps that differ from those generated by benign devices. These discrepancies can be detected by the AE as anomalies, highlighting the adaptability of the new method to various threat models.

Distributed Attacks: In distributed poisoning scenarios, where multiple adversarial devices coordinate to evade detection, our approach remains effective due to its reliance on per-device heat map analysis. The AE is designed to detect patterns in these heat maps, which allows it to isolate anomalous contributions even in coordinated attack settings.

Future work can involve evaluating GradCAM-AE against semantic targeted attacks, backdoor attacks, and other advanced poisoning techniques, ensuring robustness against a wide range of adversarial behaviors.

References

1. Blanchard P, El Mhamdi EM, Guerraoui R, Stainer J (2017) Machine learning with adversaries: Byzantine tolerant gradient descent. In: Proceedings of the 31st international conference on neural information processing systems, pp 118–128
2. Cao X, Fang M, Liu J, Gong NZ (2021) Fltrust: Byzantine-robust federated learning via trust bootstrapping. In: 28th annual network and distributed system security symposium, NDSS 2021, virtually, February 21–25, 2021. The Internet Society
3. Cao X, Gong NZ (2022) MPAF: model poisoning attacks to federated learning based on fake clients. In: Proceedings of the IEEE/CVF conference on computer vision and pattern recognition (CVPR) workshops, pp 3396–3404
4. Cao X, Zhang Z, Jia J, Gong NZ (2022) Flcert: provably secure federated learning against poisoning attacks. IEEE Trans Inf Forensics Secur 17:3691–3705
5. Chen Z, Fu A, Zhang Y, Liu Z, Zeng F, Deng RH (2021) Secure collaborative deep learning against gan attacks in the internet of things. IEEE Internet Things J 8(7):5839–5849. https://doi.org/10.1109/JIOT.2020.3033171
6. Chow JK, Su Z, Wu J, Tan PS, Mao X, Wang YH (2020) Anomaly detection of defects on concrete structures with the convolutional autoencoder. Adv Eng Inf 45:101105
7. Duan L, Li Z, Yuan X, Wang W, Liu J, Ni W (2025) Fedhydra: towards parameter-efficient and backdoor-resistant federated unlearning in human-centric metaverse service. IEEE Trans Serv Comput 1–13. https://doi.org/10.1109/TSC.2025.3620758
8. Fang M, Cao X, Jia J, Gong NZ (2020) Local model poisoning attacks to byzantine-robust federated learning. In: Proceedings of the 29th USENIX conference on security symposium, pp 1623–1640
9. Han S, Hu X, Huang H, Jiang M, Zhao Y (2022) Adbench: anomaly detection benchmark. In: Koyejo S, Mohamed S, Agarwal A, Belgrave D, Cho K, Oh A (eds) Advances in neural information processing systems, vol 35. Curran Associates, Inc., pp 32142–32159. https://proceedings.neurips.cc/paper_files/paper/2022/file/cf93972b116ca5268827d575f2cc226b-Paper-Datasets_and_Benchmarks.pdf
10. He K, Zhang X, Ren S, Sun J (2016) Deep residual learning for image recognition. In: Proceedings of the IEEE conference on computer vision and pattern recognition (CVPR)
11. Howard A, Sandler M, Chu G, Chen LC, Chen B, Tan M, Wang W, Zhu Y, Pang R, Vasudevan V, Le QV, Adam H (2019) Searching for mobilenetv3. In: Proceedings of the IEEE/CVF international conference on computer vision (ICCV)
12. Hu Y, Liang W, Wu R, Xiao K, Wang W, Li X, Liu J, Qin Z (2023) Quantifying and defending against privacy threats on federated knowledge graph embedding. In: Proceedings of the ACM web conference 2023, WWW '23. Association for Computing Machinery, New York, NY, USA, pp 2306–2317. https://doi.org/10.1145/3543507.3583450
13. Huang G, Liu Z, Van Der Maaten L, Weinberger KQ (2017) Densely connected convolutional networks. In: Proceedings of the IEEE conference on computer vision and pattern recognition, pp 4700–4708
14. Iandola FN, Moskewicz MW, Ashraf K, Han S, Dally WJ, Keutzer K (2016) Squeezenet: Alexnet-level accuracy with 50x fewer parameters and <1mb model size. CoRR arXiv:1602.07360
15. Kaissis G, Ziller A, Passerat-Palmbach J, Ryffel T, Usynin D, Trask A, Lima I Jr, Mancuso J, Jungmann F, Steinborn MM et al (2021) End-to-end privacy preserving deep learning on multi-institutional medical imaging. Nat Mach Intell 3(6):473–484

16. Krizhevsky A, Hinton G et al (2009) Learning multiple layers of features from tiny images
17. Li K, Zheng J, Yuan X, Ni W, Akan OB, Poor HV (2024) Data-agnostic model poisoning against federated learning: a graph autoencoder approach. IEEE Trans Inf Forensics Secur
18. Liu W, Lin H, Wang X, Hu J, Kaddoum G, Piran MJ, Alamri A (2023) D2MIF: a malicious model detection mechanism for federated-learning-empowered artificial intelligence of things. IEEE Internet Things J 10(3):2141–2151. https://doi.org/10.1109/JIOT.2021.3081606
19. Liu X, Li M, Yu G, Wang X, Ni W, Li L, Peng H, Ping Liu R (2025) BlockFUL: Enabling unlearning in blockchained federated learning. IEEE Trans Inf Forensics Secur 20:6635–6650. https://doi.org/10.1109/TIFS.2025.3583109
20. Lycklama H, Burkhalter L, Viand A, Küchler N, Hithnawi A (2023) Rofl: Robustness of secure federated learning. In: 2023 IEEE symposium on security and privacy (SP). IEEE, pp 453–476
21. Lyu X, Han Y, Wang W, Liu J, Wang B, Liu J, Zhang X (2023) Poisoning with cerberus: Stealthy and colluded backdoor attack against federated learning. In: Proceedings of the AAAI conference on artificial intelligence, vol 37, pp 9020–9028
22. Madry A, Makelov A, Schmidt L, Tsipras D, Vladu A (2018) Towards deep learning models resistant to adversarial attacks. In: 6th international conference on learning representations, ICLR 2018, Vancouver, BC, Canada, April 30–May 3, 2018, conference track proceedings. OpenReview.net. https://openreview.net/forum?id=rJzIBfZAb
23. McMahan B, Moore E, Ramage D, Hampson S, y Arcas BA (2017) Communication-efficient learning of deep networks from decentralized data. In: Singh A, Zhu J (eds) Proceedings of the 20th international conference on artificial intelligence and statistics, Proceedings of machine learning research, vol 54. PMLR, pp 1273–1282. https://proceedings.mlr.press/v54/mcmahan17a.html
24. Nguyen TD, Rieger P, De Viti R, Chen H, Brandenburg BB, Yalame H, Möllering H, Fereidooni H, Marchal S, Miettinen M et al (2022) {FLAME}: taming backdoors in federated learning. In: 31st USENIX security symposium (USENIX security 22), pp 1415–1432
25. Pourkabirian A, Ni W, Zhou X, Li K, Anisi MH (2025) A precoding perturbation method in geometric optimization: exploring manifold structure for privacy and efficiency. IEEE Trans Inf Forensics Secur
26. Selvaraju RR, Cogswell M, Das A, Vedantam R, Parikh D, Batra D (2017) Grad-cam: visual explanations from deep networks via gradient-based localization. In: Proceedings of the IEEE international conference on computer vision, pp 618–626
27. Shejwalkar V, Houmansadr A (2021) Manipulating the byzantine: optimizing model poisoning attacks and defenses for federated learning. In: NDSS
28. Shen S, Tople S, Saxena P (2016) Auror: defending against poisoning attacks in collaborative deep learning systems. In: Proceedings of the 32nd annual conference on computer security applications, pp 508–519
29. Shi S, Hu C, Wang D, Zhu Y, Han Z (2022) Federated anomaly analytics for local model poisoning attack. IEEE J Sel Areas Commun 40(2):596–610. https://doi.org/10.1109/JSAC.2021.3118347
30. Stallkamp J, Schlipsing M, Salmen J, Igel C (2012) Man vs. computer: Benchmarking machine learning algorithms for traffic sign recognition. Neural Netw 32:323–332. https://doi.org/10.1016/j.neunet.2012.02.016. https://www.sciencedirect.com/science/article/pii/S0893608012000457. Selected Papers from IJCNN 2011
31. Wadinger M, Kvasnica M (2024) Adaptable and interpretable framework for anomaly detection in SCADA-based industrial systems. Expert Syst Appl 246:123200
32. Wang J, Guo S, Xie X, Qi H (2022) Federated unlearning via class-discriminative pruning. In: Proceedings of the ACM web conference 2022, WWW '22. Association for Computing Machinery, New York, NY, USA, pp 622–632. https://doi.org/10.1145/3485447.3512222
33. Wang S, Tuor T, Salonidis T, Leung KK, Makaya C, He T, Chan K (2019) Adaptive federated learning in resource constrained edge computing systems. IEEE J Sel Areas Commun 37(6):1205–1221. https://doi.org/10.1109/JSAC.2019.2904348
34. Westfall PH, Henning KS (2013) Understanding advanced statistical methods, vol 543. CRC Press Boca Raton, FL

35. Xie C, Huang K, Chen PY, Li B (2020) DBA: distributed backdoor attacks against federated learning. In: International conference on learning representations
36. Yin D, Chen Y, Kannan R, Bartlett P (2018) Byzantine-robust distributed learning: Towards optimal statistical rates. In: International conference on machine learning. PMLR, pp 5650–5659
37. Yue K, Jin R, Wong CW, Baron D, Dai H (2023) Gradient obfuscation gives a false sense of security in federated learning. In: 32nd USENIX security symposium (USENIX security 23), pp 6381–6398
38. Zhang Y, Bai G, Chamikara MAP, Ma M, Shen L, Wang J, Nepal S, Xue M, Wang L, Liu J (2023) Agrevader: poisoning membership inference against byzantine-robust federated learning. In: Proceedings of the ACM web conference 2023, WWW '23. Association for Computing Machinery, New York, NY, USA, pp 2371–2382. https://doi.org/10.1145/3543507.3583542
39. Zhang Z, Cao X, Jia J, Gong NZ (2022) Fldetector: defending federated learning against model poisoning attacks via detecting malicious clients. In: Proceedings of the 28th ACM SIGKDD conference on knowledge discovery and data mining, pp 2545–2555
40. Zhang Z, Su Q, Sun X (2022) Dim-krum: backdoor-resistant federated learning for NLP with dimension-wise krum-based aggregation. In: Goldberg Y, Kozareva Z, Zhang Y (eds) Findings of the association for computational linguistics: EMNLP 2022, Abu Dhabi, United Arab Emirates, December 7–11, 2022. Association for Computational Linguistics, pp 339–354. https://doi.org/10.18653/v1/2022.findings-emnlp.25
41. Zhao Y, Deng B, Shen C, Liu Y, Lu H, Hua XS (2017) Spatio-temporal autoencoder for video anomaly detection. In: Proceedings of the 25th ACM international conference on Multimedia, pp 1933–1941
42. Zhao Y, Rossi R, Akoglu L (2021) Automatic unsupervised outlier model selection. In: Ranzato M, Beygelzimer A, Dauphin Y, Liang P, Vaughan JW (eds) Advances in neural information processing systems, vol 34. Curran Associates, Inc., pp 4489–4502. https://proceedings.neurips.cc/paper_files/paper/2021/file/23c894276a2c5a16470e6a31f4618d73-Paper.pdf
43. Zheng J, Li K, Yuan X, Ni W, Tovar E, Akan, Ö B (2025) Gradcam-ae: a new shield defense against poisoning attacks on federated learning. ACM Trans Priv Secur 28(4):1–23
44. Zhou B, Khosla A, Lapedriza A, Oliva A, Torralba A (2016) Learning deep features for discriminative localization. In: Proceedings of the IEEE conference on computer vision and pattern recognition, pp 2921–2929
45. Zhou C, Paffenroth RC (2017) Anomaly detection with robust deep autoencoders. In: Proceedings of the 23rd ACM SIGKDD international conference on knowledge discovery and data mining, pp 665–674
46. Zhou H, Yang G, Dai H, Liu G (2022) PFLF: privacy-preserving federated learning framework for edge computing. IEEE Trans Inf Forensics Secur 17:1905–1918. https://doi.org/10.1109/TIFS.2022.3174394
47. Zhu T, Guo Z, Yao C, Tan J, Dou S, Wang W, Han Z (2024) Byzantine-robust federated learning via cosine similarity aggregation. Comput Netw 254:110730
48. Zhuang H, Zhang Y, Liu S (2023) A pilot study of query-free adversarial attack against stable diffusion. In: Proceedings of the IEEE/CVF conference on computer vision and pattern recognition (CVPR) workshops, pp 2384–2391

Chapter 9
Privacy-Utility Trade-Off in Federated Learning

While preserving the privacy of FL, DP inevitably degrades the utility (i.e., accuracy) of FL due to model perturbations caused by DP noise added to model updates [11]. Existing studies have considered exclusively noise with a persistent root-mean-square amplitude and overlooked an opportunity to adjust the amplitudes to alleviate the adverse effects of the noise. This chapter presents a new DP perturbation mechanism with a time-varying noise amplitude to protect the privacy of FL and retain the capability of adjusting the learning performance. Specifically, we propose a geometric series form for the noise amplitude and reveal analytically the dependence of the series on the number of global aggregations and the (ϵ, δ)-DP requirement. We derive an online refinement of the series to prevent FL from premature convergence resulting from excessive perturbation noise. Another important aspect is an upper bound developed for the loss function of a multi-layer perceptron (MLP) trained by FL running the new DP mechanism. Accordingly, the optimal number of global aggregations is obtained, balancing learning and privacy. Extensive experiments are conducted using MLP, support vector machine (SVM), and convolutional neural network (CNN) models on four public datasets. The contribution of the new DP mechanism to the convergence and accuracy of privacy-preserving FL is corroborated, compared to the state-of-the-art Gaussian noise mechanism with a persistent noise amplitude.

9.1 Introduction

FL trains ML models at individual devices without the need to surrender any sensitive raw data of the devices to central servers [18]. It provides an effective means of model training without directly leaking private information [39]. Despite its significant potential for privacy protection, there are still risks of revealing sensitive information in the individual models uploaded to an aggregator (e.g., a central server) on each

K. Li et al., *Security and Resilience in Distributed Machine Learning*, Springer Series in Reliability Engineering, https://doi.org/10.1007/978-3-032-23959-4_9

aggregation round of FL. In particular, the local models learned from the respective local datasets can be reverse-engineered (e.g., by a curious central server) to extract private information [24, 31]. Notably, Shokri et al. [30] demonstrated that private information about local datasets can be derived from trained local models. Moreover, model inversion attacks have been shown to extract private information by using black-box attacks to predict models [12, 34].

Privacy-preserving FL is a promising method for solving the above challenges [5, 15, 19, 20, 38]. It incorporates privacy techniques into distributed ML frameworks to deliver a provable guarantee of privacy protection [1, 4, 8, 36, 37]. A Trusted Execution Environment (TEE) is a secure and isolated computing environment that uses hardware and software encryption to protect sensitive data and ensure the accuracy of computations. It guarantees the confidentiality and integrity of an individual client's application, even in an untrusted environment [26]. In FL, TEEs can be adopted by clients for local training and/or by central servers for secure aggregation of local updates to prevent attacks on the models or gradients [23, 28]. DP [22] is the *de facto* privacy mechanism that has been increasingly studied, including ϵ-DP, (ϵ, δ)-DP, Rényi DP, and (α, ϵ)-Rényi DP [2, 3, 22, 40, 42]. In [40], a time-varying noise perturbation mechanism was proposed, where a time-decaying noise was added to the model parameters. In [3], information-theoretic bounds were derived to establish a connection between information leakage and DP.

DP mechanisms have been increasingly integrated into FL, aiming to learn a secure global model while providing privacy guarantees for local datasets [29]. This allows the clients to efficiently train their local models with privacy protection according to local settings. Truex et al. [32] proposed an FL system with local DP (LDP) to ensure data privacy. The system can perform LDP-based perturbation on model parameter updating and sharing, according to the local privacy level. In [35], a framework based on the DP was proposed to prevent information leakage by injecting noise to protect the privacy of the local model parameters. In [41], three LDP mechanisms were developed to preserve privacy in different data analysis tasks. The LDP mechanisms were integrated into FL to predict traffic status, alleviate privacy threats, and reduce communications in crowd-sourcing applications. The above studies combined the LDP mechanisms with FL to address privacy issues. However, considerable noises are needed to perturb the local model parameters, reducing the efficiency and accuracy of FL.

The studies discussed above, i.e., [32, 35, 41], have considered time-invariant DP noise perturbations; in other words, the variance of the DP perturbation noise remains persistent among global aggregations. This could require a long training or convergence time and degrade the learning performance, such as validation accuracy [13]. On the other hand, there is clearly a potential to allow the variance of the DP perturbation noise to be adaptively configured and changed over different global aggregations, hence improving the learning performance of FL without compromising the privacy protection level. In particular, a small perturbation noise in the early stage of an FL process is expected to benefit convergence [7], as also observed experimentally in this chapter.

This chapter presents a new (ϵ, δ)-DP amplitude-varying perturbation mechanism with a meticulously designed time-varying root-mean-square amplitude (or amplitude for short) of the perturbation noise to strike a balance between the privacy protection and the utility (i.e., loss and/or accuracy) of FL. The DP noise can be further adjusted online to combat the potential degradation of the utility, achieving effective learning while preserving privacy.

The key contributions are listed as follows.

- We design the new (ϵ, δ)-DP perturbation mechanism, where the variance (i.e., the square of the amplitude) of the DP noise is a geometric series changing over the global aggregations of the FL to provide privacy guarantees.
- By privacy analysis, we derive the variance of the DP noise given the global aggregation number and privacy protection level of the FL. We also design the online adjustment of the variance and global aggregation rounds in the face of a model degradation of the FL.
- An upper bound is derived for the loss of a multi-layer perceptron (MLP) model trained by FL running the new DP mechanism, establishing an analytical trade-off between the loss and privacy.
- Based on the upper bound, an optimal number of global aggregations is identified to achieve the best utility of the FL and satisfy the privacy requirement.

Extensive experiments based on an MLP model show that the new DP mechanism with time-varying noise variance converges faster to better learning accuracy for a given privacy level, compared to the state-of-the-art noise perturbation with persistent variance, i.e., the Gaussian noise mechanism [35]. Moreover, the new mechanism is readily applicable to other DNN models, such as support vector machine (SVM) and CNN.

The remainder of this chapter is arranged as follows. The system and threat models are provided in Sect. 9.2. In Sect. 9.3, we elaborate on the new DP mechanism, analyze its sensitivity and privacy, and the time-varying variance of the DP noise. Section 9.4 derives the convergence upper bound of FL running the new DP mechanism, and the optimal number of global aggregations to achieve both convergence and privacy. Experimental results are discussed in Sect. 9.5, followed by concluding remarks in Sect. 9.6.

Notation: $(\cdot)^H$, $(\cdot)^\top$, and $(\cdot)^c$ are the Hermitian transpose, transpose, and conjugate of a matrix/vector, respectively. $|\cdot|$ takes element-wise absolute values. $\|\cdot\|$ denotes ℓ_2-norm. $(\mathbf{A})_{n,m}$ and $(\mathbf{A})_{\cdot,m}$ stand for the (n, m)-th element and the m-th column of the matrix $\mathbf{A}$, respectively. $\text{diag}\{a_n\}$ stands for a diagonal matrix with $a_n, \forall n$ along its diagonal. The notation used is collated in Table 9.1.

Table 9.1 List of main notations

Symbol	Definition
$\mathcal{M}$	Random differential privacy mechanism
$\mathcal{D}$, $\mathcal{D}'$	Adjacent datasets
ϵ, δ	Privacy budget parameters
$\mathcal{D}_k$	Local dataset owned by user C_k
$\nabla F(\cdot)$	Gradient operator of function $F(\cdot)$
U	Total number of users in the system
K	Number of selected users per round
t	Index of training iteration
T	Total number of training iterations
τ	Number of local update steps between two global aggregations
M	Maximum number of global aggregations
$\boldsymbol{\omega}$	Model parameter vector
$F(\boldsymbol{\omega})$	Global objective function
$f_k(\boldsymbol{\omega})$	Loss function of user k
$\boldsymbol{\omega}_k(t)$	Local model parameters of user k at iteration t
$\tilde{\boldsymbol{\omega}}_k(t)$	Noisy local model parameters of user k at iteration t
$\boldsymbol{\omega}(m)$	Aggregated model parameters at the m-th global round, $m = 0, 1, \ldots, M$
$\boldsymbol{\omega}^*$	Optimal model parameters

9.2 System and Threat Models

This section introduces the system and threat models of the considered FL system.

9.2.1 DP-Based Federated Learning

The considered FL system consists of U users collected by the set $\mathcal{U} = \{1, \ldots, U\}$ and an honest-but-curious parameter server, as illustrated in Fig. 9.1. $\mathcal{D}_i$ is the local dataset at user $i \in \mathcal{U}$. $|\mathcal{D}| = \sum_{i \in \mathcal{U}} |\mathcal{D}_i|$ is the collection of all data samples. The server wishes to train a global model on the datasets residing at the U users, by minimizing the global loss function, $F(\omega)$:

$$\boldsymbol{\omega}^* \triangleq \arg\min_{\boldsymbol{\omega}} F(\boldsymbol{\omega}), \tag{9.1}$$

where $\boldsymbol{\omega}$ stands for the model parameter, $\boldsymbol{\omega}^*$ stands for the optimal model parameter minimizing the global loss function, and $\mathcal{K} \in \mathcal{U}$ is a set of K randomly chosen users from $\mathcal{U}$ between two consecutive global aggregations. $F(\boldsymbol{\omega}) = \sum_{k \in \mathcal{K}} p_k f_k(\boldsymbol{\omega})$, with

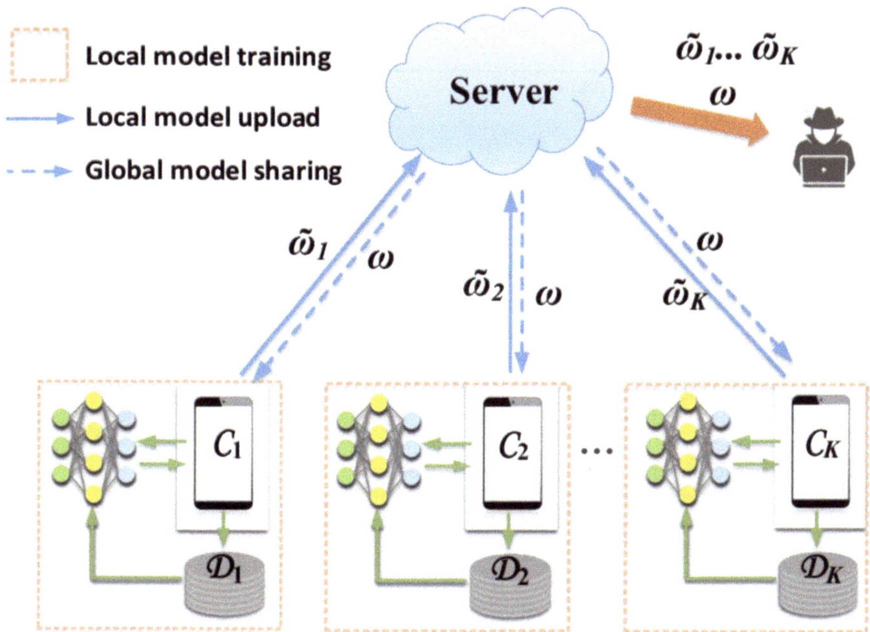

Fig. 9.1 An FL model with an honest server and an attacker trying to eavesdrop on or capture private information from both the local model upload and global model sharing process

$f_k(\boldsymbol{\omega})$ being the loss function of the k-th user in $\mathcal{K}$, and $p_k = |\mathcal{D}_k|/\sum_{k\in\mathcal{K}} |\mathcal{D}_k|$ with $\sum_{k\in\mathcal{K}} p_k = 1$.

We consider using DP to prevent the privacy leakage of a distributed gradient descent-based FL system. M is the maximum number of global aggregations. τ is the number of local update iterations between two consecutive aggregations. T is the total number of iterations. $T = M\tau$. The FL and DP operations are summarized in **Algorithm 1**, where $\boldsymbol{\omega}(m)$ denotes the global model parameter obtained at the m-th global model aggregation with the initial global model $\boldsymbol{\omega}(0)$, $\boldsymbol{\omega}_k(t)$ stands for the local model parameter of the k-th user at the t-th iteration, and q is the ratio of selected and participating users (i.e., $K = |\mathcal{K}| = qU$). $t \in [0, T]$ is the iteration index. For the k-th user, the local model is updated by

$$\boldsymbol{\omega}_k(t+1) = \boldsymbol{\omega}_k(t) - \eta \nabla f_k(\boldsymbol{\omega}_k(t)), \tag{9.2}$$

where η is the step size. The k-th user clips the local model parameter $\boldsymbol{\omega}_k(t)$ with a pre-determined threshold C, i.e., $\|\boldsymbol{\omega}_k(t+1)\| \le C$.

After every τ local updates (or iterations), e.g., the τm-th iteration, the users inject the DP noises into their local models and upload the perturbed local models, denoted by $\tilde{\boldsymbol{\omega}}_k(\tau m)$, to the parameter server for the m-th global aggregation:

Algorithm 1: Distributed gradient descent FL perturbed with time-varying DP noises

Input: τ, T, $\omega(0)$, ϵ, δ, and C.
Output: $\omega(T)$.
Initialize: $t = 0$ and $\omega(0)$;
while $t \leq T$ **do**
  % Local model update;
  while $k \in \mathcal{K}$ **do**
    Update the local parameters:
      $\omega_k(t+1) = \omega_k(t) - \eta \nabla f_k(\omega_k(t))$;
    Clip the local parameters:
      $\omega_k(t+1) \leftarrow \omega_k(t+1) / \max\left(1, \frac{\|\omega_k(t+1)\|}{C}\right)$;
  end while
  % Global model aggregation;
  if $(t+1)$ is an integer multiple of τ **then**
      $m = \lfloor \frac{t+1}{\tau} \rfloor$;
    Produce the time-varying DP noises to perturb
    the local model parameters:
      $\tilde{\omega}_k(t+1) = \omega_k(t+1) + \mathcal{N}(0, \Theta(m)\boldsymbol{I}), \ \forall k$;
    Update the global parameters:
      $\omega(m) = \sum_{k \in \mathcal{K}} p_k \tilde{\omega}_k(t+1)$;
    Update the local model parameters:
      $\omega_k(t+1) = \omega(m), \ \forall k \in \mathcal{K}$.
  end if
  $t \leftarrow t + 1$;
end while

$$\tilde{\boldsymbol{\omega}}_k(\tau m) = \boldsymbol{\omega}_k(\tau m) + \mathcal{N}(0, \Theta(m)\boldsymbol{I}), \ m \in [1, M], \tag{9.3}$$

where $\Theta(m)$ is the variance of the DP noise added to the local models for the m-th global aggregation.

The global model is the aggregation of the DP-perturbed local models. At the m-th aggregation, the global model parameter is

$$\boldsymbol{\omega}(m) = \sum_{k \in \mathcal{K}} p_k \tilde{\boldsymbol{\omega}}_k(\tau m). \tag{9.4}$$

The parameter server broadcasts $\boldsymbol{\omega}(m)$ to all users. The users start the next training iteration, i.e., $t = \tau m + 1$, by setting $\boldsymbol{\omega}_k(t) = \boldsymbol{\omega}(m)$. By executing Algorithm 1 for M global model aggregations (or T iterations), we obtain the optimal model parameter $\boldsymbol{\omega}^*$ achieving the minimum global loss function.

9.2.2 Threat Model

Suppose that the parameter server is honest but curious in the considered FL system. External attackers attempt to obtain confidential information about the users. Although the users can store and train their data locally in FL, the local model updates shared between the users and server can potentially compromise the privacy of the users, e.g., under inference attacks at the learning phase [25] and model-inversion attacks at the testing phase [12]. The attackers can hijack private information by analyzing the global model parameters broadcast by the parameter server.

Assume that the attacker has a dataset $\mathcal{D}_a$ overlapping with user k's local dataset $\mathcal{D}_k$ and attempts to find the subset of its dataset, $\mathcal{D}'_a \in \mathcal{D}_a$, that is the most likely used for the training of user k's model, i.e., $\max_{\mathcal{D}'_a \in \mathcal{D}_a} \Pr\left[\mathcal{D}'_a \in \mathcal{D}_k | \omega_k\right]$. Here, $\Pr[\mathcal{D}'_a \in \mathcal{D}_k | \omega_k]$ is the probability that the subset $\mathcal{D}'_a$ belongs to $\mathcal{D}_k$, given user k's model ω_k. It is possible that an attacker has an overlapping dataset with some participating nodes in an FL setting, which is a common assumption made in the literature when analyzing FL systems, e.g., [25].

9.3 Differential Privacy with Varying Noise Variance

This section delineates the DP mechanism with time-varying noise perturbation in the considered FL system, and analyzes the sensitivity and privacy of the mechanism.

9.3.1 Definition of Differential Privacy

For an (ϵ, δ)-DP mechanism, the privacy is parametrized by a requirement specified using ϵ and δ. Here, $\epsilon > 0$ specifies the difference beyond which the outputs on two adjacent datasets $\mathcal{D}$ and $\mathcal{D}'$ can be differentiated.[1] δ is the probability with which the ratio between the probabilities of $\mathcal{D}$ and $\mathcal{D}'$ after DP noises are added is no smaller than $\exp(\epsilon)$ [21]. The definition of the (ϵ, δ)-DP is provided below.

Definition 9.1 (*(ϵ, δ)-DP* [1]) A random mechanism $\mathcal{M}\colon \mathcal{X} \rightarrow \mathcal{R}$ that has a domain of $\mathcal{X}$ and a range of $\mathcal{R}$ meets (ϵ, δ)-DP, as long as

$$\Pr\left[\mathcal{M}(\mathcal{D}) \in \mathcal{S}\right] \leq e^{\epsilon} \Pr\left[\mathcal{M}(\mathcal{D}') \in \mathcal{S}\right] + \delta, \tag{9.5}$$

for a measurable set $\mathcal{S} \subseteq \mathcal{R}$ and adjacent datasets $\mathcal{D}, \mathcal{D}' \in \mathcal{X}$.

---

[1] Two datasets, $\mathcal{D}$ and $\mathcal{D}'$, are adjacent if $\mathcal{D}'$ can be built by inserting an example to, or discarding an example from, $\mathcal{D}$.

9.3.2 Proposed Time-Varying Perturbation Noise Variance

We propose that the variance of the DP noise, i.e., $\Theta(m)$, added to the local model parameters changes (increases or decreases), with the increasing number of global aggregations. On the one hand, for a required privacy level (e.g., (ϵ, δ) in the context of DP), adding a smaller or stronger noise at the beginning of the training can speed up convergence [7]. On the other hand, more global aggregations in the training process result in worse privacy leakage since an adversary can observe more information exposed in the global aggregations and related to the training datasets, according to [35].

We design that $\Theta(m)$ is a geometric series, and the noise variance added in the m-th global aggregation is calculated as

$$\Theta(m) = \vartheta^{m-1}\sigma^2, \; m = 1, \ldots, M, \tag{9.6}$$

where σ^2 is the initial noise variance; $\vartheta > 0$ is the scaling factor of the series. When $\vartheta = 1$, the noise variance $\Theta(m)$ remains unchanged, as it is in the existing DP schemes [35].

Apart from the specific global model aggregation m, the DP noise variance $\Theta(m)$ depends on the privacy level (ϵ, δ), the ratio of participating users q, the number of local update iterations between two global model aggregations τ, and the iteration number T (or, in other words, the number of global model aggregations M). It is of practical interest to determine the optimal values of M and τ, or their trade-off.

9.3.3 Sensitivity and Privacy Analysis

It is prudent to analyze the sensitivity and privacy performance of the proposed DP perturbation with time-varying noise variance $\Theta(m)$. We use the ℓ_2-norm to measure the sensitivity [9]

$$\Delta s = \max_{\mathcal{D}, \mathcal{D}'} \left\| s(\mathcal{D}) - s(\mathcal{D}') \right\|, \tag{9.7}$$

where $s(\cdot)$ is a general function in $\mathcal{D}$ (and $\mathcal{D}'$). Accordingly, if the batch size for training the local model is consistent with the training sample number, the sensitivity is given by $\Delta s = \frac{2C}{|\mathcal{D}_k|}$ with C being the pre-determined clipping threshold and $\mathcal{D}_k$ being the local dataset at the k-th user [35].

Given the sensitivity Δs, the amplitude (i.e., the standard deviation) of the noise injected into each global aggregation changes exponentially when $\vartheta \neq 1$. We derive the varying amplitude to meet the privacy requirement in **Theorem** 9.1.

Theorem 9.1 *To ensure the (ϵ, δ)-DP requirement of the local training dataset with M global model aggregations, the amplitude of the DP noise in the first global model aggregation of the time-varying DP perturbation mechanism is given by*

$$\sigma = \frac{\Delta s}{\epsilon}\sqrt{2q\left(\frac{\vartheta - \vartheta^{1-M}}{\vartheta - 1}\right)\ln\left(\frac{1}{\delta}\right)}, \text{ if } \vartheta \neq 1. \tag{9.8}$$

Proof See Appendix 9.7. □

Remark 9.1 Theorem 9.1 indicates that a larger σ results in a smaller ϵ, i.e., a stronger privacy guarantee, and confirms that a larger M leads to a higher likelihood of leaking private information during training, given σ. By Theorem 9.1, given ϵ, δ, and M, we adjust the DP noise variance to balance privacy preservation and the convergence of FL training.

Remark 9.2 Given a privacy budget ϵ for M global aggregations, more clients involved in the model updates, i.e., a larger q in (9.8), lead to requirements of stronger perturbation noises being added to the local model of each involved client. This indicates less privacy leakage for each client, which is consistent with the conclusion drawn in [10].

Remark 9.3 If $\vartheta = 1$, then $\sigma = \frac{\Delta s}{\epsilon}\sqrt{2qM\ln\left(\frac{1}{\delta}\right)}$ does not change over the global model aggregations for $m = 1, \ldots, M$, since $\frac{\vartheta - \vartheta^{1-M}}{\vartheta - 1} \xrightarrow{\vartheta \to 1} M$. This is consistent with the amplitude of the DP noise in a Gaussian noise perturbation mechanism developed in [35].

9.3.4 Online Adjustment of Perturbation Noise

The aim of the proposed time-varying DP noise perturbation, i.e., (9.8), is to protect user privacy and ensure reasonable learning performance. As the global aggregation increases, however, the noise added in the late stage of the model training may degrade the learning performance.

To address this issue without compromising the (ϵ, δ)-DP privacy level, we can reduce the maximum number of global aggregations, M, and accordingly reduce the noise variance σ_m^2 at the m-th global aggregation. Here, $\sigma' = \sigma'(M', m)$ depends on M' and m, as given below.

Theorem 9.2 *To reduce the loss of learning at the m-th global aggregation without compromising the (ϵ, δ)-DP privacy of the learning, we update the maximum number of global aggregations to M' ($M' < M$) and the variance of the perturbation noise to $\sigma_m^2 = \vartheta^{m-1}(\sigma')^2$ with σ' given by*

$$\sigma' = \begin{cases} \frac{\Delta s}{\epsilon}\sqrt{2q\left(\frac{\vartheta - \vartheta^{1-m}}{\vartheta - 1} + M' - m\right)\ln\left(\frac{1}{\delta}\right)}, & \text{if } \vartheta > 1; \\ \frac{\Delta s}{\epsilon}\sqrt{2qM'\ln\left(\frac{1}{\delta}\right)}, & \text{if } \vartheta = 1; \\ \frac{\Delta s}{\epsilon}\sqrt{2q\left(\frac{\vartheta^{1-m} - \vartheta + \vartheta^{m-M'}}{1 - \vartheta}\right)\ln\left(\frac{1}{\delta}\right)}, & \text{if } \vartheta < 1. \end{cases} \tag{9.9}$$

Proof See Appendix 9.8. □

Ideally, M (or M') should be as large as possible to improve the accuracy of FL training. In this sense, the online adjustment of the DP noise variance is of practical interest, as it allows for a progressive increase in M' adapting to the convergence process of FL and the remaining privacy budget. This is attractive for applications that require a balance between accuracy and privacy.

As revealed in Theorem 9.2, it is important to specify M' and then σ' based on (9.9). In practice, the aggregator may have part of the dataset for testing purposes. If the test loss function value at the aggregator stops decreasing. Then a new M' is calculated. An empirical approach is to set $M' = \lceil \alpha_d M \rceil$ at the next global aggregation, whenever it is observed at a global aggregation that the global loss function stops decreasing. α_d can be empirically determined. $\lceil \cdot \rceil$ stands for ceiling.

9.4 Convergence of Privacy-Preserving FL Under DP with Time-Varying Noise Amplitude

In this section, we establish the convergence upper bound for privacy-preserving FL (Algorithm 1) protected by the DP mechanism with time-varying noise amplitudes.

9.4.1 Definitions and Assumptions

The following definitions and assumptions facilitate analyzing the convergence of the FL under the new DP mechanism with time-varying perturbation noise variance.

Definition 9.2 (*$\mathcal{B}$-local Dissimilarity*) The local loss functions $f_k(\boldsymbol{\omega})$, $k = 1, \ldots, K$, yield $\mathcal{B}$-local dissimilarity at $\boldsymbol{\omega}$ if $\mathbb{E}_{\mathcal{D}_k}[\|\nabla f_k(\boldsymbol{\omega})\|^2] \leq \mathcal{B}^2 \|\nabla F(\boldsymbol{\omega})\|^2$, where $\mathbb{E}_{\mathcal{D}_k}[\cdot]$ takes expectation over the distribution of user k's dataset, $\mathcal{D}_k$, with the probability of $\mathcal{D}_k$ given by $p_k = |\mathcal{D}_k| / \sum_{k \in \mathcal{K}} |\mathcal{D}_k|$ and $\sum_{k \in \mathcal{K}} p_k = 1$. If $\|\nabla F(\boldsymbol{\omega})\| \neq 0$, we define $\mathcal{B}(\boldsymbol{\omega}) = \sqrt{\mathbb{E}_k[\|\nabla f_k(\boldsymbol{\omega})\|^2] / \|\nabla F(\boldsymbol{\omega})\|^2}$.

We refer to the gap in the gradient between the local and global loss functions as "gradient divergence". The gradients depend on the partition of data among users.

Definition 9.3 (*Gradient Divergence*) $\forall k, \boldsymbol{\omega}$, γ_k denotes an upper bound of the gradient divergence between the local and global loss functions, i.e., $\|\nabla f_k(\boldsymbol{\omega}) - \nabla F(\boldsymbol{\omega})\| \leq \gamma_k$. The global gradient divergence is $\gamma \triangleq \sum_k p_k \gamma_k = \frac{\sum_k |\mathcal{D}_k| \gamma_k}{|\mathcal{D}|}$.

Assumption 9.1 Consider the following assumptions: $\forall k \in \mathcal{K}$,

1. $f_k(\boldsymbol{\omega})$ is convex and L-smooth [27], that is, $\left\|\nabla f_k(\boldsymbol{\omega}) - \nabla f_k(\boldsymbol{\omega}'')\right\| \leq L \left\|\boldsymbol{\omega} - \boldsymbol{\omega}''\right\|$, $\forall \boldsymbol{\omega}, \boldsymbol{\omega}''$, with L being a constant depending on the loss function;

2. $f_k(\boldsymbol{\omega})$ is L_c-Lipschitz continuous, that is, $\left\| f_k(\boldsymbol{\omega}) - f_k(\boldsymbol{\omega}'')\right\| \leq L_c \left\|\boldsymbol{\omega} - \boldsymbol{\omega}''\right\|$, $\forall \boldsymbol{\omega}, \boldsymbol{\omega}''$;
3. The learning rate is $\eta \leq \frac{1}{L}$;
4. $f_k(\boldsymbol{\omega})$ fulfills the Polyak-Lojasiewicz requirement [16] with a positive parameter ρ, indicating that $F(\boldsymbol{\omega}) - F(\boldsymbol{\omega}^*) \leq \frac{1}{2\rho} \left\|\nabla F(\boldsymbol{\omega})\right\|^2$ and $\boldsymbol{\omega}^*$ minimizes $F(\boldsymbol{\omega})$;
5. $F(\boldsymbol{\omega}(0)) - F(\boldsymbol{\omega}^*) = \Theta$, where Θ is a constant.

9.4.2 Convergence Analysis

To analytically study the convergence of Algorithm 1, a corresponding centralized gradient descent-based learning process is typically considered, as given by

$$\boldsymbol{\omega}'(t) = \boldsymbol{\omega}'(t-1) - \eta \nabla F(\boldsymbol{\omega}'(t-1)). \tag{9.10}$$

The model parameter $\boldsymbol{\omega}'(t)$ is updated using the global loss function $F(\cdot)$ and the entire dataset $\mathcal{D}$.

The following lemma [33, Theorem 1] provides an upper bound for the gap between the global model parameter of the proposed FL process, i.e., $\boldsymbol{\omega}(m)$, and the model parameter of the centralized learning process, i.e., $\boldsymbol{\omega}'(t)$, and, in turn, an upper bound of the gap between their loss functions, $F(\boldsymbol{\omega}(m))$ and $F(\boldsymbol{\omega}'(t))$. $t = 1, \ldots, T$.

Lemma 9.1 [33, Theorem 1] *For any $t = (m-1)\tau + 1, \ldots, m\tau$, the difference of the global models between the FL process with DP perturbation and the centralized learning process in (9) is upper bounded; i.e., $\|\boldsymbol{\omega}(m) - \boldsymbol{\omega}'(t)\| \leq \mathcal{H}(t - (m-1)\tau)$. Here, $\mathcal{H}(x) \triangleq \frac{\gamma}{L}\left((\eta L + 1)^x - 1\right) - \eta\gamma x$, $x = 0, 1, 2, \ldots$. γ is the average gradient divergence over the dataset. Since $F(\boldsymbol{\omega})$ is L_c-Lipschitz, the difference of the global loss functions is also upper bounded, i.e., $\|F(\boldsymbol{\omega}(m)) - F(\boldsymbol{\omega}'(t))\| \leq L_c\mathcal{H}(t - (m-1)\tau)$.*

By Lemma 9.1, Definitions 9.2 and 9.3, and Assumption 9.1, we develop the following theorem to analyze the convergence bound of the gap between $\boldsymbol{\omega}'(t)$ and $\boldsymbol{\omega}^*$, $t = (m-1)\tau + 1, \cdots, m\tau$.

Theorem 9.3 *To satisfy the (ϵ, δ)-DP, the convergence upper bound of the FL under time-varying DP noise perturbation, i.e., Algorithm 1, after m global aggregation rounds, is obtained as*

$$\begin{aligned} F(\boldsymbol{\omega}'(t)) - F(\boldsymbol{\omega}^*) \leq & \mathcal{A}^m \Theta + \\ & \frac{qL(\Delta s)^2 \ln\left(\frac{1}{\delta}\right)\left(\vartheta^m - \mathcal{A}^m\right)\left(\vartheta - \vartheta^{1-m}\right)}{\epsilon^2 (\vartheta - \mathcal{A})(U-1)}, \end{aligned} \tag{9.11}$$

where $\mathcal{A} = 1 + 2\rho\phi$, and $\phi \triangleq \frac{\eta^2 L}{2}\left(\frac{(U-K)\mathcal{B}^2}{K(U-1)} + \frac{K-1}{UK(U-1)}\right) - \eta$.

Proof See Appendix 9.9. □

By setting $m = M$ in Theorems 9.1 and 9.3, we have the following corollary.

Corollary 9.1 *To satisfy the (ϵ, δ)-DP, the convergence upper bound of the FL under time-varying DP noise perturbation, i.e., Algorithm 1, after all M global aggregations (or, in other words, all T iterations), is obtained as*

$$F\big(\omega(M)\big) - F(\omega^*) \leq \mathcal{A}^M \Theta + \frac{qL(\Delta s)^2 \ln\left(\frac{1}{\delta}\right)\left(\vartheta^M - \mathcal{A}^M\right)\left(\vartheta - \vartheta^{1-M}\right)}{\epsilon^2 (\vartheta - \mathcal{A})(U-1)} + L_c \mathcal{H}(\tau). \tag{9.12}$$

Corollary 9.1 shows the trade-off between the learning performance and privacy protection level. In the case of a weak privacy guarantee (i.e., both ϵ and δ are large), the convergence upper bound is tighter since the second term on the right-hand side (RHS) of (9.12) approaches zero. Additionally, the convergence upper bound in (9.12) is unrestricted by the partitions of the data among the users. The data partition is captured in the gradient divergence γ, a parameter of $\mathcal{H}(\tau)$; see Lemma 9.1. Since $\frac{\partial \mathcal{H}(\tau)}{\partial M} = \frac{\partial \mathcal{H}(\frac{T}{M})}{\partial M} = -\frac{\gamma T}{M^2}(\eta L+1)^{\frac{T}{M}} \ln(\eta L+1) + \frac{\gamma T}{M^2} \leq 0$ for $M \leq \frac{T}{\ln(\eta/\ln(\eta L+1))}$, the upper bound becomes smaller when M is larger and γ is smaller under a given T.

Another interesting finding in Theorem 9.3 is that the convergence upper bound is *not* a monotonic function of either T or M, as established in the following Corollary 9.2.

Corollary 9.2 *Given ϵ and δ, the convergence upper bound is convex in the number of iterations or global aggregations, i.e., T or M, if $\vartheta \geq \mathcal{A}$ and $\tau \geq \frac{\ln(\eta/\ln(\eta L+1))}{\ln(\eta L+1)}$.*

Proof See Appendix 9.10. □

Note that the upper bound in (9.12) may not be tight, as it is obtained through the use of the triangular inequality and Jensen's inequality, which are commonly used in the derivation of convergence upper bounds for FL models [4, 5, 15, 19, 20, 38]. Despite this, the upper bound reveals that the optimality gap of an FL model under our proposed perturbation mechanism would decrease at first, and then increase as the number of global aggregations increases. This confirms that FL models can diverge, rather than converge, due to noise perturbation for DP, and highlights the existence of an optimal number of global aggregations; i.e., T or M can be optimized to minimize the loss function.

Also note that **Theorem** 9.3, and **Corollaries** 9.1 and 9.2 are based on the smoothness of the loss functions of the neural network models being trained, and applicable to the convergence bound analysis of FL systems training MLP and SVM models. On the other hand, **Theorems** 9.1 and 9.2 specify the DP noise variances to preserve the privacy of an FL training process and defer the divergence of the process caused by the DP noises. The two theorems do not rely on the smoothness of the loss functions and apply to neural networks with non-smooth loss functions, e.g., CNN.

9.4.3 Optimal Global Aggregation Rounds *M**

To improve the convergence of Algorithm 1, we optimize M to minimize the upper bound of $F(\boldsymbol{\omega}(M)) - F(\boldsymbol{\omega}^*)$ for a given T. By choosing a small enough $\eta \leq \frac{1}{L}$, Problem (9.1) can be rewritten as

$$\min_{M} \mathcal{A}^M \Theta + \frac{qL(\Delta s)^2 \ln\left(\frac{1}{\delta}\right)\left(\vartheta^M - \mathcal{A}^M\right)\left(\vartheta - \vartheta^{1-M}\right)}{\epsilon^2 (\vartheta - \mathcal{A})(U-1)}. \tag{9.13}$$

which can be solved by setting the first-order derivative of its objective, denoted by $\mathcal{G}(M)$, to zero, i.e.,

$$\begin{aligned}\frac{\partial \mathcal{G}(M)}{\partial M} =& \Theta \mathcal{A}^M \ln(\mathcal{A}) + \frac{qL(\Delta s)^2 \ln\left(\frac{1}{\delta}\right)\vartheta}{\epsilon^2 (\vartheta - \mathcal{A})(U-1)} \\ & \left(\vartheta^M \ln(\vartheta) - \mathcal{A}^M \ln(\mathcal{A}) + (\mathcal{A}/\vartheta)^M\right. \\ & \left.\ln(\mathcal{A}/\vartheta)\right) + L_c \frac{\partial \mathcal{H}\left(\frac{T}{M}\right)}{\partial M} = 0.\end{aligned} \tag{9.14}$$

The optimal number of global aggregations, denoted by M^*, can be found numerically [6], e.g., using bisection search. According to Corollary 9.2, the convergence upper bound is convex with respect to M, if $\vartheta \geq \mathcal{A}$ and $\tau \geq \frac{\ln(\eta/\ln(\eta L+1))}{\ln(\eta L+1)}$. Then, the solution to (9.14) is unique and globally optimal. The value of τ is determined by the ratio of T to M. Here, given T, we determine M. Accordingly, $\tau = T/M$ is updated.

9.5 Performance Evaluation

This section assesses the accuracy of our analysis and the effectiveness of the proposed time-varying DP noise variance for distributed gradient descent-based FL under various learning tasks, models, and real-world datasets.

9.5.1 Experimental Settings

We set the number of users to $U = 100$, the number of chosen users to $K = 10$, and the number of iterations between two consecutive global aggregations to $\tau = 5$. With reference to [35], we set the clipping threshold $C = 5$, and $\delta = 0.001$. The maximum number of global aggregations is $M = 30$ and the privacy protection level is $\epsilon = 10$; unless otherwise specified. The experiments are conducted on four datasets:

- The standard MNIST dataset comprises 60,000 training and 10,000 testing examples, which are grayscale images of handwritten digits from one to ten;

- The ADULT dataset, which contains 40,000 records extracted from census data [17] and each record has up to 58 attributes, including age, education, etc.;
- The CIFAR10 dataset, which contains 60,000 32×32 color images in ten classes (6,000 per class), 50,000 for training and 10,000 for testing; and
- The FMNIST dataset, which contains Zalando's article images (i.e., 28×28 grayscale images) in ten classes, including 60,000 examples for training and 10,000 examples for testing.

We evaluate the proposed DP mechanism with time-varying perturbation noise variance on MLP, SVM, and CNN models.

9.5.2 Evaluation of the Distributed Gradient Descent Time-Varying DP Algorithm on MLP

The MLP is a fully-connected feedforward neural network. We consider an MLP model comprising a hidden layer and 32 hidden units, and train the model on the MNIST dataset. We adopt linear activation functions and a softmax of ten classes that correspond to the ten digits. The model's error on the local training dataset is measured by a cross-entropy loss function.

9.5.2.1 Impact of Noise Scaling Factor ϑ

Figure 9.2 evaluates the impact of the proposed DP mechanism with time-varying perturbation noise variance on the convergence of FL, where the x-axis provides the index to the global aggregations m. Here, $m \leq M = 30$, since using a value of m

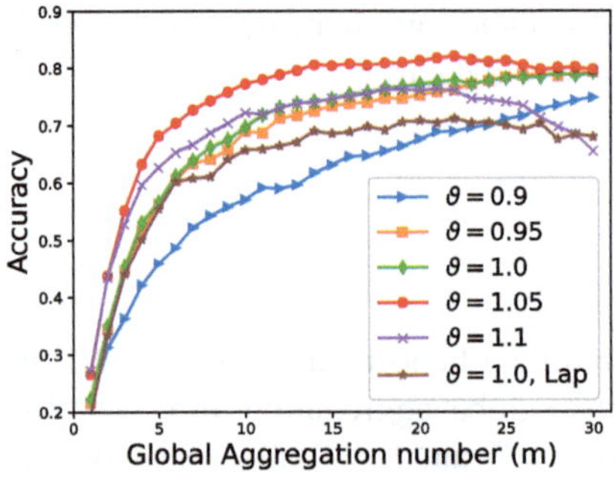

(a) Loss function value vs. m

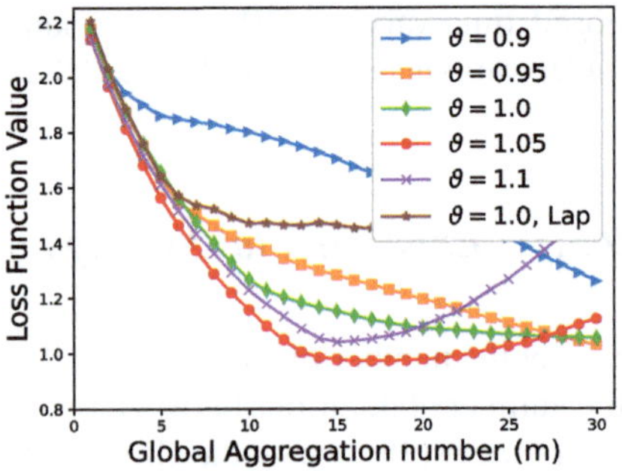

(b) Accuracy vs. m

Fig. 9.2 Comparison of loss function value and accuracy of the MLP model trained on the MNIST dataset with increasing global aggregation rounds m under different values of ϑ ('Lap' stands for the Laplacian mechanism)

larger than M would violate the privacy requirements and render the values of the training accuracy meaningless. Different ϑ values are assessed, where $\vartheta = 1$ can be viewed as the state-of-the-art design of the DP perturbation noise developed in [35]. Each curve in a figure corresponds to a standalone experiment of a training process given ϑ and ϵ.

Figure 9.2a shows that given the privacy level, the number of global aggregations needed for the (testing) loss function value to reach its minimum generally declines, as ϑ rises from 0.9 to 1.1. In the case of $\vartheta > 1$, the minimum of the loss function first declines and then increases. This is because the noise added in the early learning stage of the learning is smaller in the case of $\vartheta > 1$ than it is in the case of $\vartheta = 1$, leading to faster convergence. Moreover, the noise rises exponentially and leads to an increase in the loss function value with the growth of m. In the case of $\vartheta < 1$, the loss function value is large in the early learning stage and declines with the growth of m. The loss function value also decreases, as ϑ increases from 0.9 to 0.95. This is because the noise added in the early stage is larger for a smaller ϑ, causing slower convergence.

Figure 9.2b plots the (testing) accuracy of the proposed algorithm under different ϑ values. Similar to the loss function value, given the privacy level, the aggregation numbers required to reach the maximum accuracy decrease as ϑ increases from 0.9 to 1.1. The accuracy is more stable and remains unchanged under a smaller ϑ value. In contrast, the accuracy quickly reaches its peak and declines slowly under a larger ϑ value since a smaller noise is added in the early learning stages of the model training process and leads to faster convergence given ϵ. With the increase of global aggregations, the noises added to the model parameters grow exponentially, resulting in accuracy degradation. To this end, the value of ϑ can be adequately configured to achieve better convergence and satisfy the DP requirement in the FL process with the proposed time-varying DP perturbation noise variances.

Figure 9.3 evaluates the impact of the proposed DP mechanism with time-varying perturbation noise variance on the utility (i.e., the loss function and accuracy) of FL, as the total number of global aggregations, M, increases. Each point in a figure corresponds to a standalone experiment of a training process given ϑ and ϵ. The initial amplitude of the DP noise is calibrated for each point based on the given value of ϵ; see the second row of Table 9.2.

Figure 9.3a shows that the loss function of the MLP exhibits convex curvature with respect to M, as is consistent with Corollary 9.2. The loss functions achieve their minima under the optimal number of global aggregations, $M^* = 17$, for all considered ϑ values, validating the result in Sect. 9.4.3. The third row of Table 9.2 provides the corresponding minimum loss function values. Figure 9.3a and Table 9.2 reveal that both the time-increasing noise perturbation ($\vartheta = 1.05$) and the time-decreasing noise perturbation ($\vartheta = 0.95$) can outperform the time-invariant noise perturbation ($\vartheta = 1.0$) in terms of loss at their respective optimal numbers of global aggregations.

Figure 9.3b plots the (testing) accuracy of the proposed algorithm against M, where different values of ϑ are considered. Consistent with Fig. 9.3a, the optimal M^* values achieve the best accuracy in all experiments. We also see that when

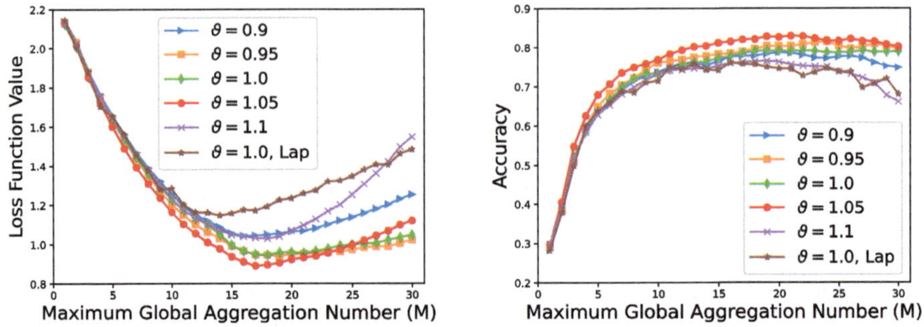

(a) Loss function value vs. M (b) Accuracy vs. M

Fig. 9.3 Comparison of loss function value and accuracy of the MLP model on the MNIST dataset with respect to the maximum number of global aggregations M under different values of ϑ with a fixed $\epsilon = 10$ ('Lap' stands for the Laplacian mechanism)

Table 9.2 Initial DP noise amplitude and minimum loss under different values of ϑ ($\epsilon = 10$)

ϑ	0.9	0.95	1.0	1.05	1.1
Initial noise σ	0.003556	0.003654	0.003749	0.003841	0.003932
Minimum loss value	1.03783	0.92706	0.94142	0.88862	1.01871

$\vartheta \neq 1$, a large number of global aggregations $M > M^*$ could overkill the learning accuracy, especially when ϑ is big, e.g., $\vartheta = 1.1$. One can potentially select the optimal ϑ value to achieve optimal learning accuracy. In the example of Fig. 9.3, the configuration of $\vartheta \geq 1.05$ allows for the smallest loss and highest accuracy of the learning. As observed in Fig. 9.3, all curves diverge, i.e., convex in Fig. 9.3a and concave in Fig. 9.3b, even when $\vartheta = 1.0$. This is because the local models are still perturbed by the DP noises with a persistent DP noise variance, when $\vartheta = 1.0$.

Figures 9.2 and 9.3 also compare our proposed DP mechanism with time-varying perturbation noise variance to the Laplacian mechanism (with a constant noise variance) [14]. Except for the DP mechanisms, all curves are under consistent experimental parameter settings in the figures. It is observed that the FL performance is worse under the Laplacian mechanism than it is under our proposed Gaussian mechanism with a time-varying DP noise variance. This is because the Laplacian mechanism adds larger noises to the data by sampling from a Laplacian distribution, which has a higher likelihood of sampling values that are farther away from the mean.

9.5.2.2 Impact of Privacy Protection Level ϵ

We evaluate the impact of ϵ on the utility of the FL protected by the proposed DP mechanism with time-varying perturbation noise variance. According to Sect. 9.5.2.1, we set $\vartheta = 1.05$. Figure 9.4 evaluates the loss of the learning with the

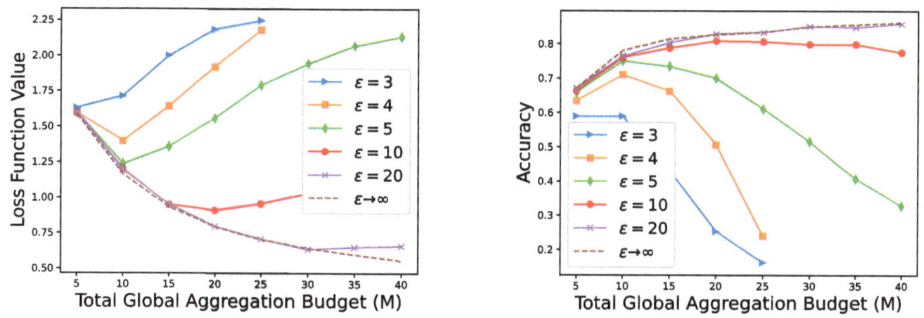

(a) Loss function value vs. M (b) Accuracy vs. M

Fig. 9.4 Loss function value and Accuracy of the MLP model on the MNIST dataset versus the maximum number of global aggregations M under different values of ϵ, where $\vartheta = 1.05$

growth of M under different settings of the privacy level ϵ. For comparison, we also plot the case with no DP noise perturbation, i.e., $\epsilon \to \infty$. Figure 9.4a and b show that the loss function value is convex and the accuracy is concave with respect to M, which is in line with Corollary 9.2. In general, the optimal number of global aggregations M^* increases with ϵ.

Figure 9.4a also shows that the loss function values decrease and approach the case with no DP noise perturbation, as ϵ increases. Figure 9.4b shows that increasing ϵ can improve the accuracy. In the case that the privacy level is larger than 20, i.e., $\epsilon \geq 20$, the convergence performance approaches the case without noise perturbation. This is because, with a lower privacy protection level ϵ, the DP noises with a larger variance are injected in the first global aggregation according to Theorem 9.1, resulting in larger loss function values. A trade-off arises between the learning performance (i.e., loss and accuracy) and privacy level, and can be adjusted through M^*.

9.5.2.3 Online Adjustment of DP Noise Variance

We evaluate the online adjustment of the aggregation number and DP noise variance proposed in Sect. 9.3.4. Figures 9.5 and 9.6 plot the updated maximum number of global aggregations, i.e., M', the loss function value, and the accuracy under different values of ϑ when $\alpha_d = 0.8$ and 0.9. As shown in Fig. 9.5, the maximum number of global aggregations M' decreases over m, resulting from the updating of σ_m to keep the global loss function decreasing, and meanwhile, the privacy protection level ϵ is satisfied. As shown in Fig. 9.6a and c, the loss function value decreases with the increase of M under all considered ϑ values. Among all four curves, $\vartheta = 1.05$ provides the best learning performance when $\alpha_d = 0.8$, as shown in Fig. 9.6a and b, and $\vartheta = 1.1$ provides the best learning performance when $\alpha_d = 0.9$, as shown in Fig. 9.6c and d. Compared to the results without online adjustment of the DP noise variance, i.e., Fig. 9.3a and b, the online adjustment of the DP noise variance can

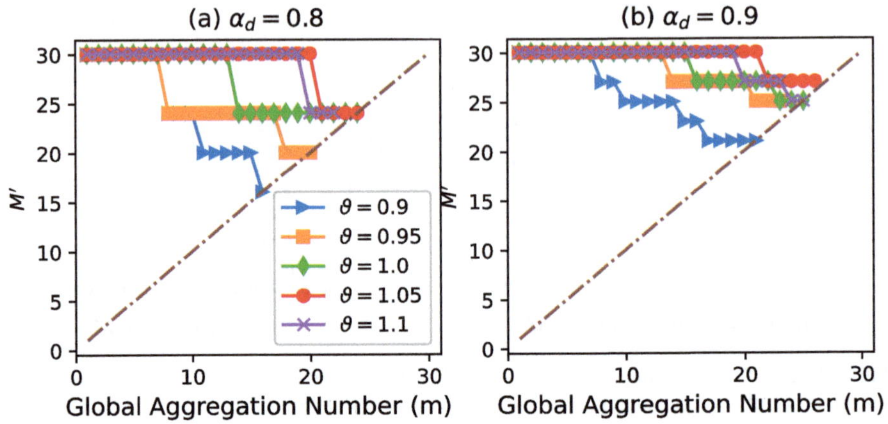

Fig. 9.5 The updated maximum number of global aggregations M' with the increasing number of global aggregations m, where M' and σ_m are updated according to Theorem 9.2 whenever the global loss function stops decreases

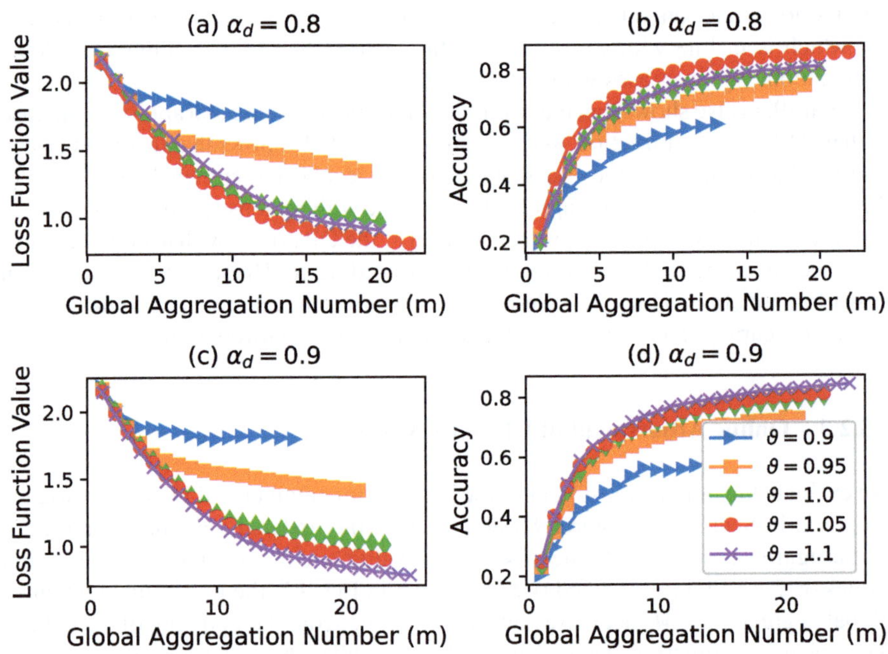

Fig. 9.6 Loss function value and Accuracy of the MLP model on the MNIST dataset versus m, where ϑ takes different values. The curves of $\vartheta = 0.9$ terminate earlier at smaller m values than the other curves because they can accommodate smaller numbers of global aggregations given the privacy level ϵ

Table 9.3 Membership inference attack success rate with and without the proposed DP mechanism (averaged over 50 independent trials, 25 shadow models)

ϵ	$\vartheta = 0.95$	$\vartheta = 1.0$	$\vartheta = 1.05$	No DP
5	0.581	0.582	0.573	0.993
10	0.599	0.587	0.583	
20	0.607	0.602	0.596	

improve the learning performance of both the loss and accuracy in the case of $\vartheta \geq 1$; see Fig. 9.6.

9.5.3 Defense Against Membership Inference Attacks

Table 9.3 evaluates the effectiveness of the proposed method in defending against membership inference attacks, which aim to determine whether a specific data point was used in training the model. The results indicate that without the mechanism, the attack success rate of the membership inference attack can be as high as 99.3%. By contrast, when the mechanism is employed, the attack success rate drops significantly, e.g., by about 40% to less than 60% when $\vartheta = 1.05$.

9.5.4 Extension to SVM and CNN Models

The proposed DP mechanism with time-varying perturbation noise can be readily applied to SVM and CNN models:

- The SVM model is trained using a standard quadratic optimization algorithm on the ADULT dataset. The loss function is $F(\omega) = \frac{\lambda_r}{2} \|\omega\|_2^2 + \max\left\{0,\ \beta_n - \omega^T \mathcal{D}_{k,n}\right\}$, where $\lambda_r > 0$ is a regularization coefficient; $\mathcal{D}_{k,n}$ is the n-th sample in $\mathcal{D}_k$, i.e., the dataset at the k-th user; and $\beta_n \in \{-1, 1\}$ for $n = 1, \ldots, |\mathcal{D}_k|$.
- The CNN model contains two convolutional layers with a kernel size of five and three fully-connected layers. The CNN model is trained separately on the CIFAR10 and FMNIST datasets. We adopt the ReLU units and softmax of ten classes for the ten classes of the CIFAR10 and the ten digits of the FMNIST. The CNN model is trained using the SGD to minimize the loss function.

Figure 9.7a and b plot the (testing) loss function and accuracy of the FL of the SVM model on the ADULT dataset under the proposed DP mechanism with time-varying perturbation noise variance. Figure 9.8a and b plot those of the CNN model on the CIFAR10 dataset. Figure 9.8c and d plot those of the CNN model on the FMNIST dataset.

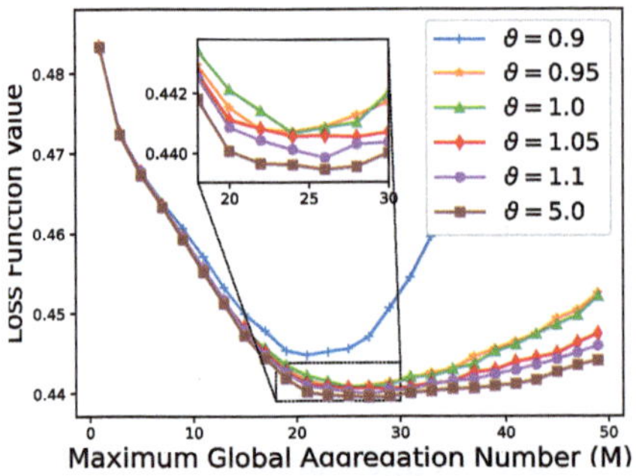

(a) Loss function value vs. M

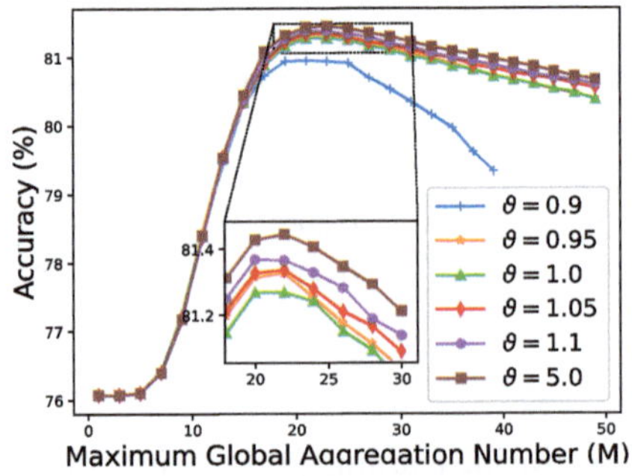

(b) Accuracy vs. M

Fig. 9.7 Loss function value and Accuracy of the SVM model on the ADULT dataset versus the maximum number of global aggregations M under different values of ϑ, where $\epsilon = 10$

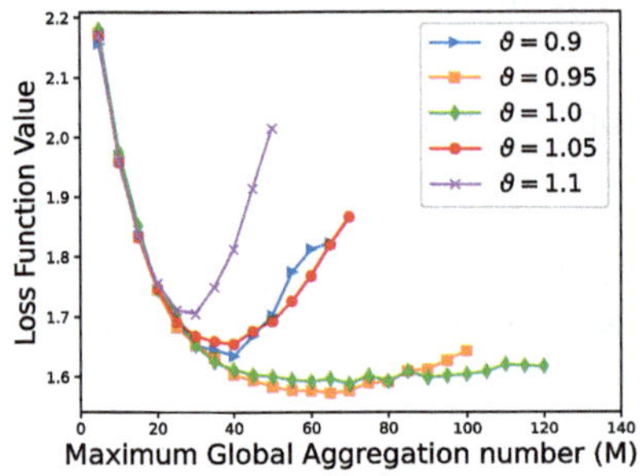

(a) Loss function value vs. M (CNN on CIFAR10)

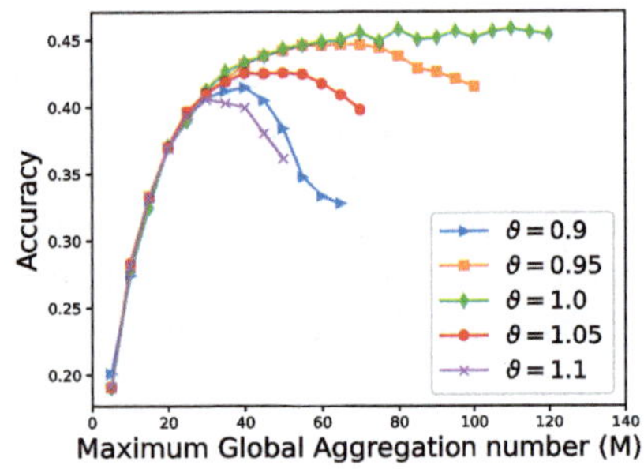

(b) Accuracy vs. M (CNN on CIFAR10)

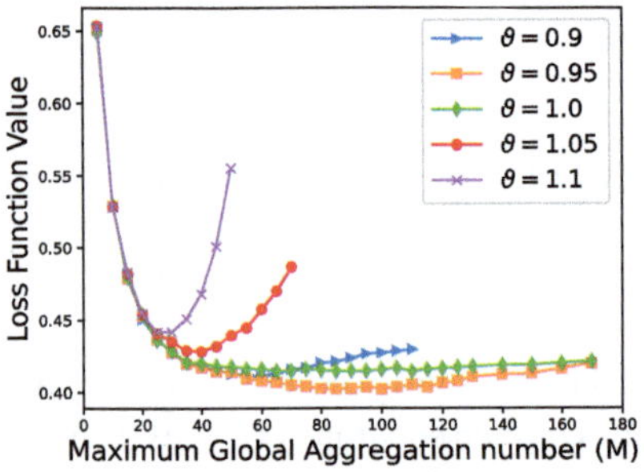

(c) Loss function value vs. M (CNN on FMNIST)

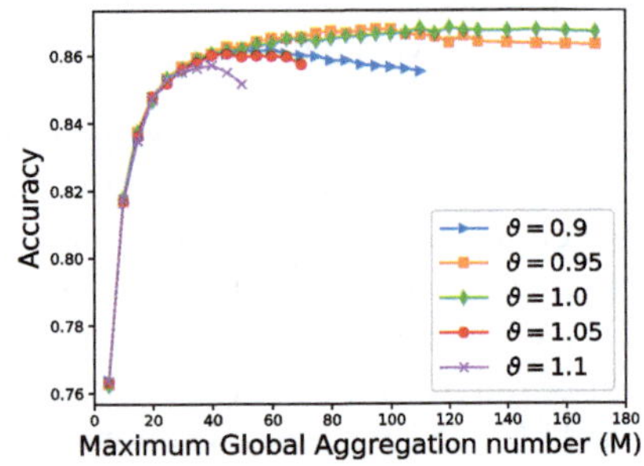

(d) Accuracy vs. M (CNN on FMNIST)

Fig. 9.8 Loss function value and Accuracy of the CNN model on the CIFAR10 and FMNIST datasets versus M under different values of ϑ, where $\epsilon = 10$

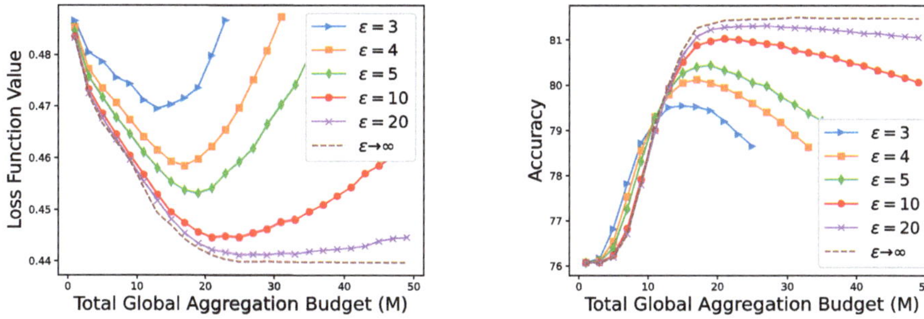

(a) Loss function value vs. $\boldsymbol{M}$ (b) Accuracy vs. $\boldsymbol{M}$

Fig. 9.9 Loss function value and Accuracy of the SVM model on the ADULT dataset versus M under different values of ϵ, where $\vartheta = 1.05$

Consistent with the observations made under the MLP model in Sect. 9.5.2, there exists the optimal number of global aggregations, M^*, that minimizes the loss function of the SVM and CNN models (and maximizes their accuracy) while satisfying the (ϵ, δ)-DP privacy level. The value of ϑ can also be configured to positively impact the utility of FL.

On the other hand, Fig. 9.7a and b show that the time-increasing noise perturbation ($\vartheta > 1$) achieves the best learning performance (i.e., the achievable smallest loss and best accuracy), followed by the time-invariant noise perturbation ($\vartheta = 1$) and then the time-decreasing noise perturbation ($\vartheta < 1$) on the SVM model. The time-decreasing noise perturbation ($\vartheta < 1$) performs the best on the CNN models under both of the considered datasets. In contrast, $\vartheta > 1$ is the best, followed by $\vartheta < 1$, and $\vartheta = 1$ is the worst on the MLP model; see Fig. 9.3a. This is due to the distinct network architectures of the MLP, SVM, and CNN models.

Figure 9.9 illustrates the effect of increasing the maximum allowed number of global aggregations, M, on the loss and accuracy of the learning process for different values of the privacy parameter, ϵ. The figure is based on the SVM model and the ADULT dataset, where $\vartheta = 1.05$ and $M = 50$. Each curve in the figure represents the results of a standalone training process with a given ϑ and ϵ. It is observed in Fig. 9.9 that as ϵ increases, the loss function values decrease and approach the case with no DP noise perturbation (i.e., $\epsilon \to \infty$). Increasing ϵ can also lead to improved accuracy. It is also observed that the FL does not diverge under $\epsilon \to \infty$, as opposed to the rest of the ϵ values. This is because when ϵ goes to infinity, no privacy is required, and the FL considered is expected to behave like regular FedAvg.

Figure 9.10 plots the (testing) loss and accuracy of the CNN models on the CIFAR10 and FMNIST datasets under different settings of the privacy level ϵ. We see that the loss functions of the CNN models are also convex with respect to M, which is in line with Corollary 9.2 and the observations made on the MLP model in Sect. 9.5.2. It is also seen that, as ϵ increases, the loss function values decrease. The optimal number of global aggregations M^* also increases. This is also consistent with the observations made on the MLP model in Sect. 9.5.2.

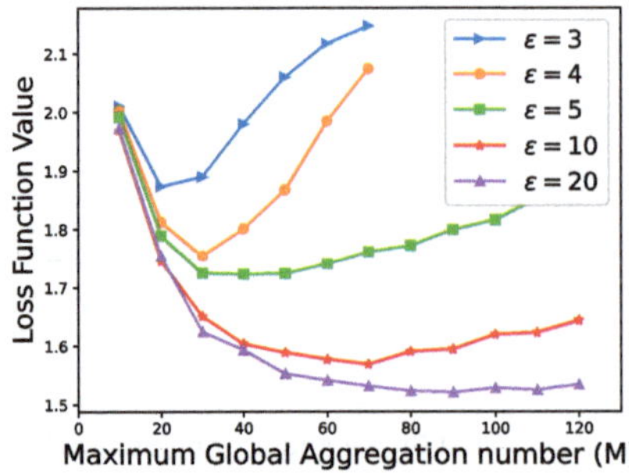

(a) Loss function value vs. M (CNN on CIFAR10).

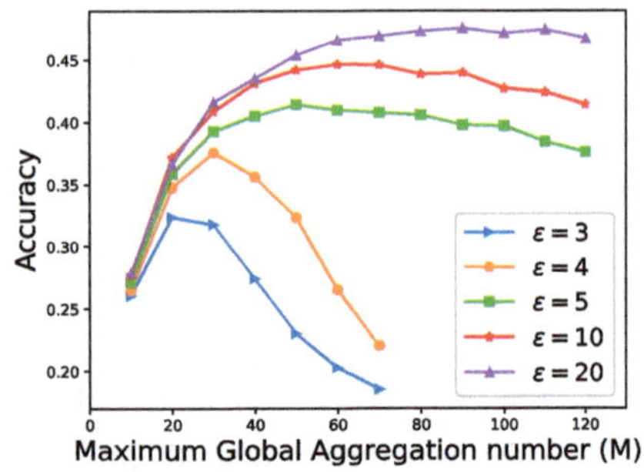

(b) Accuracy vs. M (CNN on CIFAR10).

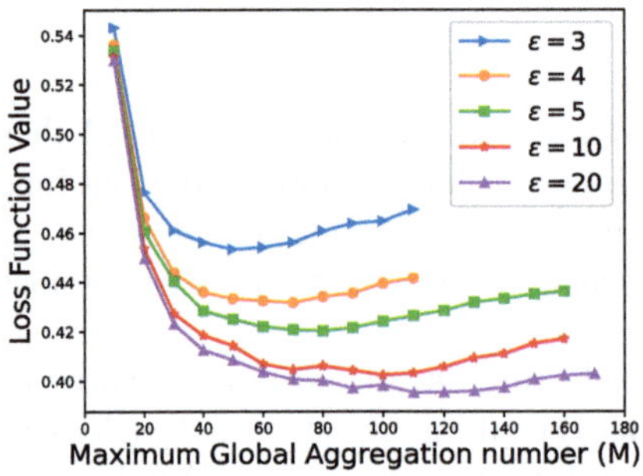

(c) Loss function value vs. M (CNN on FMNIST).

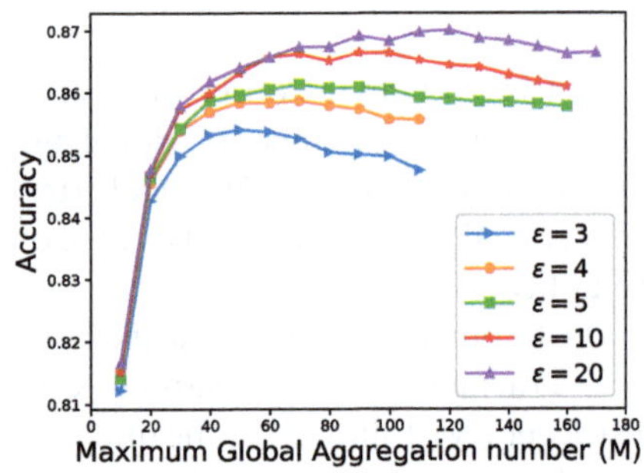

(d) Accuracy vs. M (CNN on FMNIST).

Fig. 9.10 Loss function value and Accuracy of the CNN model on the CIFAR10 and FMNIST datasets versus the maximum number of global aggregations M under different values of ϵ, where $\vartheta = 1.05$

9.6 Summary

This section presents and analyzes a DP mechanism with a time-varying noise amplitude to balance the privacy and utility of FL. The section establishes the varying amplitude as a function of the maximum number of global aggregations and the privacy protection levels. It derives a convergence upper bound for the loss function of MLP models protected by the new mechanism, revealing a trade-off between loss and privacy. The number of global aggregations is optimized based on the upper bound. Extensive experiments assess the convergence and utility of three different ML models trained using FL and protected by the new DP mechanism. The new DP mechanism with time-varying noise amplitudes exhibits faster convergence and better accuracy under given privacy protection levels, compared to existing solutions.

9.7 Proof of Theorem 9.1

The privacy loss of a mechanism $\mathcal{M}$ is defined as [1]

$$loss_p = \exp(\alpha_{\mathcal{M}}(\lambda)), \tag{9.15}$$

where $\alpha_{\mathcal{M}}(\lambda)$ is the λ-th moment, which is the logarithm of the moment generating function assessed at value λ. Based on the composability of the moment [1, Theorem 2], we have the λ-th moment for the time-varying Gaussian mechanism:

$$\begin{aligned} \alpha_{\mathcal{M}}(\lambda) &\leq \sum_{m=1}^{M} \alpha_{\mathcal{M}_m}(\lambda, \sigma_m) = \sum_{m=1}^{M} \frac{q\lambda(\lambda+1)\Delta s^2}{2\sigma_m^2} \\ &= \frac{q\lambda(\lambda+1)(\vartheta - \vartheta^{1-M})\Delta s^2}{2(\vartheta-1)\sigma^2}. \end{aligned} \tag{9.16}$$

By exploiting the tail bound of the moment [1], we can obtain the value of δ that satisfies the mechanism $\mathcal{M}$, i.e.,

$$\begin{aligned} \delta &= \min_{\lambda} \exp\left(\alpha_{\mathcal{M}}(\lambda) - \lambda\epsilon\right) \\ &= \min_{\lambda} \exp\left(\frac{q\lambda(\lambda+1)(\vartheta - \vartheta^{1-M})(\Delta s)^2}{2(\vartheta-1)\sigma^2} - \lambda\epsilon\right). \end{aligned} \tag{9.17}$$

Let $g(\lambda) = \frac{q\lambda(\lambda+1)(\vartheta-\vartheta^{1-M})(\Delta s)^2}{2(\vartheta-1)\sigma^2} - \lambda\epsilon$. Since $\exp(\cdot)$ is a monotonically increasing function, the optimization problem in (9.17) is equivalent to finding the optimum of λ, denoted by λ^*, to minimize $g(\lambda)$. We derive the first-order derivative of $g(\lambda)$, as given by

$$g'(\lambda) = \frac{q(2\lambda+1)(\vartheta - \vartheta^{1-M})(\Delta s)^2}{2(\vartheta-1)\sigma^2} - \epsilon. \tag{9.18}$$

By setting $g'(\lambda) = 0$, we obtain λ^* as

$$\lambda^* = \frac{\epsilon\sigma^2(\vartheta-1)}{q(\Delta s)^2\left(\vartheta - \vartheta^{1-M}\right)} - \frac{1}{2}. \tag{9.19}$$

As a result,

$$g(\lambda^*) = \frac{q(\Delta s)^2(\vartheta - \vartheta^{1-M})}{2\sigma^2(\vartheta-1)}\left(\frac{1}{2} - \frac{\epsilon\sigma^2(\vartheta-1)}{q(\Delta s)^2(\vartheta - \vartheta^{1-M})}\right)^2. \tag{9.20}$$

The lower bound of δ can be given by

$$\delta \geq \exp\left[-\frac{q(\Delta s)^2(\vartheta - \vartheta^{1-M})}{2\sigma^2(\vartheta - 1)}\left(\frac{1}{2} - \frac{\epsilon\sigma^2(\vartheta - 1)}{q(\Delta s)^2(\vartheta - \vartheta^{1-M})}\right)^2\right]. \tag{9.21}$$

The RHS of (9.21) provides the optimal value of δ, denoted by δ^*. By taking the logarithm on both sides of (9.21), we have

$$\ln(\delta) \geq -\frac{q(\Delta s)^2(\vartheta - \vartheta^{1-M})}{2\sigma^2(\vartheta - 1)}\left(\frac{1}{2} - \frac{\epsilon\sigma^2(\vartheta - 1)}{q(\Delta s)^2(\vartheta - \vartheta^{1-M})}\right)^2. \tag{9.22}$$

Then, we have

$$\begin{aligned} \ln\left(\frac{1}{\delta}\right) &\leq \frac{q(\Delta s)^2(\vartheta - \vartheta^{1-M})}{2\sigma^2(\vartheta - 1)}\left(\frac{1}{2} - \frac{\epsilon\sigma^2(\vartheta - 1)}{\Delta s^2(\vartheta - \vartheta^{1-M})}\right)^2 \\ &= \frac{q(\Delta s)^2(\vartheta - \vartheta^{1-M})}{8\sigma^2(\vartheta - 1)} + \frac{\epsilon^2\sigma^2(\vartheta - 1)}{2q\Delta s^2(\vartheta - \vartheta^{1-M})} - \frac{\epsilon}{2}. \end{aligned} \tag{9.23}$$

Since $\delta \in (0, 1]$ and $\ln(\delta) \leq 0$, from (9.17) we have

$$\frac{q\lambda(\lambda + 1)(\vartheta - \vartheta^{1-M})(\Delta s)^2}{2(\vartheta - 1)\sigma^2} - \lambda\epsilon \leq 0. \tag{9.24}$$

By substituting λ^* in (9.19) into (9.24), we have

$$\frac{q(\Delta s)^2(\vartheta - \vartheta^{1-M})}{8\sigma^2(\vartheta - 1)} \leq \frac{\epsilon}{4}. \tag{9.25}$$

Combining (9.23) and (9.25) leads to

$$\ln\left(\frac{1}{\delta}\right) \leq \frac{\epsilon^2\sigma^2(\vartheta - 1)}{2q(\Delta s)^2(\vartheta - \vartheta^{1-M})} - \frac{\epsilon}{4} \leq \frac{\epsilon^2\sigma^2(\vartheta - 1)}{2q\Delta s^2(\vartheta - \vartheta^{1-M})}. \tag{9.26}$$

To ensure the (ϵ, δ)-DP of the considered FL system, we choose σ that satisfies

$$\sigma^2 \leq \frac{2q(\Delta s)^2(\vartheta - \vartheta^{1-M})}{\epsilon^2(\vartheta - 1)} \ln\left(\frac{1}{\delta}\right). \tag{9.27}$$

This concludes the proof of the theorem.

9.8 Proof of Theorem 9.2

Based on the definition and composability of the λ-th moment, we have the λ-th moment for the time-varying Gaussian mechanism, as given by

$$\alpha_{\mathcal{M}}(\lambda) = \sum_{n=1}^{M'} \alpha_{\mathcal{M}_n}(\lambda, \sigma_n) \tag{9.28a}$$

$$= \sum_{n=1}^{m} \alpha(\lambda, \sigma_n) + \sum_{n=m+1}^{M'} \alpha(\lambda, \sigma_n) \tag{9.28b}$$

$$= \sum_{n=1}^{m} \frac{q\lambda(\lambda+1)(\Delta s)^2}{2\sigma_n^2} + \sum_{n=m+1}^{M'} \frac{q\lambda(\lambda+1)(\Delta s)^2}{2\sigma_n^2} \tag{9.28c}$$

$$\leq \begin{cases} \dfrac{q\lambda(\lambda+1)(\Delta s)^2}{2} \left(\displaystyle\sum_{n=1}^{m} \frac{1}{\sigma_n^2} + \sum_{n=1}^{M'-m} \frac{1}{\sigma^2} \right), \text{ if } \vartheta \geq 1; \\ \dfrac{q\lambda(\lambda+1)(\Delta s)^2}{2} \left(\displaystyle\sum_{n=1}^{m} \frac{1}{\sigma_n^2} + \frac{\vartheta^{m-M'}}{(1-\vartheta)\sigma^2} \right), \text{ if } \vartheta < 1; \end{cases} \tag{9.28d}$$

$$= \begin{cases} \dfrac{q\lambda(\lambda+1)(\frac{\vartheta - \vartheta^{1-m}}{\vartheta - 1} + M' - m)(\Delta s)^2}{2\sigma^2}, \text{ if } \vartheta > 1; \\ \dfrac{q\lambda(\lambda+1)M'(\Delta s)^2}{2\sigma^2}, \text{ if } \vartheta = 1; \\ \dfrac{q\lambda(\lambda+1)(\vartheta^{1-m} - \vartheta + \vartheta^{m-M'})(\Delta s)^2}{2(1-\vartheta)\sigma^2}, \text{ if } \vartheta < 1, \end{cases} \tag{9.28e}$$

where (9.28d) is because $\sigma_n^2 = \vartheta^{n-1}\sigma^2 \leq \sigma^2$ if $\vartheta \geq 1$, and $\sum_{n=m+1}^{M'} \frac{1}{\sigma_n^2} \leq \frac{1}{\vartheta} \frac{(\frac{1}{\vartheta})^{M'-m} - (\frac{1}{\vartheta})^{m+1}}{\frac{1}{\vartheta - 1}} \leq \frac{\vartheta^{m-M'}}{1-\vartheta}$ if $\vartheta < 1$.

By exploiting the tail bound of the moment [1], we can obtain δ that satisfies the mechanism $\mathcal{M}$, as given by

$$\delta = \min_{\lambda} \exp\left(\alpha_{\mathcal{M}}(\lambda) - \lambda\epsilon\right) \tag{9.29a}$$

$$= \begin{cases} \min_{\lambda} \exp\left[\dfrac{q\lambda(\lambda+1)(\Delta s)^2}{2\sigma^2}\left(\dfrac{\vartheta - \vartheta^{1-m}}{\vartheta - 1} + M' - m\right) - \lambda\epsilon\right], \\ \qquad \text{if } \vartheta > 1; \\ \min_{\lambda} \exp\left[\dfrac{q\lambda(\lambda+1)M'(\Delta s)^2}{2\sigma^2} - \lambda\epsilon\right], \text{ if } \vartheta = 1; \\ \min_{\lambda} \exp\left[\dfrac{q\lambda(\lambda+1)(\Delta s)^2}{2\sigma^2}\left(\dfrac{\vartheta^{1-m} - \vartheta + \vartheta^{m-M'}}{\vartheta - 1}\right) - \lambda\epsilon\right], \\ \qquad \text{if } \vartheta < 1. \end{cases} \tag{9.29b}$$

Following the steps in the proof of Theorem 9.1, we set $\lambda^* = \frac{\epsilon\sigma^2(\vartheta-1)}{q(\Delta s)^2[\vartheta - \vartheta^{1-m} + (M'-m)(\vartheta-1)]} - \frac{1}{2}$ if $\vartheta \geq 1$ and $\lambda^* = \frac{\epsilon\sigma^2(\vartheta-1)}{q(\Delta s)^2[\vartheta - \vartheta^{1-m} + \vartheta^{m-M'}]} - \frac{1}{2}$ if

$\vartheta < 1$. Then, the amplitude of the noise satisfying the (ϵ, δ)-DP of the training can be obtained in (9.9).

9.9 Proof of Theorem 9.3

The proof starts by defining

$$\boldsymbol{\omega}'(t) \triangleq \sum_{k \in \mathcal{K}} p_k \left(\boldsymbol{\omega}_k(t) + \boldsymbol{n}_k(t)\right) = \sum_{k \in \mathcal{K}} p_k \boldsymbol{\omega}_k(t) + \boldsymbol{n}(t), \tag{9.30}$$

where $\boldsymbol{n}(t) = \sum_{k \in \mathcal{K}} p_k \boldsymbol{n}_k(t)$.

Based on the L-Lipschitz smoothness of the global loss function $F(\cdot)$ and the Taylor expansion, it follows that

$$\begin{aligned} F(\boldsymbol{\omega}'(t+1)) - F(\boldsymbol{\omega}'(t)) \le\ & \nabla F\left(\boldsymbol{\omega}'(t)\right)\left(\boldsymbol{\omega}'(t+1) - \boldsymbol{\omega}'(t)\right) \\ & + \frac{L}{2}\left\|\boldsymbol{\omega}'(t+1) - \boldsymbol{\omega}'(t)\right\|^2. \end{aligned} \tag{9.31}$$

According to the gradient descent, we have

$$\boldsymbol{\omega}'_k(t+1) = \boldsymbol{\omega}'_k(t) - \eta \nabla f_k(\boldsymbol{\omega}'(t)). \tag{9.32}$$

By substituting (9.32) and (9.30) into (9.31), we obtain (9.33).

$$F\left(\boldsymbol{\omega}'(t+1)\right) - F\left(\boldsymbol{\omega}'(t)\right) \le \nabla F\left(\boldsymbol{\omega}'(t)\right)\left(\boldsymbol{n}(t+1) - \eta \sum_{k \in \mathcal{K}^m} p_k \nabla f_k\left(\boldsymbol{\omega}'(t)\right)\right) + \frac{L}{2}\left\| \sum_{k \in \mathcal{K}^m} p_k\left(\boldsymbol{n}_k(t+1) - \eta \nabla f_k\left(\boldsymbol{\omega}'_k(t)\right)\right)\right\|^2 \tag{9.33a}$$

$$= \frac{\eta^2 L}{2}\left\| \sum_{k \in \mathcal{K}^m} p_k \nabla f_k\left(\boldsymbol{\omega}'(t)\right)\right\|^2 - \eta \nabla F\left(\boldsymbol{\omega}'(t)\right) \sum_{k \in \mathcal{K}^m} p_k \nabla f_k\left(\boldsymbol{\omega}'(t)\right) + \frac{L}{2}\left\| \sum_{k \in \mathcal{K}^m} p_k \boldsymbol{n}_k(t+1)\right\|^2 \tag{9.33b}$$

Next, we take the expectation on both sides of (9.33) with respect to K randomly chosen users out of the U users at the m-th global aggregation, i.e., $\mathcal{K}^m$, and obtain

$$\mathbb{E}_{\mathcal{K}^m}\left\{F\left(\boldsymbol{\omega}'(t+1)\right)\right\} \le F\left(\boldsymbol{\omega}'(t)\right) - \eta \left\|\nabla F\left(\boldsymbol{\omega}'(t)\right)\right\|^2 \tag{9.34a}$$

$$+ \frac{\eta^2 L}{2}\mathbb{E}_{\mathcal{K}^m}\left\{\left\| \sum_{k \in \mathcal{K}^m} p_k \nabla f_k\left(\boldsymbol{\omega}'(t)\right)\right\|^2\right\} + \frac{L}{2}\mathbb{E}_{\mathcal{K}^m}\left\{\|\boldsymbol{n}(t+1)\|^2\right\}. \tag{9.34b}$$

Since $p_k = \frac{|\mathcal{D}_k|}{|\mathcal{D}|} = \frac{1}{K}$, $\mathbb{E}_{\mathcal{K}^m}\{\| \sum_{k \in \mathcal{K}^m} p_k \nabla f_k(\boldsymbol{\omega}'(t))\|^2\}$ in (9.34) is rewritten as

$$\mathbb{E}_{\mathcal{K}^m}\left\{\Big\| \sum_{k\in\mathcal{K}^m} p_k \nabla f_k\left(\omega'(t)\right) \Big\|^2\right\} = \frac{1}{UK}\left\|\sum_{k\in\mathcal{U}} \nabla f_k\left(\omega'(t)\right)\right\|^2 \tag{9.35a}$$

$$= \frac{1}{UK}\sum_{k\in\mathcal{U}} \left\|\nabla f_k\left(\omega'(t)\right)\right\|^2 + \frac{1}{UK}\frac{K-1}{U-1}\sum_{k\in\mathcal{U}}\sum_{j\in\mathcal{U}/k}\left[\nabla f_k\left(\omega'(t)\right)\right]^T \nabla f_j\left(\omega'(t)\right) \tag{9.35b}$$

$$= \frac{U-K}{UK(U-1)}\sum_{k\in\mathcal{U}} \left\|\nabla f_k\left(\omega'(t)\right)\right\|^2 + \frac{K-1}{UK(U-1)}\left\|\nabla F\left(\omega'(t)\right)\right\|^2. \tag{9.35c}$$

By the definition of $\mathcal{B}$-local dissimilarity, $\mathbb{E}_{\mathcal{K}^m}[\left\|\nabla f_k(\omega')\right\|^2] \leq \mathcal{B}^2 \left\|\nabla F(\omega')\right\|^2$, it readily follows that

$$\mathbb{E}_{\mathcal{K}^m}\left\{\left\|\sum_{k\in\mathcal{K}^m} p_k \nabla f_k\left(\omega'(t)\right)\right\|^2\right\}$$
$$\leq \left[\frac{(U-K)\mathcal{B}^2}{K(U-1)} + \frac{K-1}{UK(U-1)}\right]\left\|\nabla F\left(\omega'(t)\right)\right\|^2. \tag{9.36}$$

Since $\mathbb{E}_{\mathcal{K}^m}\{F(\omega'(t+1))\} = F(\omega'(t+1))$, substituting (9.36) into (9.34) yields

$$F(\omega'(t+1)) \leq F(\omega'(t)) + \phi\|\nabla F(\omega'(t))\|^2 + \frac{L}{2}\mathbb{E}_{\mathcal{U}^t}\{\|\boldsymbol{n}(t+1)\|^2\}, \tag{9.37}$$

where $\phi \triangleq \frac{\eta^2 L}{2}\left(\frac{(U-K)\mathcal{B}^2}{K(U-1)} + \frac{K-1}{UK(U-1)}\right) - \eta$.

Subtracting $F(\omega^*)$ from both sides of (9.37) gives

$$F(\omega'(t+1)) - F(\omega^*) \leq F(\omega'(t)) - F(\omega^*) + \phi\|\nabla F(\omega'(t))\|^2 + \frac{L}{2}\mathbb{E}_{\mathcal{K}^m}\{\|\boldsymbol{n}(t+1)\|\}. \tag{9.38}$$

Considering the Polyak-Lojasiewicz condition and $\mathbb{E}_{\mathcal{K}^m}\{\|\boldsymbol{n}(t+1)\|^2\} = \vartheta^m$ $\mathbb{E}_{\mathcal{K}^m}\{\|\boldsymbol{n}(1)\|^2\} = \vartheta^m\sigma^2$, we obtain

$$\|\nabla F(\omega'(t+1))\|^2 \leq (1+2\rho\phi)\,\|\nabla F(\omega'(t))\|^2 + \rho L\vartheta^m\sigma^2. \tag{9.39}$$

Based on the recurrence expression (9.39), we can obtain the upper bound of $\|\nabla F(\omega'(t))\|^2$, as given by

$$\left\|\nabla F\left(\omega'(t)\right)\right\|^2 \leq (1+2\rho\phi)^m\left\|\nabla F\left(\omega'(0)\right)\right\|^2 \tag{9.40a}$$

$$+ \frac{\rho L\left[\vartheta^m - (1+2\rho\phi)^m\right]}{(\vartheta - 1 - 2\rho\phi)}\sigma^2. \tag{9.40b}$$

Using the Polyak-Lojasiewicz condition again gives

$$F(\omega'(t)) - F(\omega^*) \leq \left[F(\omega'(0)) - F(\omega^*)\right](2\rho\phi + 1)^m \tag{9.41a}$$

$$+ \frac{L\left[\vartheta^m - (2\rho\phi + 1)^m\right]}{2(\vartheta - 2\rho\phi - 1)}\sigma^2, \tag{9.41b}$$

which concludes this proof.

9.10 Proof of Corollary 9.2

Define the RHS of (9.12) as $g(T)$ for the brevity of notation. The second-order derivative of $g(T)$ regarding T is

$$\begin{aligned}\frac{\partial^2 g(T)}{\partial T^2} &= \frac{\Theta}{\tau}\mathcal{A}^{\frac{T}{\tau}} \ln^2(\mathcal{A}) + \frac{qL(\Delta s)^2 \ln\left(\frac{1}{\delta}\right)\vartheta}{\tau^2\epsilon^2(\vartheta - \mathcal{A})(U-1)} \\ &\times \left(\vartheta^{\frac{T}{\tau}} \ln^2(\vartheta) - \mathcal{A}^{\frac{T}{\tau}} \ln^2(\mathcal{A}) + \left(\frac{\mathcal{A}}{\vartheta}\right)^{\frac{T}{\tau}} \ln^2\left(\frac{\mathcal{A}}{\vartheta}\right)\right) \\ &+ L\frac{\partial^2 \mathcal{H}\left(\frac{T}{M}\right)}{\partial T^2},\end{aligned} \tag{9.42}$$

where the third term on the RHS of (9.42) is the second-order derivative of $\mathcal{H}\left(\frac{T}{M}\right)$ with respect to T, i.e.,

$$\frac{\partial^2 \mathcal{H}\left(\frac{T}{M}\right)}{\partial T^2} = \frac{\gamma}{M^2}(\eta L + 1)^{T/M} \ln^2(\eta L + 1) \geq 0.$$

The first and the third terms on the RHS of (9.42) are positive, and the second term is positive when $\vartheta \geq \mathcal{A}$. As a result, $\frac{\partial^2 g(T)}{\partial T^2} > 0$ if $\vartheta \geq \mathcal{A}$, and the upper bound is a convex function of T.

By substituting $M = \frac{T}{\tau}$ into $g(T)$, $g(T)$ can be treated as a function of M, denoted by $\mathcal{G}(M)$. The second-order derivative of $\mathcal{G}(M)$ with respect to M is given by

$$\begin{aligned}\frac{\partial^2 \mathcal{G}(M)}{\partial M^2} &= \Theta\mathcal{A}^M \ln^2(\mathcal{A}) + \frac{qL(\Delta s)^2 \ln\left(\frac{1}{\delta}\right)\vartheta}{\epsilon^2(\vartheta - \mathcal{A})(U-1)} \\ &\times \left(\vartheta^M \ln^2(\vartheta) - \mathcal{A}^M \ln^2(\mathcal{A}) + \left(\frac{\mathcal{A}}{\vartheta}\right)^M \ln^2\left(\frac{\mathcal{A}}{\vartheta}\right)\right) + L\frac{\partial^2 \mathcal{H}\left(\frac{T}{M}\right)}{\partial M^2},\end{aligned} \tag{9.43}$$

where

$$\frac{\partial^2 \mathcal{H}\left(\frac{T}{M}\right)}{\partial M^2} = \frac{\gamma T^2}{M^4}\left(\eta L + 1\right)^{T/M} \ln^2\left(\eta L + 1\right) + \frac{2\gamma T}{M^3}\left(\eta L + 1\right)^{T/M} \ln\left(\eta L + 1\right) - \frac{2\eta\theta T}{M^3} \geq 0,$$

when $M \leq \frac{T \ln(\eta L+1)}{\ln(\eta/\ln(\eta L+1))}$, or $\tau \geq \frac{\ln(\eta/\ln(\eta L+1))}{\ln(\eta L+1)}$.

Like (9.42), we see that (9.43) is positive when $\vartheta \geq \mathcal{A}$ and $\tau \geq \frac{\ln(\eta/\ln(\eta L+1))}{\ln(\eta L+1)}$. We have $\frac{\partial^2 g(M)}{\partial M^2} > 0$ if $\tau \geq \frac{\ln(\eta/\ln(\eta L+1))}{\ln(\eta L+1)}$, and the upper bound is convex in M.

References

1. Abadi M et al (2016) Deep learning with differential privacy. In: Proceedings of the 2016 ACM SIGSAC conference on computer and communications security, pp 308–318
2. Acs G, Melis L, Castelluccia C, De Cristofaro E (2018) Differentially private mixture of generative neural networks. IEEE Trans Knowl Data Eng 31(6):1109–1121
3. Barthe G, Kopf B (2011) Information-theoretic bounds for differentially private mechanisms. In: 2011 IEEE 24th computer security foundations symposium. IEEE, pp 191–204
4. Bassily R, Smith A, Thakurta A (2014) Private empirical risk minimization: efficient algorithms and tight error bounds. In: Proceedings of the 55th annual symposium on foundations of computer science. IEEE, pp 464–473
5. Bhowmick A, Duchi J, Freudiger J, Kapoor G, Rogers R (2018) Protection against reconstruction and its applications in private federated learning. arXiv:1812.00984
6. Boyd S, Boyd SP, Vandenberghe L (2004) Convex optimization. Cambridge University Press
7. Cheng A, Wang P, Zhang XS, Cheng J (2022) Differentially private federated learning with local regularization and sparsification. In: Proceedings of the IEEE/CVF conference on computer vision and pattern recognition, pp 10122–10131
8. Chollet F (2017) Xception: Deep learning with depthwise separable convolutions. In: Proceedings of the IEEE conference on computer vision and pattern recognition, pp 1251–1258
9. Dwork C, Roth A et al (2014) The algorithmic foundations of differential privacy. Found Trends® Theor Comput Sci 9(3–4):211–407
10. Elkordy AR, Zhang J, Ezzeldin YH, Psounis K, Avestimehr S (2022) How much privacy does federated learning with secure aggregation guarantee? arXiv:2208.02304
11. Feng H, Wang R, Liu E, Ni W, Niyato D, Jamalipour A (2025) Over-the-air federated learning with joint privacy-accuracy optimization. IEEE Trans Inf Forensics Secur 1–1. https://doi.org/10.1109/TIFS.2025.3620108
12. Fredrikson M, Jha S, Ristenpart T (2015) Model inversion attacks that exploit confidence information and basic countermeasures. In: Proceedings of the 22nd ACM SIGSAC conference on computer and communications security, pp 1322–1333
13. Frisk O, Dormann F, Lillelund CM, Pedersen CF (2021) Super-convergence and differential privacy: training faster with better privacy guarantees. In: 2021 55th annual conference on information sciences and systems (CISS). IEEE, pp 1–6
14. Fu Y, Zhou Y, Wu D, Yu S, Wen Y, Li C (2021) On the practicality of differential privacy in federated learning by tuning iteration times. arXiv:2101.04163
15. Geyer RC, Klein T, Nabi M (2017) Differentially private federated learning: a client level perspective. arXiv:1712.07557
16. Karimi H, Nutini J, Schmidt M (2016) Linear convergence of gradient and proximal-gradient methods under the polyak-łojasiewicz condition. In: Joint European conference on machine learning and knowledge discovery in databases. Springer, pp 795–811

17. Kennedy J, Eberhart R (1995) Particle swarm optimization. In: Proceedings of ICNN'95-international conference on neural networks, vol 4. IEEE, pp 1942–1948
18. Li T, Sahu AK, Talwalkar A, Smith V (2020) Federated learning: challenges, methods, and future directions. IEEE Signal Process Mag 37(3):50–60. https://doi.org/10.1109/MSP.2020.2975749
19. Lu Y, Huang X, Dai Y, Maharjan S, Zhang Y (2019) Differentially private asynchronous federated learning for mobile edge computing in urban informatics. IEEE Trans Ind Informat 16(3):2134–2143
20. Ma C et al (2020) On safeguarding privacy and security in the framework of federated learning. IEEE Netw 34(4):242–248. https://doi.org/10.1109/MNET.001.1900506
21. McSherry F, Talwar K (2007) Mechanism design via differential privacy. In: 48th annual IEEE symposium on foundations of computer science (FOCS'07). IEEE, pp 94–103
22. Mironov I (2017) Rényi differential privacy. In: 2017 IEEE 30th computer security foundations symposium (CSF). IEEE, pp 263–275
23. Mo F, Haddadi H, Katevas K, Marin E, Perino D, Kourtellis N (2021) PPFL: privacy-preserving federated learning with trusted execution environments. In: Proceedings of the 19th annual international conference on mobile systems, applications, and services, pp 94–108
24. Nasr M, Shokri R, Houmansadr A (2019) Comprehensive privacy analysis of deep learning: passive and active white-box inference attacks against centralized and federated learning. In: 2019 IEEE symposium on security and privacy (SP). IEEE, pp 739–753
25. Nasr M, Shokri R, Houmansadr A (2019) Comprehensive privacy analysis of deep learning: passive and active white-box inference attacks against centralized and federated learning. In: 2019 IEEE symposium on security and privacy (SP). IEEE, pp 739–753
26. Nguyen J, Malik K, Zhan H, Yousefpour A, Rabbat M, Malek M, Huba D (2022) Federated learning with buffered asynchronous aggregation. In: International conference on artificial intelligence and statistics. PMLR, pp 3581–3607
27. O'Searcoid M (2006) Metric spaces. Springer undergraduate mathematics series. Springer, London. https://books.google.com.au/books?id=aP37I4QWFRcC
28. Prakash S, Hashemi H, Wang Y, Annavaram M, Avestimehr S (2022) Secure and fault tolerant decentralized learning. arXiv:2010.07541
29. Qi L, Wu B, Wang F, Li K, Ni W, Jamalipour A (2024) Differentially private energy sharing among smart grid-powered base stations. In: 2024 IEEE 100th vehicular technology conference (VTC2024-Fall). IEEE, pp 1–5
30. Shokri R, Stronati M, Song C, Shmatikov V (2017) Membership inference attacks against machine learning models. In: 2017 IEEE symposium on security and privacy (SP), pp 3–18. https://doi.org/10.1109/SP.2017.41
31. Toch E, Birman Y (2018) Towards behavioral privacy: how to understand AI's privacy threats in ubiquitous computing. In: Proceedings of the 2018 ACM international joint conference and 2018 international symposium on pervasive and ubiquitous computing and wearable computers, pp 931–936
32. Truex S, Liu L, Chow KH, Gursoy ME, Wei W (2020) LDP-Fed: federated learning with local differential privacy. In: Proceedings of the third ACM international workshop on edge systems, analytics and networking, pp 61–66
33. Wang S et al (2019) Adaptive federated learning in resource constrained edge computing systems. IEEE J Sel Areas Commun 37(6):1205–1221
34. Wang Y, Si C, Wu X (2015) Regression model fitting under differential privacy and model inversion attack. In: IJCAI, pp 1003–1009
35. Wei K et al (2020) Federated learning with differential privacy: algorithms and performance analysis. IEEE Trans Inf Forensics Secur 15:3454–3469. https://doi.org/10.1109/TIFS.2020.2988575
36. Wei K et al (2022) Low-latency federated learning over wireless channels with differential privacy. IEEE J Sel Areas Commun 40(1):290–307. https://doi.org/10.1109/JSAC.2021.3126052

37. Wei K et al (2022) User-level privacy-preserving federated learning: analysis and performance optimization. IEEE Trans Mobile Comput 21(9):3388–3401. https://doi.org/10.1109/TMC.2021.3056991
38. Wu M et al (2021) Incentivizing differentially private federated learning: a multi-dimensional contract approach. IEEE Internet Things J
39. Yang Q, Liu Y, Chen T, Tong Y (2019) Federated machine learning: concept and applications. ACM Trans Intell Syst Technol 10(2):1–19
40. Yu L, Liu L, Pu C, Gursoy ME, Truex S (2019) Differentially private model publishing for deep learning. In: 2019 IEEE symposium on security and privacy (SP), pp 332–349. https://doi.org/10.1109/SP.2019.00019
41. Zhao Y et al (2020) Local differential privacy based federated learning for internet of things. IEEE Internet Things J 1–1. https://doi.org/10.1109/JIOT.2020.3037194
42. Zhu T, Li G, Zhou W, Philip SY (2017) Differentially private data publishing and analysis: a survey. IEEE Trans Knowl Data Eng 29(8):1619–1638

Part IV
Emerging Trends and Future Directions

Chapter 10
Integrating Graph Neural Networks with Federated Learning

The integration of GNNs with FL has emerged as a compelling paradigm for building resilient, secure, and adaptive distributed learning systems. While FL addresses privacy and scalability challenges by enabling decentralized model training without direct data sharing, it struggles with issues, such as non-IID data distributions, system heterogeneity, and vulnerability to adversarial attacks. GNNs, on the other hand, are well-suited to capture relational structures and higher-order dependencies [21], making them a natural complement to FL in contexts where clients, data, and updates exhibit graph-like interactions [9]. By embedding FL into graph-structured frameworks, researchers can leverage structural insights to improve robustness against poisoning and Byzantine attacks, enhance personalization for heterogeneous clients, and optimize communication efficiency in large-scale, dynamic environments [11]. This chapter explores the convergence of GNNs and FL, examining how graph-based techniques can enrich FL with new capabilities for resilience, adaptability, and trustworthiness across diverse application domains, such as IoT, healthcare, autonomous systems, and the Metaverse.

10.1 Centralized and Decentralized FL-GNNs

Based on the availability of explicit graph information, the taxonomy of integrating GNN and FL can be broadly divided into two primary categories: GNN-assisted FL and FL-assisted GNNs. Each category can be broken down into specific scenarios depending on system configurations and deployment settings. Within these scenarios, different integration strategies are proposed, along with tailored solutions to address critical challenges. For clarity, we first provide a concise introduction to the two overarching categories of this taxonomy.

GNN-assisted FL can be categorized into two main scenarios: centralized FL-GNNs and decentralized FL-GNNs. In the centralized case, where both a server and

K. Li et al., *Security and Resilience in Distributed Machine Learning*, Springer Series in Reliability Engineering, https://doi.org/10.1007/978-3-032-23959-4_10

clients are present, two subscenarios arise depending on the placement of the implicit graph: (1) the graph is maintained on the server [14], or (2) the graph is constructed locally by the clients [12]. In contrast, decentralized FL-GNNs operate without a central server, meaning the implicit graph has to be distributed solely among the clients. For each subscenario, we begin by outlining the underlying assumptions and general integration strategies, followed by a presentation of the key challenges along with the techniques proposed to address them.

10.1.1 Centralized FL-GNNs

In centralized FL-GNNs, a central server is responsible for coordinating client participation. Depending on where the implicit graph resides, GNN training can take place directly on the server. In this setup, the server constructs an inter-client graph under the assumption that neighboring clients share similar local models or feature embeddings. Model parameters uploaded by clients, collected in the same manner as standard FL, are represented as node features within this graph [17]. The server then trains a GNN using these features to guide the aggregation process, after which the refined parameters are redistributed to the clients. The inter-client graph may be predefined or dynamically learned through mechanisms such as self-attention during training. A key difficulty in this approach lies in bilevel optimization, as the server needs to jointly manage the training of both local client models and the global GNN, making convergence more challenging.

10.1.2 Decentralized FL-GNNs

Decentralized FL-GNNs operate without a central server, meaning coordination among clients is achieved solely through peer-to-peer communication. This setting involves a single subscenario in which the implicit graph is distributed across the clients, representing only the inter-client relationships. Under this assumption, each client can exchange information exclusively with its neighbors, making decentralized aggregation a key challenge [5]. Efficient neighbor-to-neighbor communication is also essential, as repeated exchanges may otherwise cause excessive overhead. To address these issues, prior work has introduced two main strategies for graph-based decentralized aggregation [20]: (a) updating model parameters through weighted summation of neighboring updates according to the underlying graph structure, and (b) leveraging graph regularization to refine the aggregation process.

In weighted-summation approaches, clients iteratively share parameters with their neighbors and update their local models by averaging these inputs based on graph topology. Because this requires communication with all neighbors, efficiency becomes a pressing concern. For instance, Rizk and Sayed [13] proposed a multi-server FL-GNNs framework that improves scalability by deploying multiple servers arranged in a fixed graph topology rather than relying on a single central coordinator. Within each server, clients follow a standard centralized FL process, and the servers subsequently aggregate updates among themselves. In addition, PSO-GFML [4] reduces communication cost by allowing only partial parameter exchanges with servers. Beyond static graph assumptions, recent studies have explored learning the inter-client graph dynamically during training, drawing inspiration from GATs. In this approach, edge weights are computed adaptively, often based on similarity between unlabeled embeddings or hidden parameters from participating clients.

10.2 Potential for Enhancing Resilience and Robustness

The integration of GNNs into FL frameworks offers transformative potential for enhancing resilience and robustness in distributed ML systems. FL was originally conceived to preserve privacy and reduce communication overhead by training local models on client data and aggregating updates at a central server. However, the inherent heterogeneity of data across clients, adversarial threats, such as MP or backdoor attacks, and the instability caused by dynamic participation in large-scale networks introduce significant challenges to its reliability [1]. GNNs provide a natural structural lens to address these issues, as they are designed to capture relationships and dependencies in graph-structured data. By constructing inter-client graphs where clients represent nodes and their similarities (in terms of data distributions, parameter updates, or communication patterns) form edges, GNNs can exploit relational information that traditional distance-based defenses overlook. This ability to model higher-order dependencies allows FL systems to better distinguish between benign and malicious updates, as adversarial manipulations often fail to preserve complex structural correlations.

Beyond security, graph-based modeling also addresses the non-IID data problem by enabling clients with similar distributions to share knowledge more effectively through localized graph neighborhoods. This creates a more balanced trade-off between personalization and global generalization, supporting robust performance across diverse applications such as smart healthcare, autonomous driving, and metaverse ecosystems [7]. By leveraging GNNs as an overlay framework for FL, resilience is not only about filtering out anomalies but also about reinforcing collaborative strength through the structural properties of graphs.

10.2.1 Malicious Models Detection

A major potential of GNN-assisted FL lies in strengthening security mechanisms to counter adversarial behaviors, including poisoning, inference, and Byzantine attacks [2]. Traditional defenses in FL often rely on statistical outlier detection, such as Euclidean distance [16], cosine similarity [24], or median-based aggregation [10], which are effective only under simplistic threat models. GNNs extend this defense landscape by enabling graph-based anomaly detection, where local updates are analyzed not in isolation but in relation to their neighborhoods. For example, a malicious client attempting to craft poisoned gradients that mimic benign updates may succeed against distance-based metrics but will struggle to reproduce consistent structural relationships across multiple neighbors in the inter-client graph. This relational context increases detection accuracy and reduces false positives. Furthermore, GNNs can support dynamic graph construction, where attention mechanisms recalibrate edge weights during training to emphasize reliable clients and diminish the influence of potentially adversarial ones.

In addition to detecting malicious updates, GNNs enhance robustness through graph regularization, where structural priors constrain model updates to remain consistent with observed relational patterns [6]. This not only mitigates attacks but also improves fault tolerance in noisy environments by smoothing updates across similar clients. Privacy preservation also benefits from graph-based defenses: by embedding sensitive data distributions into anonymized graph structures, direct access to raw data or local embeddings can be avoided, limiting opportunities for inference or reconstruction attacks. Importantly, explainable AI (XAI) techniques, such as GradCAM [22] or LayerCAM [23] visualizations integrated into GNN pipelines, further strengthen trust by offering interpretable evidence for why certain updates are flagged as malicious. This transparency is vital in domains like healthcare or finance, where resilience should coexist with accountability. Therefore, the security potential of integrating GNNs with FL lies in moving beyond point-based defenses to context-aware resilience, where adversarial manipulations are countered not only statistically but also structurally.

10.2.2 Heterogeneity, Scalability, and Adaptability

Beyond security, the synergy between GNNs and FL holds substantial promise for tackling system-level challenges in heterogeneity, scalability, and adaptability. One of the persistent difficulties in FL is the heterogeneity of client models and data distributions, especially in cross-device environments such as IoT, where clients vary widely in hardware capacity, connectivity, and dataset composition [8]. GNNs offer a unifying mechanism to abstract this diversity into graph representations. For instance, by modeling neural architectures as graphs where layers represent nodes and computational flows form edges, graph hypernetworks can align heterogeneous models into a shared latent space, enabling aggregation without requiring identical client architectures. This innovation facilitates broader participation across diverse devices while preserving training stability.

Scalability is also enhanced, as hierarchical or partitioned graph structures can be used to cluster clients into subgraphs for localized aggregation, reducing synchronization delays and communication costs. Empirical studies have already demonstrated that hierarchical aggregation combined with graph regularization can lower synchronization latency by 30–40% while sustaining accuracy under non-IID distributions [15], a critical advantage for latency-sensitive applications like autonomous vehicles or AR/VR metaverse services [18, 19].

Moreover, GNN-based peer-to-peer FL settings allow serverless architectures, where clients exchange updates directly with their neighbors based on graph topologies [3]. While communication efficiency and robustness remain challenges in these settings, GNNs can adaptively prune or re-weight edges to maintain efficiency and resilience. Importantly, graph-based mechanisms support continual learning in dynamic environments: when new clients join or leave, graph attention networks can reconfigure relationships without requiring complete retraining, thereby maintaining adaptability in real-world deployments. Thus, the integration of GNNs with FL extends beyond enhancing security; it can provide a scalable, flexible foundation for resilient learning across diverse and evolving ecosystems.

References

1. Bi S, Yuan X, Hu S, Li K, Ni W, Hossain E, Wang X (2025) Resilience and failure analysis in next-generation communication networks: a contemporary survey. IEEE Trans Netw Sci Eng
2. Biadgligne Y, Baghoussi Y, Li K, Jorge A (2025) Resilience under attack: Benchmarking optimizers against poisoning in federated learning for image classification using CNN. In: International work-conference on artificial neural networks. Springer, pp 27–40
3. Chen G, Li K, Abdelmoniem AM, You L (2024) Exploring representational similarity analysis to protect federated learning from data poisoning. In: Companion proceedings of the ACM web conference 2024, pp 525–528
4. Gogineni VC, Werner S, Huang YF, Kuh A (2022) Decentralized graph federated multitask learning for streaming data. In: 2022 56th annual conference on information sciences and systems (CISS). IEEE, pp 101–106
5. He C, Ceyani E, Balasubramanian K, Annavaram M, Avestimehr S (2022) Spreadgnn: Decentralized multi-task federated learning for graph neural networks on molecular data. In: Proceedings of the AAAI conference on artificial intelligence, vol 36, pp 6865–6873
6. Kasyap H, Tripathy S (2024) Sine: similarity is not enough for mitigating local model poisoning attacks in federated learning. IEEE Trans Dependable Secure Comput 21(5):4481–4494
7. Li K, Li C, Yuan X, Li S, Zou S, Ahmed SS, Ni W, Niyato D, Jamalipour A, Dressler F et al (2025) Zero-trust foundation models: a new paradigm for secure and collaborative artificial intelligence for internet of things. IEEE Internet Things J
8. Li K, Liang Y, Yuan X, Ni W, Crowcroft J, Yuen C, Akan OB (2025) A novel framework of horizontal-vertical hybrid federated learning for edgeIoT. IEEE Netw Lett
9. Li K, Zhang Z, Pourkabirian A, Ni W, Dressler F, Akan OB (2025) Towards resilient federated learning in cyberedge networks: recent advances and future trends. arXiv:2504.01240
10. Li X, Qu Z, Zhao S, Tang B, Lu Z, Liu Y (2021) Lomar: a local defense against poisoning attack on federated learning. IEEE Trans Dependable Secure Comput 20(1):437–450
11. Liu R, Xing P, Deng Z, Li A, Guan C, Yu H (2024) Federated graph neural networks: overview, techniques, and challenges. IEEE Trans Neural Netw Learn Syst
12. Ma X, Hu J, Liang S, Wu Y (2025) Federated learning and resource-aware graph neural network for intrusion detection in 6G-IoT driven healthcare system. IEEE Internet Things J
13. Rizk E, Sayed AH (2021) A graph federated architecture with privacy preserving learning. In: 2021 IEEE 22nd international workshop on signal processing advances in wireless communications (SPAWC). IEEE, pp 131–135

14. Sarkar A, Sen T, Roy AK (2021) Grafehty: graph neural network using federated learning for human activity recognition. In: 2021 20th IEEE international conference on machine learning and applications (ICMLA). IEEE, pp 1124–1129
15. Su L, Zhou R, Wang N, Chen J, Li Z (2023) Low-latency hierarchical federated learning in wireless edge networks. IEEE Internet Things J 11(4):6943–6960
16. Tian P, Liao W, Yu W, Blasch E (2022) WSCC: a weight-similarity-based client clustering approach for non-IID federated learning. IEEE Internet Things J 9(20):20243–20256
17. Wang C, Chen B, Li G, Wang H (2023) Automated graph neural network search under federated learning framework. IEEE Trans Knowl Data Eng 35(10):9959–9972
18. Wen Y, Zou S, Sun Y, Gao H (2024) Adaptive network function chain orchestration strategy for 6g inclusive intelligent services. In: Proceedings of the 19th workshop on mobility in the evolving internet architecture, pp 7–12
19. Wen Y, Zou S, Sun Y, Ni W, Li K, Gao H (2025) Adaptive network function chain orchestration strategy for 6g inclusive intelligent services. IEEE Trans Cogn Commun Netw 11(5):3515–3528. https://doi.org/10.1109/TCCN.2025.3532304
20. Yu G, Wang X, Sun C, Wang Q, Yu P, Ni W, Liu RP (2025) IronForge: an open, secure, fair, decentralized federated learning. IEEE Trans Neural Netw Learn Syst 36(1):354–368. https://doi.org/10.1109/TNNLS.2023.3329249
21. Yu X, Lv T, Li W, Ni W, Niyato D, Hossain E (2025) Multi-task semantic communication with graph attention-based feature correlation extraction. IEEE Trans Mob Comput 24(5):4371–4388. https://doi.org/10.1109/TMC.2025.3525477
22. Zheng J, Li K, Yuan X, Ni W, Tovar E (2024) Detecting poisoning attacks on federated learning using gradient-weighted class activation mapping. In: Companion proceedings of the ACM web conference 2024, pp 714–717
23. Zheng J, Li K, Yuan X, Ni W, Tovar E, Crowcroft J (2024) Exploring visual explanations for defending federated learning against poisoning attacks. In: Proceedings of the 30th annual international conference on mobile computing and networking, pp 1596–1598
24. Zhu T, Guo Z, Yao C, Tan J, Dou S, Wang W, Han Z (2024) Byzantine-robust federated learning via cosine similarity aggregation. Comput Netw 254:110730

Chapter 11
Towards Fully Explainable Federated Learning

This chapter outlines a practical path toward fully explainable FL: we connect client-level attributions and visual explanations (e.g., LayerCAM-based heat maps) with representation-level diagnostics and lightweight, server-side screening that informs aggregation, fairness, and privacy decisions. Emphasizing deployment-ready tools with minimal overhead, we show how explainability improves robustness to poisoning, accelerates convergence under Non-IID data, and supports energy- and carbon-aware orchestration, thereby turning transparency from a reporting afterthought into a core mechanism for trustworthy FL.

11.1 Importance of Interpretability in Defense Mechanisms

FL has recently emerged as a privacy-preserving distributed ML paradigm that allows multiple clients to collaboratively train a shared model under the orchestration of a server without sharing their raw data [7], as shown in Fig. 11.1. While FL is able to enhance user data privacy protection, the distributed architecture of FL makes it particularly susceptible to poisoning attacks [2]. Clients compromised by adversarial actors can alter model update parameters, subsequently contaminating the global FL models [4]. Existing countermeasures, using Euclidean distance to distinguish the deviations between malicious and benign models, have been proven to effectively resist such attacks [1, 6]. However, attackers can carefully fabricate malicious local model updates such that the Euclidean distances to benign counterparts are below a specified threshold, thereby evading defense detection that relies on this metric.

MP is a central threat in FL: compromised clients can submit updates that subtly steer the global model toward adversarial objectives while remaining statistically inconspicuous. Classical defenses-distance-based outlier filters, pairwise similarity tests, or black-box anomaly detectors, operate primarily in parameter space. Although useful under benign or IID-like conditions, these mechanisms degrade in realistic

K. Li et al., *Security and Resilience in Distributed Machine Learning*, Springer Series in Reliability Engineering, https://doi.org/10.1007/978-3-032-23959-4_11

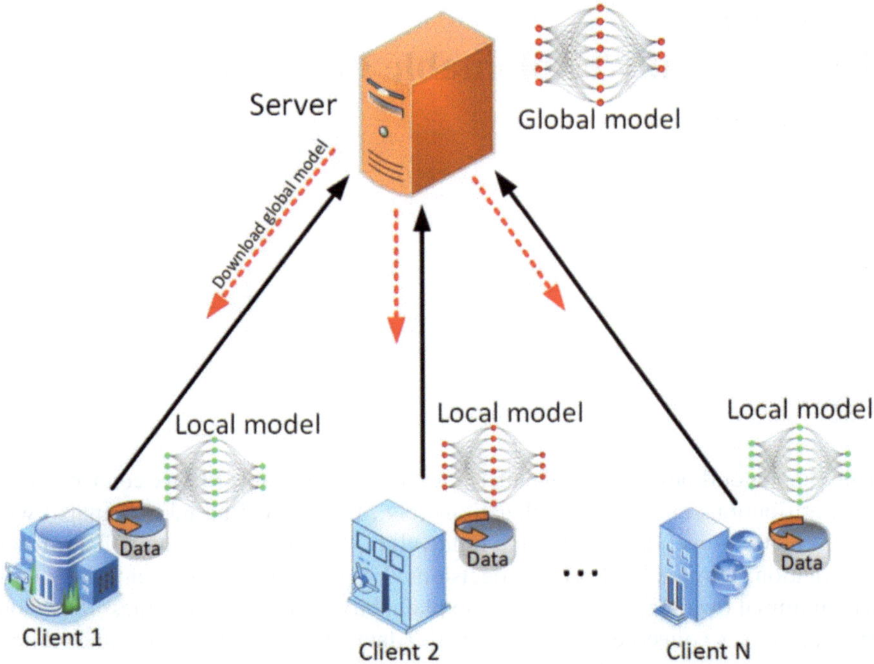

Fig. 11.1 An application of FL in banking, where multiple clients jointly train a global ML model without exposing their private and sensitive data

deployments where (i) attackers craft perturbations that stay within preset Euclidean or cosine thresholds, (ii) heterogeneity induces natural dispersion that overlaps with attack manifolds, and (iii) high dimensionality and class imbalance erode the reliability of generic detectors. A more reliable lens is *interpretability*: assess *what* evidence a model uses and *where* it attends, not merely how far its weights moved.

We adopt an interpretability-first stance through **FedCAMAE**, a visual explanation-guided defense that interrogates the *representational behavior* of each uploaded update using a consistent, server-held probe, as shown in Fig. 11.2. Rather than inferring intent from weight deltas, the server renders per-update class activation patterns via LayerCAM, producing fine-grained heat maps that reveal the spatial evidence the update would rely on for a fixed input. These maps serve as standardized, model-agnostic explanations: benign updates tend to preserve salient regions and relational structure, whereas poisoned updates often shift or diffuse attention in ways that are hard to conceal with small norm changes. To make this signal operational at scale, FedCAMAE embeds an AE that learns the manifold of benign explanations and flags deviations by reconstruction error, thereby converting human-auditable cues into a quantitative screening criterion. Because the probe is server-held and the computation is a single forward pass per update, the approach adds negligible overhead and does not weaken privacy guarantees.

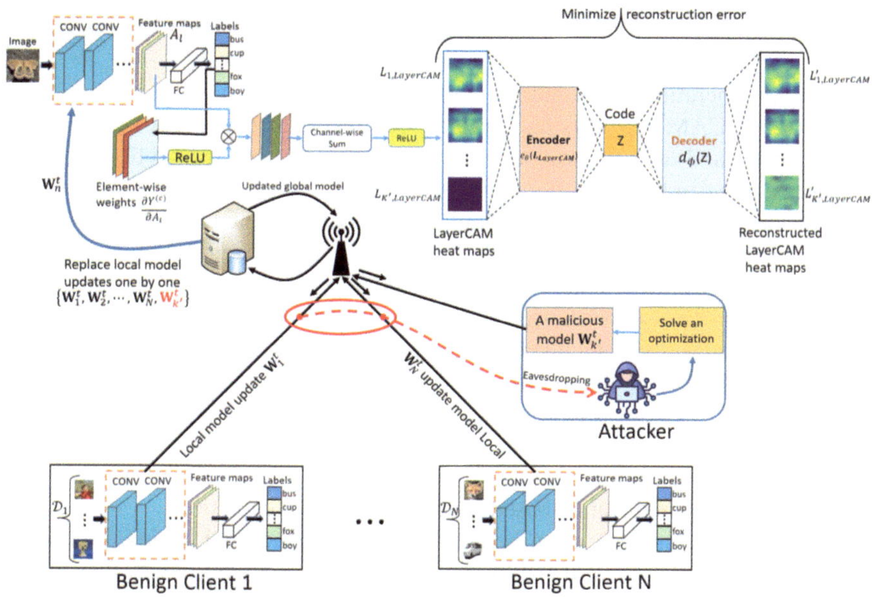

Fig. 11.2 Illustration of the proposed FedCAMAE architecture, where the server arbitrarily selects an image (e.g., an image with the label "butterfly") from the global model testing dataset to create a heat map for every uploaded model update. These heat maps flow into an AE for anomaly detection. The heat map corresponding to the local model update exhibits a large reconstruction error and is deemed malicious

The payoff is practical and multifold. First, explanation-level comparisons are inherently more robust to non-IID variation: heterogeneous but honest clients can differ in parameters yet converge on similar evidence patterns for canonical probes. Second, the method scales across architectures (e.g., ResNet-50, RegNetY-800MF) and datasets (SVHN, CIFAR-100) without redesigning distance metrics or retraining task-specific detectors. Third, the resulting screening improves both security and systems outcomes: by excluding low-fidelity, adversarial updates before aggregation, FedCAMAE preserves test accuracy, accelerates convergence, and prevents wasteful remedial rounds, benefiting fairness, privacy budgeting, and energy/carbon objectives alike. In short, interpretability is not merely a reporting aid but a *decision-making primitive* for trustworthy FL, transforming explanations into actionable defenses that withstand adaptive, low-norm poisoning in heterogeneous environments.

11.2 Visual and Analytical Techniques

FedCAMAE operationalizes interpretability through a two-stage visual-analytical pipeline: LayerCAM produces fine-grained, class-focused heat maps for each local model update, and an AE analyzes these heat maps to amplify subtle but telling differences between benign and malicious behavior.

On the device side, each of the K benign clients utilizes the DNN tailored specifically for image classification tasks, as shown in Fig. 11.2. The DNN model extracts relevant features from an input image (e.g., a butterfly) through convolution (CONV) and pooling operations and subsequently maps them to the corresponding classes through fully connected (FC) layers. The attacker overhears the local model updates uploaded by the benign clients. In this work, we aim to defend against the GAE-based MP attacks [4]. Specifically, the attacker creates and uploads a malicious local model update to progressively contaminate the global model and subsequently the local models of the benign devices while remaining imperceptible by the server, which uses Euclidean distance-based or similarity-based detection.

The server randomly picks an image from the global model testing dataset that encompasses all categories of the clients' dataset as input and passes through the convolutional layers with weight and bias parameters that are replaced by each local model update $\mathbb{W}_l$, obtains the feature maps A_l, which are fed into a fully connected layer for final classification, as shown in Fig. 11.2. The heat map is computed as follows:

$$L_{l,LayerCAM}^{(c)}(i,j) = ReLU\Big(\sum_{k=1}^{K} \alpha_{l,k}^{(c)}(i,j) \cdot A_l^k(i,j)\Big) \tag{11.1}$$

with the importance of element-wise weights $\alpha_{l,k}^{(c)}(i,j) = ReLU\Big(\frac{\partial Y^{(c)}}{A_l^k(i,j)}\Big)$, where $A_l^k(i,j)$ is the activation of node in the target layer of the model at position (i,j), $Y^{(c)}$ is the model output score for class c before softmax, and $k \in [1, K]$ is the index of the channels of the feature maps [3].

Each heat map $L_{l,LayerCAM}^{(c)}$ is flattened into a vector and concatenated with the vectors of other heat maps as the input to the encoder. During the AE training, the encoder compresses the heat maps from a high-dimensional space to a low-dimensional space Z, also called the code or the latent space. The code learns the underlying features or representation of the heat maps, which are input to the decoder for reconstructing the original heat maps. The reconstructed heat map with a significant error compared to the original heat map is flagged as abnormal.

11.3 Performance Evaluation

We use the SVHN and CIFAR-100 datasets to evaluate the performance of our proposed FedCAMAE. Two existing schemes of **Multi-Krum** [1] and **FAA-DL** [5] are compared. **Multi-Krum**: computes a score of the sum of each local model update to its neighbor's Euclidean distance. The score is the sum of its Euclidean distance from its neighbors; those with high scores are excluded as malicious model updates. **FAA-DL** [5]: a lightweight, unsupervised anomaly detection method based on a one-class SVM-based method, support vector machine, which utilizes an appropriate kernel function and soft margins to estimate a nonlinear decision boundary and separate the benign and malicious local model updates. There are 21 benign clients and 3 attackers.

Figure 11.3 plots the test accuracy of ResNet-50 on Non-IID SVHN and Non-IID CIFAR-100 datasets, demonstrating the superiority of FedCAMAE. Under the Non-IID SVHN setting, FedCAMAE achieves the highest test accuracy (0.95) of the FL global model and converges quickly (around the 20th communication round) as it involves more benign clients in FL training. This indicates that FedCAMAE can accurately filter malicious model updates.

Multi-Krum directly detects millions or even more parameters of local model updates based on Euclidean distance. The Euclidean distance between carefully designed malicious local model updates and benign local model updates is within the threshold specified by the server. This suggests that malicious local model updates can evade server detection and participate in the FL training process through multiple rounds, leading to the corruption of the global model.

FAA-DL also directly classifies local model updates collected by servers as benign and malicious. However, there are two key reasons why FAA-DL failed in the experiments. Firstly, local model updates have the characteristics of high-dimensional feature spaces. The curse of dimensionality can lead to data sparsity, making it difficult for FAA-DL to find a suitable boundary to distinguish between benign and malicious local model updates. Secondly, FAA-DL is sensitive to class imbalance (21 benign local model updates and 3 malicious local model updates), which means that FAA-DL tends its decision boundary towards the majority class, making it struggle to effectively detect updates of malicious local models. We obtain the same trend even if we replace the SVHN dataset with the CIFAR-100 dataset.

We replaced the ResNet-50 model with RGENTY-800MF, and the trend of FL global model testing accuracy with communication rounds is consistent with Fig. 11.3, as shown in Fig. 11.4. The experimental results indicate that although the complexity of local data plays a crucial role, the complexity of the DNN model itself has a limited impact on the malicious detection of FedCAMAE.

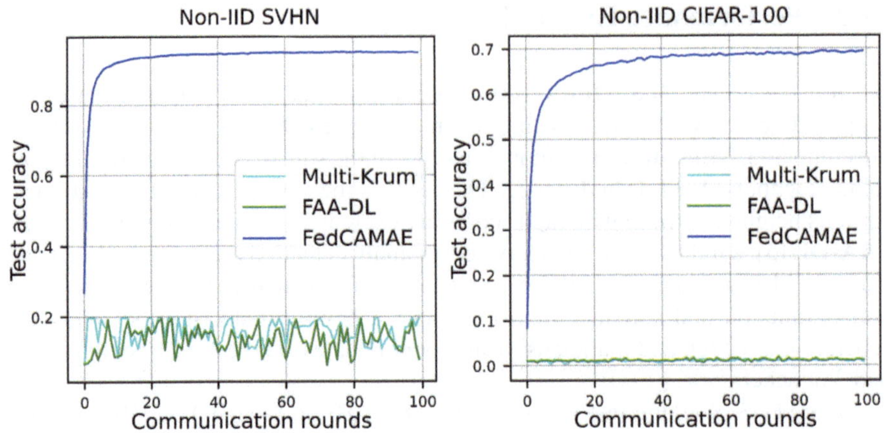

Fig. 11.3 Test accuracy versus communication rounds for **ResNet-50** on Non-IID SVHN and CIFAR-100

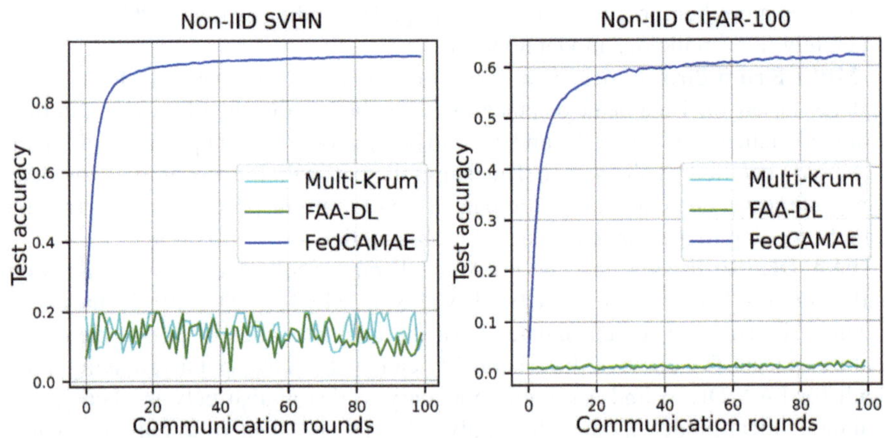

Fig. 11.4 Test accuracy versus communication rounds for **RGENETY-800MF** on Non-IID SVHN and CIFAR-100

11.4 Summary

In this chapter, we propose FedCAMAE, a novel and robust shield defense against poisoning attacks on FL. LayerCAM was leveraged to process the received local model updates with a selected image from the test dataset, generating the corresponding heat maps. An AE was seamlessly embedded in FedCAMAE to fortify the hidden features of the heat maps. It was demonstrated experimentally that FedCAMAE significantly outperforms the existing defense schemes under the same setting tested.

References

1. Blanchard P, El Mhamdi EM, Guerraoui R, Stainer J (2017) Machine learning with adversaries: Byzantine tolerant gradient descent. In: Proceedings of the 31st international conference on neural information processing systems, pp 118–128
2. Fang M, Cao X, Jia J, Gong NZ (2020) Local model poisoning attacks to byzantine-robust federated learning. In: Proceedings of the 29th USENIX conference on security symposium, pp 1623–1640
3. Jiang PT, Zhang CB, Hou Q, Cheng MM, Wei Y (2021) Layercam: exploring hierarchical class activation maps for localization. IEEE Trans Image Process 30:5875–5888. https://doi.org/10.1109/TIP.2021.3089943
4. Li K, Zheng J, Yuan X, Ni W, Akan OB, Poor HV (2024) Data-agnostic model poisoning against federated learning: a graph autoencoder approach. IEEE Trans Inf Forensics Secur
5. Shi S, Hu C, Wang D, Zhu Y, Han Z (2022) Federated anomaly analytics for local model poisoning attack. IEEE J Sel Areas Commun 40(2):596–610. https://doi.org/10.1109/JSAC.2021.3118347
6. Zhang Z, Cao X, Jia J, Gong NZ (2022) Fldetector: defending federated learning against model poisoning attacks via detecting malicious clients. In: Proceedings of the 28th ACM SIGKDD conference on knowledge discovery and data mining, pp 2545–2555
7. Zheng J, Li K, Mhaisen N, Ni W, Tovar E, Guizani M (2022) Exploring deep-reinforcement-learning-assisted federated learning for online resource allocation in privacy-preserving EdgeIoT. IEEE Internet Things J 9(21):21099–21110

Chapter 12
Sustainability in Federated Learning

FL can reduce data-transfer volumes and mitigate privacy risks by pushing computation to the edge; however, it may also increase energy use and carbon emissions if naively orchestrated across heterogeneous devices and networks. This chapter formalizes sustainability objectives for FL, connects them to the United Nations Sustainable Development Goals (UN SDGs), and develops concrete, measurable strategies for energy- and carbon-aware FL pipelines. We conclude with a robustness module based on *Representational Similarity Analysis* (RSA) that prevents wasteful training under data poisoning, thereby improving both accuracy and sustainability.

12.1 Aligning with UN SDGs

FL advances multiple UN SDGs when designed with resource awareness and robustness in mind. At its core, FL decentralizes computation and data, which can lower privacy risk and reduce wide-area data transfer, but it also introduces new orchestration challenges that affect energy and carbon outcomes. The following mapping clarifies how concrete design choices in FL contribute to specific SDGs and why they must be considered jointly.

SDG 7 (Affordable and Clean Energy)

Distributing training across heterogeneous edge devices can unintentionally raise aggregate energy use if participation is not coordinated. Carbon- and energy-aware client selection and pacing address this by deciding *when* and *where* to train to minimize energy intensity while meeting utility targets.

K. Li et al., *Security and Resilience in Distributed Machine Learning*, Springer Series in Reliability Engineering, https://doi.org/10.1007/978-3-032-23959-4_12

Table 12.1 Sustainability levers across the FL stack

Layer	Representative levers
Model	Pruning, quantization, low-rank adapters, distillation
Client	Adaptive batch/epochs, early stopping, on-device validation
Server	Energy/carbon-aware selection & pacing; partial participation
Comm.	Sparse/sketched updates; error-feedback; delta encoding
Robustness	RSA-based filtering to avoid wasteful poisoned rounds
Operations	Time-of-use/geo carbon scheduling; renewable-aligned windows

SDG 9 (Industry, Innovation, and Infrastructure)

By reducing raw-data centralization, FL enables resilient, privacy-preserving analytics over constrained networks. A sustainable FL stack reinforces this with efficient edge infrastructure, such as low-power accelerators and on-device quantization, and with robust aggregation that limits rework due to noisy or adversarial updates.

SDG 12 (Responsible Consumption and Production)

Responsible use of compute and network resources is operationalized through model compression, sparsity, adaptive participation, and communication-efficient aggregation. These measures reduce both the embodied and operational footprints of training and update exchange.

SDG 13 (Climate Action)

A carbon-aware FL controller, incorporating grid mix, time-of-use carbon intensity, and renewable availability, directly lowers per-round CO_2e. Coupling this with robustness to poisoning prevents wasted rounds and rollbacks, further cutting avoidable emissions. Overall, these levers, as summarized in Table 12.1, frame sustainability as a first-class objective in FL: utility gains are pursued alongside explicit targets on energy and carbon, with robustness ensuring that progress is not undone by ineffective or adversarial training.

12.2 Strategies for Energy-Efficient Systems

Designing sustainable FL requires a clear view of where energy and carbon are spent, followed by orchestration policies that trade utility for cost in a principled way. We first introduce simple accounting models, then formulate round-level scheduling as a multi-objective problem, and finally outline practical levers on the model, client, server, and communication stack to reduce both energy and emissions.

12.2.1 Energy and Carbon Accounting Models

Consider an FL process running for T rounds over clients $\mathcal{N}$. In round t, the server selects $S_t \subseteq \mathcal{N}$. For client $i \in S_t$, we decompose total energy into on-device training, communication, and overhead:

$$E_{i,t} \;=\; E_{i,t}^{\text{train}} + E_{i,t}^{\text{comm}} + E_{i,t}^{\text{over}}. \tag{12.1}$$

Training energy scales with computational work and device efficiency:

$$E_{i,t}^{\text{train}} \approx \frac{\text{FLOPs}_{i,t}}{\eta_i^{\text{hw}}}, \qquad \text{FLOPs}_{i,t} = \kappa\, B_{i,t}\, E_{i,t}^{\text{loc}}\, \text{MACs}(w_t), \tag{12.2}$$

where $B_{i,t}$ is local batch size, $E_{i,t}^{\text{loc}}$ the number of local epochs, $\text{MACs}(w_t)$ per-example multiply–accumulate operations for the current global model, κ a proportionality constant, and η_i^{hw} the device's energy efficiency.

Communication energy depends on payload size and radio characteristics:

$$E_{i,t}^{\text{comm}} \;\approx\; \gamma_i^{\uparrow}\, \text{bits}_{i,t}^{\uparrow} \;+\; \gamma_i^{\downarrow}\, \text{bits}_{i,t}^{\downarrow}, \tag{12.3}$$

with $\gamma_i^{\uparrow}, \gamma_i^{\downarrow}$ capturing uplink/downlink energy per bit.

Carbon emissions follow from local or regional carbon intensity $I_{i,t}$ (kg CO_2e/kWh):

$$C_{i,t} \;=\; \xi\, I_{i,t}\, E_{i,t}, \qquad C_{\text{total}} \;=\; \sum_{t=1}^{T} \sum_{i \in S_t} C_{i,t}, \tag{12.4}$$

where ξ converts Joules to kWh. If only a grid-level signal is available, set $I_{i,t} \equiv I_t$.

12.2.2 Multi-objective Round Orchestration

Given these accounts, round t selection balances expected utility against energy and carbon costs. Let $\Delta A_t(S_t)$ denote the expected accuracy gain from clients S_t. A per-round objective with tunable weights $\lambda_E, \lambda_C \geq 0$ is

$$U_t(S_t) \;=\; \Delta A_t(S_t) \;-\; \lambda_E \sum_{i \in S_t} E_{i,t} \;-\; \lambda_C \sum_{i \in S_t} C_{i,t}. \tag{12.5}$$

Selection then becomes a budgeted knapsack:

$$\max_{S_t \subseteq \mathcal{N}} U_t(S_t) \quad \text{s.t.} \quad \sum_{i \in S_t} E_{i,t} \leq \bar{E}_t, \quad \sum_{i \in S_t} C_{i,t} \leq \bar{C}_t, \quad \Phi(S_t) \leq \bar{\epsilon}_t, \tag{12.6}$$

where $\bar{E}_t$, $\bar{C}_t$ are energy/carbon budgets and $\Phi(S_t)$ encodes privacy (e.g., Rényi DP) or fairness constraints. In practice, ΔA_t is approximated by proxy signals, i.e., gradient norms, loss plateaus, or historical contribution scores, and a greedy or bandit policy selects clients under the budgets.

12.2.3 Communication-Efficient Updates

Most cost arises from local compute and radio use; both can be reduced with complementary levers:

- **Model & update compression:** width/depth scaling, pruning, low-rank adapters, post-training quantization (8/4-bit), and sparsified or sketch-based updates (e.g., top-k, error-feedback) to reduce bits$^{\uparrow}$ and on-device FLOPs.
- **Adaptive local work:** per-client tuning of $(B_{i,t}, E_{i,t}^{\mathrm{loc}})$ to device efficiency and energy availability, with early stopping via on-device validation.
- **Partial participation with stratification:** more frequent sampling of energy-efficient clients while preserving statistical coverage to protect the utility.

12.2.4 Carbon-Aware Time and Place

Let I_t be a (forecast) time-of-use carbon intensity. A carbon-aware controller defers or advances rounds to greener windows while meeting service and privacy targets:

$$\min_{\{S_t\}} \sum_{t=1}^{T} \sum_{i \in S_t} I_t \, E_{i,t} \quad \text{s.t.} \quad A_T \geq A_{\min}, \quad \text{deadline}, \quad \text{privacy/fairness}. \tag{12.7}$$

Geo-diversity further helps by prioritizing clients in low-intensity regions when representativeness permits. In Async-FL, *staleness-aware* aggregation accommodates carbon-aware arrival patterns without sacrificing convergence.

12.2.5 A Practical Controller

At each round t, the server forecasts a per-client utility signal $\widehat{\Delta A}_{i,t}$ from recent contribution histories, estimates energy and carbon costs $\widehat{E}_{i,t}$ and $\widehat{C}_{i,t}$ using device/link profiles and time-of-use carbon intensity, and then selects a participant set by approximately maximizing the round objective U_t under energy, carbon, privacy, and fairness budgets (e.g., via a greedy or bandit policy). Before aggregating, it performs robustness screening to exclude low-consistency or suspicious updates, and finally updates the controller weights (λ_E, λ_C) online so that cumulative energy and carbon remain within targets while maintaining model utility.

12.3 Protecting FL with Representational Similarity

Robustness and sustainability are tightly coupled in FL. Every poisoned aggregation that slips through quality control tends to trigger compensatory rounds, rollbacks, or even full retraining, each of which consumes additional energy and increases carbon emissions. This section presents a practical, server-side defense based on *RSA* that filters adversarial contributions before aggregation by comparing how local and global models *encode relations* among a small, shared set of stimuli. The approach is simple to implement, compatible with common FL pipelines, and adds negligible overhead when the stimulus set is kept small.

12.3.1 Threat Model and Design Rationale

We focus on cross-device FL where a fraction of clients perform *data poisoning* (e.g., label flipping attack (LFA)), and the server cannot inspect raw client data. Under non-IID partitions, defenses that rely on parameter- or gradient-level proximity often confound distribution shift with adversarial behavior. RSA sidesteps this by evaluating the *geometry of representations* on a common reference set: if a client's model has learned a target-shifted boundary, the pairwise relations among stimulus embeddings will be distorted relative to those of the global model, even when parameter distances remain inconclusive.

12.3.2 Representational Consistency

In an FL system with N clients and a central server, the overall design of RSim-FL contains the workflow of clients and server, respectively.

In general, clients in the FL system cooperate with the server to train global models. However, malicious clients lurked in the system result in a chaotic situation comprising (1) benign clients that comply with the server, train, and share local models using standard datasets, and (2) malicious clients that attempt to interfere with the system by launching divergent data poisoning attacks (e.g., LFA and backdoor attacks). It is worth noting that since malicious clients behave similarly to a benign client, except for uploading nasty models based on poisoned local datasets, they are of high invisibility and are hard to discover.

The global server coordinates the overall training. At the beginning of each round, it broadcasts the current global model w_G to clients and, after local training, receives their updated models. Once the uploads are collected, the server initiates an identification procedure to reduce the risk of aggregating poisoned updates. Concretely, for each local model $w_n \in \{w_1, w_2, \ldots, w_N\}$, the server performs *RSA* using a small, server-held stimulus dataset $s = \{s_1, s_2, \ldots, s_h\}$. Note that the stimulus dataset consists of several samples from each category and is prepared by the server before the global training.

For model w_n, let $O(s_i)$ denote the representation (e.g., logits or a penultimate-layer feature) for stimulus s_i. The representational (dis)similarity matrix $\mathrm{RDM}_n \in \mathbb{R}^{h\times h}$ is computed with cosine similarity:

$$\mathrm{RDM}_n[i, j] = \frac{O(s_i)\cdot O(s_j)}{\|O(s_i)\|\,\|O(s_j)\|}, \qquad i, j \le h. \tag{12.8}$$

In parallel, the server forms the global model's matrix RDM_G by applying the same procedure to w_G.

Assuming malicious models can be distinguished by their representational geometry relative to the global model, the server measures the alignment between each local model and the global model via the Pearson correlation coefficient ρ. Let vec$(\cdot)$ denote vectorization of the upper triangle (excluding the diagonal). The *representational consistency* is

$$\mathrm{RC}_{n,G} = \left[\rho(\mathrm{vec}(\mathrm{RDM}_n),\ \mathrm{vec}(\mathrm{RDM}_G))\right]^2. \tag{12.9}$$

Collect the scores $\{\mathrm{RC}_{1,G}, \mathrm{RC}_{2,G}, \ldots, \mathrm{RC}_{N,G}\}$ and rescale them to $[0, 1]$ using a Min–Max normalizer.

To separate benign from malicious updates, the server runs a lightweight clustering algorithm (e.g., k-means with k=2) on the normalized RC scores. Under the common assumption that fewer than half of the clients are malicious [4, 8, 9], the cluster containing the majority of models is treated as benign. The server then updates the global model by aggregating the benign set, e.g., with simple (possibly data-size–weighted) averaging.

Keep the stimulus set small, often, a single exemplar per class suffices, and prefer logits or a stable penultimate layer for classification tasks. Because cost scales with the number of stimuli rather than with dataset size, the additional server work (one forward pass on h inputs per client and an $h \times h$ similarity/correlation) is typically negligible relative to local training. To avoid oscillations near the decision boundary, a simple hysteresis rule can be used: require two consecutive rounds below $\tau_{\min}$ before blacklisting, and two above before reinstatement. Finally, periodically refreshing a fraction of the stimulus set or varying class-conditional mini-batches helps guard against adaptive adversaries, while privacy remains unaffected since the stimuli are server-held and no extra client data are exposed.

12.3.3 Complexity and Overhead

Per the client model, RSA requires a single forward pass on h inputs and forming an $h \times h$ similarity matrix. Embedding costs $O(h\,d)$; forming and correlating upper triangles costs $O(h^2)$. With h in the single digits, the server's extra work scales linearly in $|S_t|$ and is dominated by standard inference. Communication volume and privacy exposure are unchanged relative to a baseline FL round.

12.3.4 Coupling Robustness with Sustainability

Screening reduces the probability of harmful aggregation, thereby lowering the expected number of remedial rounds and the associated energy/carbon overhead. To align this explicitly with the orchestration policies, incorporate RC into the client selection score:

$$\mathrm{score}_{i,t} = \alpha\,\widehat{\Delta A}_{i,t} + \beta\,\mathrm{RC}_{i,G} - \gamma\,\widehat{E}_{i,t} - \delta\,\widehat{C}_{i,t}, \tag{12.10}$$

where $\widehat{\Delta A}_{i,t}$ is a proxy for the client's expected utility (e.g., from historical contribution signals), and $\widehat{E}_{i,t}$, $\widehat{C}_{i,t}$ are energy and carbon estimates from the accounting models. This single scoring rule prioritizes clients who are simultaneously useful (high $\widehat{\Delta A}$), trustworthy (high RC), and economical (low E, low C).

12.3.5 Experimental Evaluation

12.3.5.1 Experimental Settings

Datasets. We evaluate on **MNIST** [3] and the **German Traffic Sign Recognition Benchmark (GTSRB)** [2]. To simulate Non-IID scenarios, local datasets are generated via a Dirichlet partition with concentration $\alpha = 0.5$. For GTSRB, we use the ten most frequent categories.

Baselines. To showcase the advantages of RSA-based screening (RSim-FL), we compare against **FedAVG** [6], **Median** [9], **PEFL** [5], **LFR-PPFL** [7], and **FoolsGold** [1].

System Setup. We instantiate an FL system with 30 clients and 1 server. Each client holds 5,000 training and 1,000 test samples. Local models are trained with a five-layer CNN, learning rate 0.001, batch size 128, and 500 iterations per round. Unless otherwise stated, the stimulus set for RSA uses one image per class.

Attack Setting. A subset of clients (40%) is malicious and launches a *LFA*. Following [7], MNIST flips "9→7", and for GTSRB the label "No passing (3.5t)" is flipped to "Keep right".

Evaluation Metrics. We report (i) Top-1 test accuracy (**Acc**) and (ii) attack success rate (**ASR**). With TP, TN, FP and FN denoting true/false positives/negatives, accuracy is

$$\text{Acc} = \frac{\text{TP} + \text{TN}}{\text{TP} + \text{TN} + \text{FP} + \text{FN}}. \tag{12.11}$$

Let N_S be the number of test samples whose source label is S (the flipped-from class) and N_T the number misclassified as the target class T; then

$$\text{ASR} = \frac{N_T}{N_S}. \tag{12.12}$$

12.3.5.2 Performance Comparison

Table 12.2 summarizes results on the MNIST and GTSRB datasets under IID and Non-IID partitions.

Table 12.2 Comparison of RSim-FL with baselines under IID and Non-IID settings. Best is **bold**; second best is underlined. "–" indicates non-convergence

Method	MNIST				GTSRB			
	IID		Non-IID		IID		Non-IID	
	Acc↑	ASR↓	Acc↑	ASR↓	Acc↑	ASR↓	Acc↑	ASR↓
RSim-FL (Ours)	**0.989**	**0.002**	**0.985**	**0.002**	**0.950**	**0.000**	**0.944**	**0.005**
FedAVG	0.972	0.102	0.934	0.402	0.872	0.207	0.851	0.335
Median	0.979	0.084	0.973	0.136	<u>0.943</u>	0.005	0.922	0.191
PEFL	0.974	0.012	–	–	0.884	0.004	–	–
LFR-PPFL	0.981	0.024	0.968	0.104	0.908	0.023	0.901	0.093
FoolsGold	<u>0.984</u>	<u>0.004</u>	0.949	0.338	0.902	<u>0.003</u>	0.876	0.179

IID setting. All defenses mitigate LFAs to some extent. RSim-FL achieves the highest accuracy with the lowest ASR on both datasets, outperforming the baselines by correctly separating benign and malicious updates.

Non-IID setting. Differences are amplified under data heterogeneity. Most baselines degrade (e.g., FoolsGold's ASR increases on GTSRB), while RSim-FL maintains high Acc and low ASR. PEFL fails to converge in some non-IID cases because it treats Median-like statistics as representative despite the lack of typicality.

12.3.5.3 Robustness Analysis

We vary source-target flip pairs on GTSRB (Non-IID) to assess robustness. RSim-FL consistently yields ASR between 0.00% and 5.82% (mean $\approx$ 0.60%), demonstrating resilience across divergent LFAs. In contrast, an undefended FedAVG shows ASR ranging from 2.76% to 91.67% (mean $\approx$ 49.45%), underscoring the necessity of RSA-based screening.

RSim-FL improves both security and utility. Relative to distance- or similarity-based baselines, it raises global accuracy by about 2.76% (IID) and 3.61% (non-IID) and substantially reduces ASR, including an average reduction of 43.956% compared to an unshielded FedAVG in robustness sweeps.

12.3.6 Summary

We leverage *RSA* to detect malicious updates in FL and propose **RSim-FL**, which computes representational consistency between each local model and the global model on a small server-held stimulus set, followed by lightweight clustering. Experiments validate its advantages under non-IID data and diverse label-flip attacks. Promising directions include broadening to other poisoning paradigms (e.g., backdoors), accelerating RSA via efficient sketches, and adapting thresholds and screening to Async-FL.

References

1. Fung C, Yoon CJM, Beschastnikh I (2020) The limitations of federated learning in sybil settings. In: 23rd international symposium on research in attacks, intrusions and defenses (RAID 2020), pp 301–316
2. Houben S, Stallkamp J, Salmen J, Schlipsing M, Igel C (2013) Detection of traffic signs in real-world images: the German traffic sign detection benchmark. In: 2013 international joint conference on neural networks (IJCNN), pp 1–8. https://doi.org/10.1109/IJCNN.2013.6706807
3. LeCun Y, Bottou L, Bengio Y, Haffner P (1998) Gradient-based learning applied to document recognition. Proc IEEE 86(11):2278–2324. https://doi.org/10.1109/5.726791

4. Liu B, Ding M, Shaham S, Rahayu W, Farokhi F, Lin Z (2021) When machine learning meets privacy: a survey and outlook. ACM Comput Surv (CSUR) 54(2):1–36
5. Liu X, Li H, Xu G, Chen Z, Huang X, Lu R (2021) Privacy-enhanced federated learning against poisoning adversaries. IEEE Trans Inf Forensics Secur 16:4574–4588
6. McMahan B, Moore E, Ramage D, Hampson S, y Arcas BA (2017) Communication-efficient learning of deep networks from decentralized data. In: Proceedings of the 20th international conference on artificial intelligence and statistics, Proceedings of machine learning research, vol 54, pp 1273–1282. PMLR
7. Shen X, Liu Y, Li F, Li C (2023) Privacy-preserving federated learning against label-flipping attacks on non-iid data. IEEE Internet of Things J. Early Access
8. Sun P, Shen S, Wan Y, Wu Z, Fang Z, Gao XZ (2024) A survey of iot privacy security: architecture, technology, challenges, and trends. IEEE Internet Things J 11(21): 34567–34591
9. Yin D, Chen Y, Kannan R, Bartlett P (2018) Byzantine-robust distributed learning: towards optimal statistical rates. In: Proceedings of the 35th international conference on machine learning, Proceedings of machine learning research, vol 80, pp 5650–5659. PMLR

Chapter 13
Summary and Outlook

The advancement of FL marks a transformative step toward building resilient, secure, and scalable distributed intelligence. However, this emerging paradigm introduces new challenges in adversarial robustness, privacy protection, and system efficiency. This chapter presents key research frontiers that define the next generation of resilient FL systems, focusing on *security and resilience*, *personalization*, and the integration of XAI to enhance transparency and trust.

13.1 Robust and Trustworthy Graph Construction

One of the foundational challenges in graph-based FL lies in constructing robust inter-client graphs. Most existing methods rely on similarity measures between model updates or feature embeddings to infer relationships among clients. Such measures are often noisy and vulnerable to adversarial manipulation. Future research should explore secure graph construction mechanisms that combine *secure multi-party computation*, *homomorphic encryption*, and *DP*, thereby ensuring that similarity estimation does not expose sensitive information or allow poisoning of graph topology. Dynamic graph construction offers a promising direction for resilience. By allowing edge weights or connectivity to evolve according to the temporal reliability and behavioral history of clients, adaptive graphs can detect and mitigate long-term poisoning campaigns. Integrating *temporal GNNs* could further enable FL to recognize shifting trust patterns and respond autonomously to emerging adversarial behaviors.

K. Li et al., *Security and Resilience in Distributed Machine Learning*, Springer Series in Reliability Engineering, https://doi.org/10.1007/978-3-032-23959-4_13

13.2 Adversarial Robustness and Zero-Trust Architectures

While GNNs can be leveraged to detect malicious participants or anomalous updates, adversaries are likely to adapt their strategies to exploit graph dependencies. Attacks such as node injection, edge manipulation, or structural poisoning can compromise both convergence and fairness. To address these challenges, future work should design *robust graph defense mechanisms* that explicitly model and defend against structural adversaries.

Integrating *zero-trust principles* within graph-based FL architectures offers a compelling pathway toward systemic resilience. In such settings, no client, server, or communication link is inherently trusted. Instead, decentralized verification, continuous authentication, and distributed anomaly detection operate within the graph structure to ensure secure collaboration. This paradigm represents a shift from reactive defenses to *inherent resilience by design*, where trust must be earned dynamically rather than assumed.

13.3 Explainable Graph Defenses

Transparency is critical in high-stakes domains such as healthcare, finance, and autonomous systems. Integrating *XAI* into graph-based FL defenses can bridge this gap. By coupling GNNs with interpretable techniques–such as visual explanations, attribution mapping, or symbolic reasoning–researchers can reveal why certain nodes or edges are flagged as malicious or unreliable. Explainable graph defenses enhance trust among stakeholders, support compliance with regulatory frameworks, and enable human-in-the-loop oversight for sensitive applications.

13.4 Personalization and Adaptability

Traditional FL often struggles to balance global model generalization with local personalization, especially when clients differ in data distributions, architectures, or computational capabilities. Graph-based FL provides a natural mechanism for *personalized learning* by embedding clients into relational neighborhoods and selectively propagating information among similar participants.

Future research could explore *hierarchical personalization*, where subgraphs of related clients perform localized learning before merging updates into the global model. Similarly, *graph hypernetworks* may facilitate aggregation across heterogeneous architectures by mapping diverse model types into a shared latent space. Such approaches can enable *cross-silo collaboration* among institutions that operate under different constraints while maintaining data sovereignty and system resilience.

13.5 Communication Efficiency and System Optimization

The structural richness of graph-based FL comes at a cost–additional communication and computation overhead. Large or dynamic graphs amplify this challenge. To address it, researchers may employ *graph compression*, *sparsification*, or *hierarchical partitioning* to maintain robustness while reducing message complexity. Adaptive protocols that regulate communication frequency based on convergence metrics or trust dynamics could further optimize resource use in bandwidth-constrained settings. Combining *asynchronous optimization* or *gossip-based updates* with graph-enhanced aggregation can make distributed learning more scalable and resilient to stragglers, enabling real-time adaptation to fluctuating network conditions.

13.6 Domain-Specific Applications and Explainability

Graph-based FL is poised to transform multiple application domains:

- **Healthcare:** Patient and institution graphs can support privacy-preserving collaborative diagnosis using medical ontologies.
- **Autonomous Vehicles:** Mobility-based graphs can enable resilient, low-latency perception and control across vehicular networks.
- **Finance:** Transaction networks can harness FL for fraud detection while preserving cross-institutional data confidentiality.
- **Metaverse and IoT:** Dense, dynamic graphs can support adaptive, personalized interactions across immersive environments.

In all these domains, embedding explainability into model decisions–both at the node (client) and edge (relationship) level–will be essential to achieving accountable and trustworthy distributed intelligence.

13.7 Benchmarking and Evaluation Frameworks

Despite the growing body of literature, the evaluation of graph-based FL remains fragmented. Current benchmarks largely focus on accuracy under non-IID conditions, often overlooking graph structural dynamics and adversarial robustness. There is a pressing need for *standardized evaluation protocols* and *graph-aware metrics* that measure not only performance but also resilience, fairness, and communication efficiency. Developing dynamic benchmark datasets that simulate real-world behaviors, e.g., node churn, adversarial injection, and temporal evolution, will be crucial. Similarly, resilience indicators (e.g., degradation under attack) and fairness metrics (e.g., inter-client performance disparity) should be incorporated into future frameworks to provide a comprehensive understanding of system reliability.

13.8 Summary

The convergence of *GNNs*, *FL*, and *XAI* establishes a powerful foundation for building secure, transparent, and adaptive distributed AI ecosystems. Resilience, in this context, evolves from a passive defense mechanism into a *structural property* of the learning process–emerging from the interplay of trust, explainability, and graph-based intelligence. By embedding FL within graph and zero-trust frameworks, the next generation of distributed AI systems can operate safely in dynamic, adversarial, and resource-limited environments. These concepts not only strengthen federated architectures but also pave the way for broader applications in *Large Language Models (LLMs)* and *agentic AI*, where decentralized reasoning, privacy preservation, and continual adaptation are equally critical.

MIX
Papier aus verantwortungsvollen Quellen
Paper from responsible sources
FSC® C105338

If you have any concerns about our products,
you can contact us on
ProductSafety@springernature.com

In case Publisher is established outside the EU,
the EU authorized representative is:
Springer Nature Customer Service Center GmbH
Europaplatz 3, 69115 Heidelberg, Germany

Printed by Libri Plureos GmbH
in Hamburg, Germany